International Research on Poverty Reduction

Series Editors

Xiaolin Wang, Fudan University, Shanghai, China

Qianqian Liu, Finance Center for South-South Cooperation, Hong Kong, China

This book series covers poverty reduction practices and experience, knowledge mobilization, poverty reduction strategy analysis, and comparative research of different poverty reduction models. By analyzing China's poverty reduction approaches and experiences, and comparing them with those of other countries, this book series provides policy makers, researchers and practitioners in poverty reduction with theoretical and technical support, as well asproviding a new channel for sharing experiences as well as facilitating collaboration and communication in poverty reduction development fields between China and the rest of the world. Poverty remains a crucial development issue for the world, and China has become an important player in the global endeavor in poverty reduction and development. China has made a substantial contribution to the achievement of the Millennium Development Goals (MDGs) by reducing half of its poverty population.

Weizhong Zeng · Fan Yang

Poverty Alleviation Via Forest Carbon Sequestration

Theory, Empirical Evidence, and Policy Implications

SSAP

Springer

Weizhong Zeng
School of Management
Sichuan Agricultural University
Chengdu, Sichuan, China

Fan Yang
School of Public Administration
Sichuan University
Chengdu, Sichuan, China

ISSN 2195-9781 ISSN 2195-979X (electronic)
International Research on Poverty Reduction
ISBN 978-981-99-8190-8 ISBN 978-981-99-8191-5 (eBook)
https://doi.org/10.1007/978-981-99-8191-5

Jointly published with Social Sciences Academic Press
The print edition is not for sale in China (Mainland). Customers from China (Mainland) please order the print book from: Social Sciences Academic Press.
ISBN of the Co-Publisher's edition: 978-7-5201-5707-0

Translation from the Chinese language edition: "森林碳汇扶贫: 理论、实证与政策" by Weizhong Zeng and Fan Yang, © Social Sciences Academic Press (China) 2019. Published by Social Sciences Academic Press (China). All Rights Reserved.

Sponsored by the Chinese Fund for the Humanities and Social Sciences

This Springer imprint is published by the registered company Springer Nature Singapore Pte Ltd.
The registered company address is: 152 Beach Road, #21-01/04 Gateway East, Singapore 189721, Singapore

Paper in this product is recyclable.

Preface

Forest carbon sequestration/sink (FCS) originates from marketized ecological compensation and aims to achieve a win-win outcome by addressing climate change and poverty reduction through economic incentives. The international FCS trading market and its industrial development began to take shape following the ratification of global climate change protocols such as the Kyoto Protocol (1997), Bali Roadmap (2007), Copenhagen Accord (2009), and Paris Agreement (2016). In China, afforestation and reforestation (AR) pilot projects for carbon sequestration have been successively implemented in poverty-stricken areas. The Panda Standard (PS),[1] which was proposed alongside the establishment of China's unified carbon trading market, was China's first voluntary carbon emission reduction standard designed to provide transparent and reliable carbon credits for dealers in the Chinese carbon market. Hence, FCS has gradually become the focus and practical means in China to meet emission reduction commitments, relieve poverty by providing more income-earning options for farmers, and establish marketized eco-compensation mechanisms. However, it cannot be overlooked that—although existing FCS projects are mainly conducted in poverty-stricken areas—those projects are unable to precisely target poverty-stricken farmers because they are market-driven and centred on carbon trade.

In response to the two main issues attracting global attention, e.g., climate change and poverty, China is actively implementing strategies targeting poverty reduction, poverty elimination, rural revitalisation, and constructing an ecological and prosperous country, which—in China—is formally called ecological civilisation. Against this macroscopic background, it is important to determine how to formulate appropriate governmental policies to regulate market operations and compensate for market failure. To ensure FCS projects are focused on poverty alleviation, it is essential to formulate scientific and feasible guidance, motivation, safeguarding, and normative

[1] The PS was developed by the Beijing Environmental Exchange in December 2009. As China's first voluntary emission reduction standard, it establishes emission reduction testing standards and principles in a narrow sense, while specifying processes, assessment bodies, and rules in a broad sense. It was established to meet the needs of Chinese domestic companies and individuals to take action on climate issues.

policies during the process of planning, design, authentication, registration, organisation, construction, monitoring, and assessment of such projects. The above actions ensure that farmers targeted for poverty alleviation can truly benefit from the projects and that the actions taken can prevent the externalisation of natural resources in poverty-stricken areas.

The ethnic regions of southwest China—dominated by ethnic minorities and poverty-stricken populations—are the preferred areas for FCS pilot projects. These regions are primarily inhabited by extremely poverty-stricken populations and are defined as restricted or forbidden development zones, which are excluded from development-prioritised zones. Nevertheless, these ethnic regions enjoy exceptional advantages for FCS project implementation and have great potential for poverty alleviation via forest carbon sequestration (PAFCS) as they have abundant resources and suitable climate and soil conditions for forest growth. Therefore, in-depth scientific research into PAFCS is needed from the governmental macro-control perspective to determine how related policies can be better promoted in the ethnic regions of southwest China to make better use of their exceptional advantages and huge PAFCS potential. The research endeavour discussed above has great significance and practical value for the achievement of ecological civilisation, political stabilisation, national solidarity, and social harmony.

Based on the basic principle of *theory–practice–theory*, this book adopts the methodology of *theoretical research –empirical analysis–discussion–policy suggestions* and *macroscopic analysis–case study–microscopic survey*. Using relevant research findings within and outside China, the researchers analyse the effect of FCS project exploitation on regional and individual poverty alleviation. The PAFCS stakeholders and their basic appeals for benefits are clarified, and the constituents of the PAFCS system are identified. The dynamic mechanism that combines FCS project exploitation and poverty alleviation is uncovered, and a theoretical framework of PAFCS is established. Based on the analyses of the current situations and challenges of PAFCS in the ethnic regions of southwest China, this book sampled FCS projects implemented in these regions where afforestation or reforestation has been partially completed. Theories of economics, management science, ethnology, ecology, and sociology have been drawn on, and multiple econometric methods, including the logistic model, ordered probit model, analytic hierarchy process (AHP), and structural equation modelling (SEM), have been employed.

From the microscopic perspective, farmers involved in the projects are selected as samples for the purpose of empirically studying their willingness to participate sustainedly, barriers to their behaviour selection and participation, elite drive and elite capture, compatibility between the ethnic tradition and the commercialisation of FCS, farmers' perception of poverty alleviation effects, and other key factors. A PAFCS performance evaluation index system is established, and the periodical comprehensive poverty alleviation performances of some projects are quantitatively measured and validated. Based on the research findings, some implications for future development, policy frameworks, and policy suggestions for PAFCS are proposed. The main conclusions of the research are as follows.

First, PAFCS is a novel form of poverty alleviation proposed by China as part of its participation in global joint endeavours to address climate change. It is characterised by a definite poverty alleviation target, the involvement of multiple stakeholders, multidimensional effects, and policy guidance. The resulting endogenous dynamics include the driving force of emission reduction obligation, economic benefits, and the promotion of corporate responsibilities. The exogenous dynamics include the constraints of emission reduction standards, governmental policy regulations, and public demands. To ensure the success of PAFCS, it will be far from sufficient to rely exclusively on the responses to climate changes and policies of the FCS industry. Appropriate governmental interventions are required, specifically the integration of various poverty alleviation and eco-developmental policies. The goals of poverty alleviation can be achieved only by encouraging and enhancing cooperation among PAFCS stakeholders, integrating poverty alleviation resources, and using appropriate poverty alleviation methods. The fundamental principle of "government guidance, enterprise leadership, farmer participation, and third-party evaluation" must be established as a requisite framework to improve the mechanism of PAFCS to benefit poverty-stricken people continually. A marketized system subject to the intervention of governmental policies should be established to achieve a booming FCS market, sustainable industrial development, and sustainable project management.

Second, the formulation and implementation of FCS voluntary emission reduction rules aiming for poverty reduction form the basis of PAFCS development. Surveys in this book show that FCS pilot projects have made great progress; however, in many poverty-stricken areas, PAFCS is simply carried out in the same way as other FCS projects and is usually regarded by locals as no different from an ordinary industrial poverty alleviation project. Generally, poverty alleviation goals are almost not integrated into the various procedures that compare FCS projects, including planning and design, authentication and registration, organisation and construction, and monitoring and assessment. PAFCS is still at the exploratory stage and is led by the government. The present policies introduced by local governments are implemented in a mandatory but temporary manner to meet any urgent requests for targeted poverty alleviation and to solve the practical problems of FCS projects. However, these policies are ad hoc and short term.

Hence, macroscopic management policies are required to promote sustainable PAFCS development. Normative regulations should be immediately introduced, including FCS standards and methods, preferential policies concerning FCS-related mortgages or subsidised loans, FCS insurance, and prevention of deforestation in FCS. Additionally, it is essential to integrate fiscal taxation, finance, investment, forest ecological compensation, technical assistance, and other general preferential policies. Government authorities should formulate and implement Chinese FCS voluntary emission rules that are tailored to local conditions and highlight the goal of *alleviating the truly poor—truly alleviating the poor*.

Third, highlighting poverty alleviation functions should be an important choice for transitioning FCS regulations in China. Empirical results in this book show that the large-scale AR carbon sequestration projects implemented in the poverty-stricken ethnic regions of southwest China have achieved significant poverty alleviation in multiple ways. On the temporal scale, the poverty alleviation performances of FCS projects are on the rise, even though the same FCS project has significantly different effects in different areas. About 72.56% of sampled farmers reported that the benefits of PAFCS outweigh the costs. This perceived value has a significantly positive impact on farmers' willingness to support the subsequent operation of the projects, showing a coefficient of 0.61. This result indicates that highlighting poverty alleviation improves the attractiveness of the FCS market and the market shares of FCS projects and decreases the uncertainty of FCS industrial development. As a guarantee, it is important to reduce the transaction costs and risks of the projects to secure support from farmers and achieve the sustainable operation of FCS projects.

Fourth, it is crucial to foster active participation among poverty-stricken populations and enhance the latter's sense of gain to ensure the long-term sustainability of FCS projects. Empirical results show that farmers, including poverty-stricken farmers, are hesitant to continue their participation in FCS projects: about 46.35% are neutral, and 24.69% express unwillingness to do so. Many factors significantly affect farmers' willingness to participate sustainedly, including age, land area, household income, degree of mixed occupation, mode of project organisation, satisfaction with previous earnings, expectation of subsequent earnings, degree of governmental support, difficulty in acquisition of forestry information, and road conditions. As such, adequate attention should be paid to opportunities to participate, degree of participation, and the risk that poverty-stricken families will have lower income levels and smaller enrolled forest lands and depend more heavily on traditional farming for their livelihood. These families should be equally assisted in boosting their earnings, as doing so could reduce the risk of their withdrawing from the project.

Fifth, given the benefits some groups could derive from elite capture, community elites are surprisingly unwilling to participate in FCS projects. Results show that the overall degree of elite capture is 0.254. Specifically, the capture degrees of institutional elites, economic elites, and traditional elites are 0.095, 0.201, and 0.092, respectively; thus the capture degree of economic elites is the most significant. Meanwhile, only half (55.3%) of community elites are willing to encourage, organise, and guide farmers to participate in PAFCS projects. The willingness of rural elites, in particular, is significantly affected by factors including age, prior employment in forestry, the main source of family income, perceived incomes from FCS projects, perceived difficulty in project completion, relationships with farmers, knowledge of FCS policies, and type of elite. The result implies that PAFCS is affected not only by the behaviours of intra-institutional elites, but also by the impact of extra-institutional ones, such as respected elders in the community, educated farmers, and the heads of patriarchal clans. In this regard, PAFCS designers and implementers should not exclude ensuring that economic elites receive reasonable individual benefits considering their important role in leading and organising community participation in FCS projects; they should simultaneously prevent elite capture as much as possible.

Sixth, enhancing the functions of informal institutions and farmer autonomy is a critical strategy for promoting PAFCS. Empirical results indicate that farmers in Yi communities of Liangshan identify more with traditional cultures than with the modern business culture brought by FCS projects. Farmers' average identification scores with traditional and business cultures are 4.05 and 3.48, respectively. Integration and separation are farmers' two principal choices for adapting traditional and business cultures. Female farmers, middle-aged farmers, better-educated farmers, and farmers with larger areas of land have higher cultural adaptiveness. This result suggests that in remote poverty-stricken areas such as ethnic areas, the implementation of market-dominated FCS projects should rely not only on contract norms, economic incentives, and administrative means, but also on local cultures and traditional customs. In particular, the ecological philosophy and ecological consciousness of traditional agriculture cultures should be followed, and informal institutions and their former autonomy should be respected.

Seventh, public administration of the community guarantees the development of PAFCS. This book demonstrates that the community played a more important role in 2016 than it did in 2013. The values of the publicity function, income allocation function, and project maintenance function of all communities slightly fluctuated from 2013 to 2016 but were on the rise. In particular, the income allocation function and project maintenance function significantly improved. This result suggests that the community's functions gradually switch from organisation and promotion to project maintenance and that communities influence the organisation, implementation, subsequent maintenance, and income allocation of FCS projects either directly or indirectly.

Eighth, preventing potential natural and market risks is the long-term driving force that ensures a trickle-down effect of FCS projects on regional economic development and poverty alleviation. Results show that FCS projects make a remarkable contribution to regional economic growth. However, due to the length of such projects, this effect is not significant in the short term. Local economic development is mainly manifested in optimised industrial structure, increased deposits of local residents, and increased financial income and expenses of local governments. Hence, efforts should be made to expand the coverage areas of FCS projects, increase special investments, and implement the projects in ecologically fragile and severely poverty-stricken areas. Project performance assessment should not be limited to short-term effects but should be more focused on long-term ones. Farmers should be guided to pay attention to the long-term benefits of FCS projects, which will also enhance their willingness to participate. Moreover, the financial environment at the regional level should be improved.

Ninth, reducing situational risks is key. Results show that the perceived benefits and risks of farmers directly influence their perceptions of the value of PAFCS. Specifically, the perceived economic, social, and ecological benefits of farmers significantly positively affect such perceptions, with coefficients of 0.36, 0.35, and 0.22, respectively. The results imply that economic benefits are farmers' major concern, followed by social and ecological benefits. The perceived economic, mental, and situational risks of farmers have significantly negative impacts on their perceptions of the value of PAFCS, with path coefficients of −0.28, −0.25, and −0.16, respectively. Thus, the economic risk most affects farmers' perception of value, followed by perceived mental and situational risks. Obviously, ensuring the maximisation of economic benefits obtained by farmers and actively decreasing the negative effects due to forestry eco-construction are important strategies to improve farmers' sense of gain from project participation.

Limitations of the research include researchers' knowledge structure and research capacity, difficulties resulting from the scattered habitation of farmers, language barriers, poor transport infrastructure, and high research costs. As a result, the researchers cannot achieve a larger sample size or longer time-series data from different areas. Instead, statistical analyses based on data at a certain time point or from a specific area were used to represent overall conditions, and data relating to currently filed and registered poverty-stricken families were used to represent all poverty-stricken families. Moreover, the global institutional arrangements in response to climate change are changing, and so are China's national and regional carbon emission trading systems and low-carbon policies. As the construction of a carbon trading market and the practice of AR for carbon sinks in China are still in the developing stage, much attention has been paid to emerging topics, such as REDD and wood products. More work needs to be done regarding how to trace the latest advances in international climate negotiation and institutional transitions, how to resolve the relative poverty problems in the post-well-off era, namely the time after 2020, and how to align with the implementation of rural revitalisation strategies in China. Lastly, rarely discussed in the literature is how PAFCS programmes are planned and how innovation within related institutions occurs. Hence, interested researchers in multidisciplinary areas, policymakers, and practitioners are encouraged to increase their endeavours in response to climate change and poverty alleviation across China and the world.

Chengdu, China

Weizhong Zeng
Fan Yang

Principal Researchers

Zeng Weizhong is a professor of anti-poverty at the School of Management, Sichuan Agricultural University, Chengdu, China
Yang Fan is an associate professor at the School of Public Administration, Sichuan University, Chengdu, China
Zhuang Tianhui is a professor at the School of Economics, Sichuan Agricultural University, Chengdu, China
Yang Hao is a professor at the Southwestern Poverty Reduction and Development Research Centre, Sichuan Agricultural University, Chengdu, China
Cheng Rongzhu is a Ph.D. candidate at the School of Public Administration, Zhejiang University, Hangzhou, China
Gong Rongfa is a Ph.D. candidate at the School of Economics, Sichuan Agricultural University, Chengdu, China
Hu Yuan is an associate professor at the School of Management, Sichuan Agricultural University, Chengdu, China
Qiu Lingling is an assistant professor at the School of Public Administration and Policy, Dalian University of Technology, Dalian, China

Translation Team

Dr. Long Chengzhi is a professor of the Brand Sustainability Research Centre at the School of Management, Guangdong University of Finance, Guangzhou, China
Dr. Lin Jin is an assistant professor at the School of Foreign Language and Cultures, Guangdong University of Finance, Guangzhou, China
Zeng Weizhong is a professor of anti-poverty at the School of Management, Sichuan Agricultural University, Chengdu, China
Dr. Yang Fan is an associate professor at the School of Public Administration, Sichuan University, Chengdu, China

Dr. Liz Dixon is an associate professor of management communication at the Marriott School of Business, Brigham Young University, Provo, UT, USA

Contents

Chapter 1
Introduction

This book is the first academic monograph on PAFCS in China. In this book, the researchers construct a theoretical framework of PAFCS; review the current situation; investigate farmers' intentions, constraints, and willingness to participate sustainedly in FCS projects; analyse farmers' cultural adaptability, community participation, elite participation, and farmers' returns; and propose PAFCS performance evaluation methods. As an introduction, this chapter will help readers establish a general understanding of the book. First, the researchers introduce the research background and state the salient implications. Second, the book's research objectives and main contents are outlined. Next, research ideas, methodologies, methods, and data sources are introduced. Finally, the originality and possible limitations of the research are stated.

1.1 Background and Significance

Forests are the largest carbon reservoir in terrestrial ecosystems and play a unique role in absorbing carbon dioxide, reducing the concentration of greenhouse gases in the atmosphere, and slowing global warming. AR, sustainable forest management, and reducing deforestation have become internationally recognised and advocated as climate change adaptation and mitigation strategies. At the same time, they are crucial elements of China's National Climate Change Programme (China State Council, 2007) and the Action Plan for Carbon Dioxide Peaking Before 2030 (China State Council, 2021a, 2021b). In the post-Kyoto era since 1997, and given the development of AR projects in poverty-stricken areas and the official opening of China's national carbon emission trading market in 2021, the intrinsic linkage between FCS and anti-poverty work has increasingly emerged as an issue of concern in research and practice. The researchers elaborate on the background of this book and its practical and theoretical significance in this section.

W. Zeng and F. Yang, *Poverty Alleviation Via Forest Carbon Sequestration*, International Research on Poverty Reduction, https://doi.org/10.1007/978-981-99-8191-5_1

1.1.1 *Background of the Selected Topic*

1.1.1.1 FCS is a Pivotal Element of the Win–Win Strategy for Global Climate Change and Poverty Reduction

Combating climate change and poverty is a crucial mission for today's world, as these issues not only constrain the survival and development of human beings but also determine the harmony and prosperity of the world. As one of the three flexible emission reduction mechanisms under the framework of the Kyoto Protocol (1997), the Clean Development Mechanism (CDM) aligns with the FCS philosophy that developed countries should meet their emission reduction commitments by investing in emission reduction or carbon sequestration projects in developing countries and providing funds and advanced technology to assist the latter in alleviating poverty as agreed by the international community. In the post-Kyoto era, with the drafting or signing of the Bali Roadmap in 2007, Copenhagen Accord in 2009, Paris Agreement in 2016, and other international documents, FCS, which is more economical and efficient than other emission reduction methods, has gradually become an alternative way of reducing carbon dioxide emissions. Along with the progress of carbon sequestration measurement technology and market trading rules, the international FCS industry and trading market have developed rapidly, and their irreplaceable role in addressing climate change and anti-poverty work has been highly praised globally.

1.1.1.2 FCS is a Prominent Measure for China to Meet Its Commitment to Reducing Emissions, Benefitting Farmers, and Combating Poverty

According to the principle of common but differentiated responsibilities established by the Kyoto Protocol and the United Nations (UN) Framework Convention on Climate Change (FCCC), China is not obligated to reduce greenhouse gas emissions. As a responsible developing country and a major emitter of greenhouse gases, however, China is committed to reducing emissions and mitigating global warming, focusing on the strategic position and unique role of forestry in addressing climate change, and has actively supported FCS development, AR, and related work.

The Chinese central government and various ministries have promulgated a series of documents and policies:

- China's National Climate Change Programme (China State Council, 2007),
- the 12th Five-Year Plan for Controlling Greenhouse Gas Emissions (China State Council, 2011),
- the 13th Five-Year Plan for Addressing Climate Change in Forestry (China State Council, 2012),
- China Forestry Adaptation to Climate Change Action Plan (2016–2020) (China State Forestry Administration and Grassland Bureau, 2016),

- the Guidance of the State Forestry Administration on Promoting Forestry Carbon Exchange (China State Forestry Administration and Grassland Bureau, 2014),
- the Establishment of Market-oriented and Diversified Ecological Protection Compensation Mechanism Action Plan (China Development and Reform Commission, 2018),
- the Carbon Emissions Trading Management Measures (for Trial Implementation) (China Ministry of Ecology and Environment, 2021),
- the Suggestions on Comprehensively Implementing the New Development Concept Within the Complete Sovereignty and Doing a Good Job of Carbon Neutral Work (China State Council, 2021c),
- China National Adaptation Strategy for Climate Change 2035 (China Ministry of Ecology and Environment et al., 2022).

With the support of these policies, China's FCS market and industry are developing rapidly, and the number of projects based on both Kyoto rules and non-Kyoto rules is increasing. In particular, with the implementation of the China Certified Emission Reduction (CCER) mechanism based on the domestic market and the PS, the China carbon emission trading market was officially launched on 16 July 2021, based on the seven carbon emission trading pilots that started nationwide in 2017. At the end of 2021, the first compliance cycle of the national carbon emission trading market was completed, with a cumulative turnover of 179 million tons of carbon emission allowances and a cumulative turnover of 7.66 billion RMB (about 1.11 billion USD), making it the largest carbon emission trading market in the world. All these efforts demonstrate China's active participation in international cooperation on climate change, fulfilling its commitment to increase sequestrations and reduce emissions and addressing the issue of benefiting farmers and alleviating poverty.

Meanwhile, in January 2018, six ministries and commissions, namely China Development and Reform Commission (CDRC), China State Forestry Administration, China Ministry of Finance, China Ministry of Water Resources, China Ministry of Agriculture, and China State Council's Poverty Alleviation Office, jointly issued China Ecological Poverty Alleviation Programme, which set out not only to develop the national carbon emission trading market but also to reform the CDM and voluntary greenhouse gas emission reduction trading mechanism, develop FCS projects to obtain carbon emission reduction compensation, increase support for poverty-stricken areas, and elevate PAFCS to an unprecedented level.

1.1.1.3 FCS Creates an Important Opportunity to Promote Ecological Civilisation and Green Poverty Reduction in the Ethnic Areas of Southwest China

The ecological location of the ethnic regions of Southwest China is extremely important. The world's primary natural carbon bank, with abundant forestable land and other resources, its climate, and soil conditions are suitable for the growth of trees. It is also prioritised as a major pilot area for FCS projects. As early as 2017,

China registered five CDM projects with the Executive Board of the United Nations Clean Development Mechanism (UN CDM-EB), and Southwest China accounted for two: the Novartis *Southwest Sichuan Forestry Carbon Sequestration, Community and Biodiversity Project* (henceforth, *Novartis Sichuan FCS Project*), the Reforestation of Degraded Land in Northwest Sichuan (henceforth, Northwest Sichuan FCS Project), *Yunnan Tengchong Small-scale Reforestation Landscape Restoration Carbon Sequestration Project* (henceforth, *Tenchong FCS Project*), *the Bamboo Reforestation Project in Xishuangbanna, Yunnan under PS*, the Walt Disney Panda Habitat Restoration Project in Southwest China, and China Green Foundation's Public Welfare Carbon Sequestration Project.

The projects listed above are concentrated in remote and poverty-stricken mountainous areas in Sichuan, Yunnan, and Guizhou. Most poverty-stricken areas with restricted or prohibited development zones are in areas inhabited by ethnic minorities and poverty-stricken populations. These projects not only bring economic income, employment opportunities, and new technologies to these remote and poverty-stricken areas but also take the lead in implementing a system of resource reimbursement and ecological compensation, thus accelerating the development of green poverty reduction in ethnic areas in Southwest China.

In summary, in the context of the two global issues of climate change and poverty alleviation, China has announced ecological poverty alleviation, ecological civilisation, and rural revitalisation strategies. In this vein, this study conducted empirical research on pre-forestation or reforestation projects within and outside the Kyoto rules in ethnic areas of Southwest China. The research findings are convincing, and the policy recommendations will help China and the world formulate strategies and sustainable poverty alleviation strategies that align with local economic, social, and cultural characteristics. The study has important theoretical reference and practical inspiration value.

1.1.2 Significance of the Study

1.1.2.1 Practical Significance

Chinese FCS projects are currently developed in remote and poverty-stricken areas inhabited by ethnic minorities. However, it must be recalled that FCS projects, led by a market mechanism to achieve carbon trading, will not automatically focus on and reach the goal of poverty alleviation by creating benefits and development opportunities for poverty-stricken people. Therefore, it is necessary to determine how to choose and formulate corresponding guiding and incentive policies, guarantee and regulate the market operation, and guide and intervene in FCS project development to promote poverty alleviation in the planning and design, certification and registration, organisation and construction, and monitoring and evaluation of the project. In other words, this research will help enhance the accuracy with which poverty alleviation is targeted, find a poverty reduction path, raise the poverty benefit effect, and improve monitoring performance. Thus, poverty-stricken farmers will benefit from FCS projects, and the externalisation of natural resources in poverty-stricken areas will be prevented.

The results of this study are of particular and important practical significance for the development of FCS projects in the ethnic areas of Southwest China to better exploit their unique advantages and potential for carbon sequestration for poverty alleviation. Additionally, this study offers inspiration to promote green development and to build a beautiful, stable, harmonious, and united China.

1.1.2.2 Theoretical Significance

Poverty alleviation is the most usual goal of ecological compensation projects, including FCS projects. As ecological compensation projects consider social equity and the objective effect of poverty alleviation, research on the relationship between ecological compensation and poverty alleviation has received much attention globally. A need remains, however, for in-depth research on the trade-off between the two and how to more effectively use the poverty alleviation potential of ecological compensation (Wang & Zhong, 2009; Milder et al., 2010; Xu et al., 2015).

The researchers consider PAFCS to be an intersection of ecological compensation and poverty research in this study. They aim to explore how to make the FCS industry more beneficial to and create more development opportunities for poverty-stricken people, not only a key issue of poverty research but also an important branch of ecological compensation research, inclusive growth research, and related fields. In this book, the researchers not only provide new research perspectives, approaches, and methods for poverty reduction theory innovation and ecological compensation theory development but also further the design of poverty reduction tools, path selection, and institutional arrangements.

1.2 Research Objectives and Subjects

In this section, the researchers introduce the research objectives and subjects.

1.2.1 Research Objectives

The main objective of this study is to reveal the willingness of farmers and community elites to participate in FCS projects, their behavioural choices, the perception of poverty alleviation effects, and the influencing factors of these variables and to quantitatively measure project performance at different stages. It aims to provide scientific evidence and policy recommendations to promote the transformation from a focus on poverty-stricken areas alone to one that considers both regional and personal (i.e., of the farmers) factors.

Specifically, the objectives of this study are fourfold. The first is to analyse the impact of the projects on regional and individual poverty reduction, clarify the stakeholders and their basic interests, reveal the dynamic mechanism of combining project development and poverty alleviation, clarify the basic elements and essential characteristics of PAFCS, and build a theoretical analysis framework for future research.

The second is to investigate the current situation of PAFCS in ethnic areas of Southwest China; identify farmers' willingness to participate, barriers to participation, and behavioural choices; and examine elite driving and elite capture, the performance of project poverty alleviation, and farmers' perceptions of poverty alleviation effects.

The third is to design a performance evaluation index system for PAFCS to conduct a comprehensive evaluation, an analysis, and empirical tests on the poverty alleviation performance of the typical project in China.

The fourth is to propose ideas for future development and policy frameworks based on theoretical analysis and empirical studies.

1.2.2 Research Subjects

This study investigated AR FCS projects located in the ethnic regions of Southwest China, specifically Yunnan, Guizhou, Sichuan, and Chongqing provinces, but not Tibet. The projects considered were at the pre-forestation or reforestation stage.

1.3 Research Framework and Structure

In this section, the researchers present the research framework and structure.

1.3.1 Research Framework

In the vein of the currently active global response to climate change and poverty, this study focuses on four main issues in China: targeted poverty alleviation, green poverty reduction and development, ecological civilisation, and rural revitalisation strategies. It follows the principles of *theory–practice–theory, theoretical research–empirical analysis–solution exploration–policy recommendation, and macro analysis–case analysis–micro investigation.* The theoretical analysis framework of PAFCS follows the combined methodology of historical literature review, qualitative analysis, sampling and data analysis, comparative analysis, and case study. Based on the in-depth analysis of the status quo of PAFCS in Southwest China, this study selects

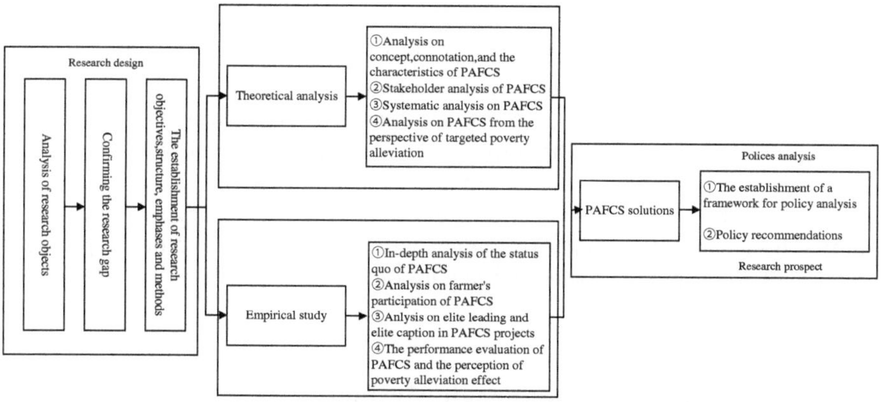

Fig. 1.1 Research schema

the Kyoto rules and non-Kyoto rules FCS projects in that area and focuses on four frontier issues: farmer participation, elite leading, farmers' perceptions of the poverty alleviation effect, and the monitoring and evaluation of poverty alleviation performance. Lastly, the researchers explore the development ideas, policy framework, and supporting policies. The research framework is shown in Fig. 1.1.

1.3.2 Research Structure

This study consists of 17 chapters, which can be divided into four parts.

The first part, in this chapter, is the introduction. It principally introduces the research background, significance, objectives, subjects, structures, data, and methods and explains the research gaps and limitations.

The second part is the theoretical framework, presented in Chaps. 2–3, in which core concepts are introduced, the literature on FCS and poverty alleviation is reviewed, and two basic scientific questions are proposed. It emphasises the combination of poverty alleviation theory and ecological compensation theory, clarifies the essential characteristics of PAFCS, analyses the dynamic mechanism of combining FCS projects and poverty alleviation, dissects the mechanism of the impact of project development on regional and individual poverty alleviation, and discusses poverty alleviation targets, resources, and methods as well as the closely related dynamics, participation, and environment systems. In particular, it scaffolds a theoretical framework for targeted poverty alleviation.

The third part is the empirical research presented in Chaps. 4–16. Focusing on the theoretical and practical frontier issues raised in the second part, from the practical perspective and from macro to micro, it provides an in-depth analysis of opportunities and challenges of PAFCS in ethnic areas in Southwest China and reveals the relationship between FCS and poverty alleviation. On this basis, first-hand survey data are analysed from the perspectives of farmer and elite participation, farmers' perception of the effect of poverty alleviation, and performance evaluation to provide empirical support and scientific evidence for policy innovation.

The fourth part, Chap. 17, presents solutions and recommendations. Based on the unique characteristics of the poverty-stricken ethnic areas in Southwest China in terms of ecological type, resource environment, ethnic culture, and economic and demographic characteristics and given the problems found in the study, it proposes a policy framework and policy recommendations to promote the development of PAFCS and points out the issues that require further study.

1.4 Data Sources and Research Methods

In this section, the researchers explain the data sources and research methods.

1.4.1 Data Sources

This study draws on two data sources, the first of which is surveys. Based on the project, *Policy Research on Promoting Forest Carbon Sequestrations for Poverty Alleviation in Ethnic Areas of southwest China,* funded by the National Social Science Foundation of China, the research team conducted an on-site investigation of five carbon sequestration AR projects, namely *the* Northwest Sichuan FCS Project, Novartis Sichuan FCS Project, Tengchong FCS Project, *Yunnan Xishuangbanna Bamboo Reforestation Project, and Guizhou Zazuo FCS Afforestation Demonstration Project.*

The team obtained data by visiting 11 counties (districts), 35 townships (towns), 72 poverty-stricken villages, and more than 110 AR areas (communities) and by holding in-depth interviews with or disseminating questionnaires to more than 1,200 community farmers and 150 relevant government departments, enterprises and institutions, and non-governmental organisation (NGO) leaders and staff at all levels. Second, some data are taken from official documents:

- the China Rural Poverty Monitoring Report (2012–2017),
- the China Statistical Yearbook (2010–2022),
- official statistics,
- government work reports,
- research reports,
- newspapers and magazines, and
- websites from the four southwestern provinces of
  - Sichuan
  - Yunnan
  - Guizhou

  - Chongqing
- websites from the ethnic autonomous prefectures of
  - Ganzi Prefecture
  - Tengchong County of Baoshan City
  - Aba Prefecture
  - Xishuangbanna, and
- the planning reports and annual work summaries (2010–2020) of the foregoing afforestation projects.

1.4.2 Research Methods

Based on theoretical analyses in the disciplines of economics, management, ethnology, ecology, sociology, and others, this study focuses on empirical research, macroscopic (regional factors) and microscopic (farmers) research, qualitative analysis with quantitative analysis, and theoretical research with practical application. Specifically, the following research methods are principally used.

The first is a **literature review**. By assessing progress made in the literature on FCS and poverty alleviation, the researchers identify the gaps in current research, put forward two basic scientific issues, and then provide the basis for the formulation of the research ideas. The theoretical analysis framework is built on stakeholder theory, system theory, and targeted poverty alleviation theory. Specifically, to lay a theoretical foundation for the study, it considers the types of project stakeholders and their basic interests and the components of the PAFCS system and its operation mechanism.

The second is a **field survey.** The research team adopted a combination of Delphi, group discussion, and Likert-scale methods to design the questionnaire, which was piloted before being carried out and revised in several subsequent rounds. The team then conducted focus-based interviews and administered a questionnaire with industry-related government departments, enterprises, and institutions, and NGO heads of staff in Sichuan, Yunnan, and Guizhou. The team combined typical sampling and stratified sampling in the household questionnaire surveys and semi-structured interviews with farmers in the communities.

The third is **statistical analysis**. Along with the descriptive analysis, this study employs SPSS, AMOS, and other software programs to model, analyse the process, and empirically study the collected data using various econometric methods, such as logistic, ordered probit, AHP, and SEM. The specific methods are explained in detail in the methodology section of each chapter.

The fourth is **comparative analysis.** This method is used in many parts of this book, including a comparative analysis of Sichuan, Yunnan, and Guizhou Provinces and between different projects and the same project as carried out in different regions, and the difference between the barriers to farmer participation and effects of poverty alleviation under different rules and operation modes.

1.5 Research Originality and Limitations

In this section, the researchers state the originality and limitations of this research.

1.5.1 Originality

This study has original cross-disciplinary features, perspectives, opinions, and empirical results, which are summarised below.

First, the research topic is exploratory. Although FCS has been increasingly linked to anti-poverty work, a relative lack of research focusing on FCS from the perspective of poverty alleviation means the topic of PAFCS is still under-researched. This study focuses on how to better exploit the poverty alleviation potential of FCS and promote the transformation of FCS projects currently underway so they are more beneficial to and create more development opportunities for poverty-stricken people. The topic is not only a core issue in the field of poverty research but also an important part of the research on FCS, ecological compensation, and related fields. In this sense, the topic is forward-looking and exploratory.

Second, the original research perspective expands the theoretical basis of PAFCS research. It emphasises that FCS projects in poverty-stricken areas are not the same as PAFCS, highlights the combination of poverty alleviation theory and ecological compensation theory, and considers PAFCS as an intersection of poverty research and ecological compensation research. It is among the few studies to propose the trade-offs between the benefit to be derived and creation of development opportunities for poverty-stricken populations and the trade-offs between FCS enhancement and the function of poverty alleviation, while the two trade-offs are the fundamental issues for in-depth research on PAFCS. It systematically clarifies the essential characteristics of PAFCS, builds an analytical framework, helps to solve the problem that theoretical research lags behind its practical development, and pioneers work in this field.

Third, the research content is novel, and the research methods are a breakthrough. The study emphasises that FCS projects are not general poverty alleviation projects, highlights the need for full respect to be granted to the subjective will and value judgement of farmers—including poverty-stricken farmers—takes the lead in focusing on the win–win goal of combating climate change and reducing poverty, and advances research on key PAFCS issues that need to be solved in the ethnic areas of Southwest China. It bridges the literature gap, which contains far fewer empirical studies than theoretical discussions, and offers a qualitative elaboration rather than a quantitative analysis. Therefore, it provides innovative ideas and scientific evidence for promoting the practice of climate change and anti-poverty in China. For example, it focuses on economic, social, and cultural conditions in the ethnic areas of Southwest China, develops a scale to measure farmers' identification with traditional culture and modern business culture, such as FCS trading, and identifies farmers' adaptation strategies and their main influencing factors. It establishes a performance evaluation index system for PAFCS. It is among the first studies to introduce and validate situational risk into the theoretical model of farmers' perceptions of the value of poverty alleviation effect. The modelling reveals the relationship between farmers' perception of the poverty alleviation effect value and their willingness to support the subsequent operation of the project.

In addition, the research has produced abundant results. As the first academic monograph on PAFCS, some of the results have not only been published in the *Journal of Forest Economics*, *International Forestry Review*, and other SSCI journals, but also have had a direct impact on policy-making. For example, with the support of the Sichuan Forestry and Grassland Bureau, the Sichuan Science and Technology Department, the Sichuan Poverty Alleviation and Migration Bureau, other relevant government departments, and Novartis Group (China), Sichuan has initiated PAFCS projects and, in turn, promoted this practice.

News

Sichuan Province Launches PAFCS Pilot Project

On the morning of 16 May 2018, the launch event and seminar of the Sichuan Province PAFCS pilot project was held in Chengdu. The meeting was hosted by the Sichuan Forestry Department, Sichuan Agricultural University, and Sichuan Dadu River Afforestation Bureau. The attendees came from the Sichuan Forestry Department, the Sichuan Provincial Science and Technology Department, the Sichuan Poverty Alleviation and Migration Bureau, the Sichuan Agricultural University, the Sichuan Rural Development Research Centre, the Sichuan Greening Foundation, the Sichuan Forestry Society, Liangshan Forestry Bureau, The Nature Conservancy, Zhaoge County, Yuexi County, and Leibo County.

As a crucial part of the world's response to climate change and poverty, FCS is an important initiative to fulfil China's commitment to increase sequestration, reduce emissions, and reduce poverty. Sichuan is the first province and prioritised area where China's national FCS project piloted, with the unique advantage of developing FCS projects and huge potential for carbon sequestration poverty alleviation. In the context of the world's active response to global warming, including China's strategy to realise ecological civilisation, targeted poverty alleviation, and rural revitalisation, Sichuan has seized the opportunity to take advantage of the national carbon trading market and initiated the PAFCS pilot project. The pilot project will provide a replicable case and practical experience in Sichuan to promote targeted poverty alleviation employing FCS.

(Editor: Zhang Huawei, Gao Hongxia).

Source: *People's Daily Online*, 18-05-2018.

http://sc.people.com.cn/n2/2018/0518/c345509-31597650.html.

1.5.2 *The Limitations of the Study*

This study is constrained by the researchers' knowledge structure and research capacity, coupled with a variety of insurmountable difficulties such as the scattered nature of farmers' residences in southwest ethnic areas, language communication barriers, poor transportation, and high research costs. Therefore, obtaining large samples or time-series data for each project was difficult; in some cases, cross-sectional data from a small region was substituted for the actual overall data. Thus, the research results are inevitably inconsistent with the overall and actual situation in the ethnic areas of Southwest China, and the research conclusions need to be tested in practice.

Other limitations exist; for example, the international arrangements for addressing climate change challenges are still evolving; the domestic or regional carbon trading systems and low-carbon policies established by relevant countries or regions vary significantly; the construction of China's carbon trading market and its carbon sequestration, AR, and improved forest management (IFM) practices are still in their infancy; and sustainable management of forests and wood products has not yet developed. Some important issues still need to be considered by future studies, such as how to track the latest progress in international climate negotiations or institutional changes; how to deal with the relative poverty, which continues to exist in the post-prosperity era (i.e., after 2020) in China; and how to explore innovation in solutions and institutional arrangements of PAFCS under different FCS mechanisms. Efforts are needed from international experts and scholars, policymakers, and practitioners from multiple disciplines to continuously deepen the research and share the results to make positive contributions to China and the world, leading to the win–win situation of addressing both climate change and poverty reduction.

References

China State Council. (2007). China's National Climate Change Programme. Retrieved 24 December, 2021, from https://sthj.gansu.gov.cn/__local/7/59/08/A6C1A2FE8E0CBE469250D221D2F_AA1EE0A1_2FAD7.pdf?e=.pdf

China State Council. (2008). China's National Programme to Address Climate Change. Retrieved 28 March, 2008, from http://www.gov.cn/zhengce/content/2008-03/28/content_5743.htm,published

China State Council. (2011). The 12th Five-Year Plan for Controlling Greenhouse Gas Emissions. Retrieved 24 December, 2021, from http://www.gov.cn/zwgk/2012-01/13/content_2043645.htm

China State Council. (2012). The 13th Five-Year Plan for Addressing Climate Change in Forestry. Retrieved Retrieved 23 June, 2016, from https://www.gov.cn/xinwen/2016-06/22/content_5084284.htm

China State Council. (2021a). Opinions on the Comprehensive Implementation of the New Development Concept within the Complete Sovereignty for Better Carbon Neutral Work. Retrieved 24 December 2021, from http://www.gov.cn/zhengce/2021-10/24/content_5644613.htm

China State Council. (2021b). Action Plan to Reach the Carbon Peak by 2030. Retrieved 26 October, 2021, from http://www.gov.cn/zhengce/content/2021-10/26/content_5644984.htm

China State Council. (2021c). The Suggestions on Comprehensively Implementing the New Development Concept Within the Complete Sovereignty and Doing a Good Job of Carbon Neutral Work. Retrieved 19 December, 2021, from https://www.gov.cn/gongbao/content/2021/content_5649728.htm.

China State Forestry Administration and Grassland Bureau. (2014). The guidance of the state forestry administration on promoting forestry carbon exchange work. Retrieved 7 May, 2014, from https://www.gov.cn/xinwen/2014-05/05/content_2671841.htm

China State Forestry Administration and Grassland Bureau. (2016). China forestry adaptation to climate change action plan (2016–2020). Retrieved 22 July, 2016, from https://www.gov.cn/xinwen/2016-07/22/content_5093660.htm

China Ministry of Ecology and Environment. (2021). Carbon Emissions Trading Management Measures (for Trial Implementation). Retrieved 1 February 2021, from http://www.gov.cn/zhengce/zhengceku/2021-01/06/content_5577360.htm

China Ministry of Ecology and Environment et al. (2022). China National Adaptation Strategy for Climate Change 2035. Retrieved 8 May, from https://www.cma.gov.cn/2011xwzx/2011xmtjj/202207/t20220709_4969154.html

China State Council & China Development and Reform Commission Notice on the Issuance of the Work Programme on Ecological Poverty Alleviation. Retrieved 24 January 2018, from http://www.gov.cn/xinwen/2018-01/24/content_5260157.htm

Milder, J. C., Scherr, S. J., & Bracer, C. (2010). Trends and future potential of payment for ecosystem services to alleviate rural poverty in developing countries. *Ecology and Society, 15*(2), 4.

China Development and Reform Commission. (2018). The Establishment of Market-oriented and Diversified Ecological Protection Compensation Mechanism Action Plan. Retrieved 28 December 2018, from http://www.ndrc.gov.cn/zcfb/zcfbtz/201901/t20190111_925436.html

Wang, L. A., & Zhong, F. L. (2009). Research progress on the relationship between ecological compensation and poverty alleviation. *Forestry Economic Issues, 9*(3), 201–205.

Xu, J. Y., Liu, X. X., Feng, L., et al. (2015). Advances in research on ecological compensation trade-off relationships. *Journal of Ecology, 35*(20), 6901–6907.

Chapter 2
Definitions of Core Concepts and Literature Review

This chapter presents a literature review of PAFCS. First, core concepts are defined, including forest, FCS, FCS projects, poverty, poverty-stricken populations, community, poverty alleviation, poverty alleviation performances, and eco-compensation. Second, the literature is summarised within the following categories: development of FCS industry and poverty alleviation, PAFCS project design and implementation, poverty alleviation performances, and challenges in exploiting PAFCS potential. Finally, research suggestions and prospects are highlighted.

2.1 Core Concepts of PAFCS

In this section, the researchers introduce and define many of the concepts related to PAFCS research, including forest, FCS, FCS project, poverty, poverty-stricken, community, poverty alleviation, poverty alleviation performance, and ecological compensation.

2.1.1 Forests

Forests—biotic communities dominated by woody plants—are the ecosystem formed from the mutual dependence of and mutual restriction between concentrated trees and other plants, animals, microbes, and soils. As the lungs of the earth, forests possess diverse species and complex structures and perform myriad functions. Forests have rich natural and social values: as the modulator of nature, they modulate air and water circulation—affecting climate change—and they also protect soils from wind or rain invasion and alleviate environmental pollution hazards affecting humans. Hence, forests are a buffer that controls global warming. Regarding social values, forests

W. Zeng and F. Yang, *Poverty Alleviation Via Forest Carbon Sequestration*, International Research on Poverty Reduction, https://doi.org/10.1007/978-981-99-8191-5_2

can improve human living environments and provide necessary resources for life and production.

Forests are also the habitats of diverse animals and plants, harbouring the most active biological reproduction on earth and protecting biodiversity resources, making forests a natural species and gene bank (Li et al., 2017). Forests are the largest carbon pool of all terrestrial ecosystems. Thus, enlarging forest coverage areas is a low-cost and economically feasible way to address climate change over the next 30 to 50 years. This huge carbon-dioxide-collecting function of forests is called carbon sequestration.

Conversely, forest destruction, forest land deterioration, forest fires, plant diseases, and insect pests can release the stored carbon dioxide into the atmosphere, generating a carbon source. Effective measures must be taken to enhance the carbon-storing and carbon-sequestration functions of forest ecosystems and to decrease their carbon emissions. These efforts have become the global consensus to address climate change and is one of the goals set by China in its two commitments to voluntary emission reduction (Li & Su, 2018).

2.1.2 FCS

FCS, short for forest carbon sequestration and sometimes known as forest carbon sink in the natural sciences, is a process by which forest ecosystems absorb atmospheric carbon dioxide and fix it in vegetation and soils, decreasing atmospheric carbon dioxide concentrations (Lv, 2014). This concept originates from the definition of carbon sequestration given in *the Kyoto Protocol (1997),* which was signed by the contracting parties of *the FCCC*. The concepts of ocean carbon sequestration, grassland carbon sequestration, and forest carbon sequestration were gradually introduced in subsequent research and practice.

Forestry carbon sequestration is a concept coined by Chinese scholars. In the literature outside China, forestry carbon sequestration and forest carbon sequestration/sink (FCS) are regarded as the same concept so are both described as FCS. To differentiate that part of FCS used for carbon trading, Chinese researchers started to use the concept of forestry carbon sequestration. Currently, one clear definition of forestry carbon sequestration is that it refers to the processes, activities, or mechanisms by which atmospheric carbon dioxide is absorbed through AR, forest management, carbon sink trading, and the decrease of forest destruction. A primary component of FCS, forestry carbon sequestration has both natural and socioeconomic attributes. Since the scope of FCS is larger than that of forestry carbon sequestration, the former is widely accepted in international climate negotiations for its contributions to carbon sequestration AR, REDD, and IFM. Thus, FCS and forestry carbon sequestration are not significantly different in research and practice, but FCS is preferred.

2.1.3 FCS Projects

FCS projects which are intended for forest carbon credit transactions aim to absorb atmospheric carbon dioxide through AR, forest management, and forest conservation and to reduce or prevent the emissions of forest-stored carbon dioxide into the atmosphere. There are generally two categories, namely Kyoto rules FCS projects and potential *non-Kyoto rules* ones, depending on the operational legal basis, standards, and procedures (Li et al., 2006a, 2006b, 2007). Globally, the CDM acknowledges three types of carbon emission reduction, namely AR, REDD, and IFM. But the CDM trade system has not yet formally included REDD or REDD+projects actually, which are progressing very slowly. Thus, FCS projects in this book refer to carbon sequestration AR projects based on Kyoto or non-Kyoto rules.

The *Kyoto Protocol* defined strict standards and procedures, for example, regarding the land eligibility[1] of CDM-AR projects, plot boundary, date of inclusion, identification of baseline situations, additionality[2] check, confirmation, registration, and recording of leakage investigation, certified emission reduction (CER), and non-permanence.

In China, depending on the entities of exploitation and implementation, FCS projects are generally practised in two models, independent exploitation and cooperative exploitation, and three types, namely unilateral, bilateral, and multilateral. Depending on different registration, documenting, and certification standards, there are four classes of FCS projects: CDM projects; international verified carbon standard (VCS) projects; CCER projects; and other FCS projects, such as climate, community, and biodiversity standards (CCB), PS, and the China Green Carbon Sink Fund (CGCF) (He et al., 2016; Lv, 2014), as shown in Fig. 2.1.

The two key types of CCER projects in China are AR projects and the management of existing forests. Both types of projects are applied as per the exploitation procedures provided in the *Interim Measures for Management of Greenhouse Gas Voluntary Emission Reduction* (CDRC 2012). All projects must be examined and checked by six institutions approved and authorised by CDRC to authenticate voluntary greenhouse gas reduction projects, after which the CDRC documents and registers the voluntary greenhouse gas reduction of each project.

The implementation of these two types of projects is based on significantly different methodologies. First, forest management carbon sequestration projects refer

---

[1] Land eligibility is the basic premise of whether an FCS project can be registered as a CDM-AR project. One definite provision is that the plots must have been afforestation lands for over 50 years or AR lands, which is forestless, deteriorated, bare hills since 31 December 1989, when not all three strict requirements are met: closure degree ≥ 0.2, continuous area $\geq$ 1 mu, or standing tree height ≥ 2 m.

[2] Additionality has two basic meanings: first, the decrease of greenhouse gas emissions or increase in amount of carbon fixed is added to the decrease or increase that would occur if no project were implemented. That is, the natural carbon storage is counted as the baseline. Second, the emission-reducing behaviours or carbon sequestration behaviours are additional, including technology, fund, investment, environment, and policy additionality.

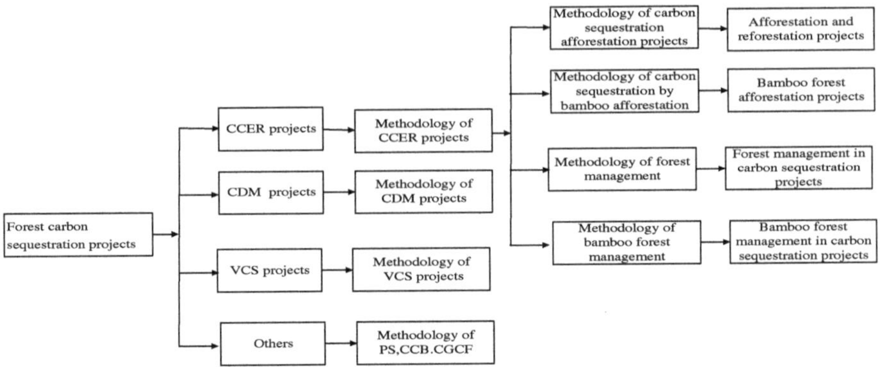

Fig. 2.1 Classification of FCS projects in China

to the operating activities conducted in grown forests whereas AR projects are principally aimed at afforestation. Second, they differ as regards the land eligibility provision. Forest management projects are mainly conducted on arbour lands that conform to national regulations; AR projects, in contrast, are carried out on lands that have been forestless since 31 December 1989 and exclude arbour forests. Third, AR projects are the first of the two types to be implemented, with pilot projects having been conducted in Sichuan, Yunnan, and Guangxi since 2003. Forest management projects were started in 2005 and are mostly still at the stage of planning and designing. Finally, in China, a total of 99 CCER projects were documented in CDRC by 2017, the majority being AR projects (68 projects, 68.69%).

In summary, given the prominent role of AR projects in China, all FCS projects mentioned in this research are AR projects based on *Kyoto or non-Kyoto rules* (Table 2.1).

2.1.4 Poverty

Poverty is a global, historical, dynamic, and regional concept that refers to the lack of social, material, and spiritual life. Poverty evolves continually alongside the transformation of time, space, and human thought and philosophy. Dynamic definitions of poverty have moved from simple to complex and from specific to generalised. The process of definitional change can be divided into many stages, including the proposal of the concept of relative poverty as compared to absolute poverty, the development of the concept of subjective poverty versus objective poverty, and the introduction of the concept of income poverty versus capacity poverty (Yang et al., 2012).

To date, there is no unified definition of poverty. Though many researchers have explicated poverty from different perspectives, critical development is lacking. From the regional and individual perspectives, poverty involves two aspects: regional

Table 2.1 Comparison between and AR projects and forest management projects

Comparison of contents	AR projects	Forest management projects
Methodology	Methodology of afforestation projects	Methodology of forest management projects
Content of projects	Afforestation activities mainly aimed to enhance carbon sequestration	Forest operating activities mainly aimed to enhance carbon sequestration
Land requirements	Forest-free lands since 31 December 1989	Arbour forest lands in accordance with national provisions
Main activities	Afforestation activities: determination of provenance, seedling raising, soil preparation, complementary planting, weeding; forest culture activities: fostering, intermediate logging, fertilisation	Forest management: complementary planting, complementary forestation, fostering and logging, rejuvenation, comprehensive measures

poverty and individual poverty. Individual poverty can usually be summarised as one of three types: income poverty, which refers to lacking necessities for daily life; capacity poverty, which entails being unable to obtain life necessities; and rights poverty, which refers to lacking political rights (Guo, 2005). Currently, one consensus is that poverty is not a singular problem related to low income but a multidimensional problem resulting from lack of human rights; low-risk resistance; frailty and ease of returning to poverty; and unevenness in housing, possessions, resources, abilities, and opportunities (Xian et al., 2016).

The poverty issues in the ethnic regions of Southwest China are also characterised by the two perspectives and three basic types mentioned above. Specifically, poverty is not limited to families with low incomes and expenditures but involves the structural poverty of the whole region. In addition to the specific concept that a deficit of living necessities leads to below-standard living conditions, poverty refers to the multidimensional state induced by the lagged development of the regional economy and cultural education.

2.1.5 *Poverty-Stricken Populations*

Poverty-stricken populations, sometimes known as the poor, refer to individuals or families living under the poverty-stricken status and are usually divided into populations in absolute poverty and those in relative poverty. Populations in absolute poverty comprise individuals or families who cannot meet their basic survival needs through their earnings and other legal forms of income. In most countries, populations in absolute poverty are defined as those whose living standards cannot reach the socially acceptable lowest standard, otherwise known as being below the poverty line. The absolute poverty line differs among countries and across periods. Populations in relative poverty encompass those individuals or families whose resources are able to meet their basic life needs but who cannot reach the average living standard in society, and these populations remain at a level far below the average living standard (Guo & Luo, 2005).

Generally, a certain proportion of the total community population is classified as relatively poverty-stricken people. According to the World Bank, social members whose incomes are only or less than one-third of the average income can be considered relatively poverty-stricken people. Some countries define a population with less than 40% of the average income as relatively poverty-stricken people.

In this research, the term poverty-stricken population mainly refers to those who live below the rural poverty standard of China. Specifically, the rural population in absolute poverty is defined as those whose family annual net income per capita is lower than 2300 RMB (about 333 USD) in constant price of 2010. In our empirical research, the poverty-stricken population (families) consists of those registered and filed in the Targeted Poverty Alleviation and Targeted Poverty Relief project in China during 2015–2020.

2.1.6 Community

The Chinese word *She Qu* was translated by Fei Xiaotong in the 1930s from the English word *community*, but no single global definition of the latter term has, as yet, been devised. The relatively authoritative concept put forward by the World Health Organisation (WHO) in 1974 is that a community comprises members of a social group in a certain geographical space who share common interests and mutually know and communicate. These members conduct social functions, create social norms, constitute special value systems, and undertake social welfare. Each member is integrated through families, close neighbours, and communities into a larger community. This concept continues to evolve alongside modern socioeconomic development, especially in regard to changes in community scale, structure, and functions. Nevertheless, a community, that consists of members who live in the same geographical region and have the same language and customs as well as similar values and common interests, is still basically defined as a social group with relatively independent economic or cultural identity systems.

A community, which is the simplest social form, can be categorised into many types according to political, natural, social, economic, or cultural factors (Liu, 2013). The community mentioned in FCS projects is a unique one. Some researchers consider a community to be a group of members involved in an FCS project (Arun et al., 2012), but others regard it as a group of members who have the right and opportunity to participate in a project. The communities mentioned in this book refer to the rural natural communities located within the outreach of carbon sequestration projects.

A community can acquire the largest benefit and promote sustainable project management through collective negotiation, collective actions, or participation in project management. Depending on the size of project areas, a community may be a certain natural geographical zone of one or more administrative villages. The community includes farmers living around the lands on which projects are being implemented who are stakeholders participating either directly or indirectly, temporarily or long-term, and are either beneficiaries or victims of the FCS project.

2.1.7 *Poverty Alleviation*

Poverty alleviation is the social task of supporting and helping poverty-stricken areas and poverty-stricken people to increase their income and move out of poverty. Poverty alleviation is aimed at changing the status of poverty-stricken families or areas (Fan, 2011). There are three aspects to its basic content and characteristics (Fan, 2011). First, it has both short-term and long-term plans as well as definite goals, concrete plans, steps, and measures. It integrates temporary solutions and permanent solutions, with a focus on the latter. Second, it helps poverty-stricken people to solve difficulties in life by increasing income and, more importantly, aligning poverty-stricken areas with modern development trends, thereby fundamentally eliminating poverty and becoming self-sufficient. Third, relevant governmental departments and social forces are mobilised to cooperate and effectively help with the development of poverty-stricken families and poverty-stricken areas.

Since the Reform and Opening-up in 1978, poverty alleviation in China has experienced four stages: pure poverty relief focusing on certain regions (1978–1985), development of poverty relief projects and industries focusing on poverty-stricken counties (1986–2000), comprehensive poverty alleviation focusing on poverty-stricken villages (2001–2010), and targeted poverty alleviation focusing on poverty-stricken families (2011–present). Therefore, rural poverty alleviation in China has switched from extensive poverty alleviation to targeted poverty alleviation (Lu & Huang, 2016).

2.1.8 *Poverty Alleviation Performance*

Usually measured by achievement, result, and benefit, performance is a combination of achievement, effect, and efficiency. Referring to the good or bad result of a task, performance reflects the efficiency with which an organisation, team, or individual achieves its goals under certain resource and external conditions (Zhuo, 2004). Measuring performance was first adopted in investment project management and then gradually applied to human resource management, enterprise management, and governmental activity management. From the management perspective, performance is an organisation's expected result and effective output at different levels to achieve its goals. It has two aspects: individual performance and organisational performance. In the field of economic management, performance is the result and achievement of economic activities (Zhuo, 2004).

Generally, poverty alleviation performance refers to the achievement, effect, result, and efficiency of poverty alleviation activities. In poverty alleviation, performance evaluation has been adopted in regional poverty alleviation, poverty alleviation via tourism, poverty alleviation projects, poverty alleviation funds, targeted poverty alleviation, and poverty alleviation via financing. Generally, there are objective and subjective, static and dynamic, macroscopic and microscopic, qualitative and quantitative, and one-dimensional and multidimensional performance evaluations. The performance of PAFCS in this book includes the objective performance of FCS projects at the macroscopic scale and the subjective effects perceived by community farmers at the microscopic scale.

2.1.9 Eco-compensation

Eco-compensation is generally considered as an important way to coordinate environmental protection and economic development through market mechanisms and economic incentives. So far, no unified definition of eco-compensation has emerged. The concept originates from ecology theories and refers in particular to the scope of natural eco-compensation. Since the 1990s, eco-compensation has been introduced into the field of socioeconomy and understood as an economic stimulus for environmental protection (Lai et al., 2008).

There are general and specific definitions of eco-compensation (Wang & Zhuang, 2006). General eco-compensation includes, first, the gains acquired from the protection of the ecosystem and natural resources and the penalties imposed for damages done to these and, second, the fees charged by those responsible for environmental pollution. Specific eco-compensation mainly refers to the gains and penalties from eco-compensation. Eco-compensation can be divided into longitudinal and horizontal compensation according to fields and into fund, material, policy, and intelligence compensation according to methods. It can be classified as government-dominated compensation (fiscal transfer payment, policy compensation, and eco-compensation funds) and market-dominated compensation (carbon sink trading, pollution right trading, water right trading, environment tax, eco-product, and eco-service logos). It can further be separated into four modes: global, interregional, regional, and project compensation (Lai et al., 2008; Zhang et al., 2012). In practice, the core goal of eco-compensation is to supply ecosystem services and relieve poverty through economic incentives. Hence, eco-compensation has been implemented worldwide (Xu et al., 2015).

The FCS discussed in this book is among the most important ways to achieve marketised compensation and highlight the benefits and values of forest ecology. The researchers focus on the socioeconomic benefits of FCS eco-compensation, the theoretical and practical progress of PAFCS in pro-poor perspective, and management optimisation and institutional arrangements of PAFCS.

2.2 Literature Review

FCS is an important research topic of global climate change (Zeng et al. 2017). Traced back to the 1990s, discussion of FCS and anti-poverty is rooted in projects with the double goals of poverty alleviation and emission reduction. FCS, as one of the three emission reduction mechanisms of the CDM frame in the Kyoto *Protocol*, has as its primary philosophy that developed countries, while meeting their commitments to reduce and limit emissions through carbon sequestration projects, must provide additional funds and advanced techniques and actively promote sustainable development in developing countries (Montagnini & Nair, 2004; Nair et al., 2009; Noble & Scholes, 2001). Some researchers stress that poverty elimination is one of the core contents of sustainable development. CDM FCS projects offer a brand-new model for developed countries to help developing countries, where alleviating poverty is highly expected (Scurlock & Hall, 1998).

In the twenty-first century, *the Bonn Political Agreement (2001)* and *Marrakesh Agreement* (2001) both appointed AR carbon sequestration projects as the only qualified type of CDM carbon sequestration project in the *first commitment phase*. The poverty alleviating function and utility of carbon sequestration by AR projects are universally valued by academics. The mainstream view holds that developing FCS projects is a win–win choice that helps developed countries achieve emission reduction tasks at low cost and promotes sustainable development in developing countries, thus combating both climate change and poverty (Antle & Stoorvogel, 2009; Chen & Song, 2006; Gong & Li, 2006; Perez et al., 2007). In the post-Kyoto era, the effects of FCS have become increasingly evident (Huang, 2008). As the number of carbon sequestration AR pilot projects rises, academics have further associated FCS directly with the development of poverty-stricken areas, the sustainable livelihood of local farmers, and the participation and benefit-sharing of poverty-stricken populations. A new concept, namely PAFCS, was put forward in China (Ding & Ma, 2013; Huang, 2008; Ma, 2012).

2.2.1 Development of FCS Industry and Poverty Alleviation

FCS is a result of the innovation of multi-level (international, regional, national, and local) eco-compensation systems under the market mechanism. The formation and sustainable industrial development of the FCS market is undoubtedly the premise on which poverty alleviation can be achieved. Researchers generally believe that the standpoint and concrete measures for climate change still differ among countries and that the *Paris Agreement,* signed in 2016, is not perfect. Nevertheless, FCS is more economical and efficient than other emission reduction methods, and it has gradually substituted other methods to become the main means employed to reduce carbon dioxide emissions. As the econometric techniques of carbon sequestration and market trading rules are increasingly deployed, the industrial development of FCS has given rise to huge commercial opportunities, showing the enormous potential of poverty alleviation (Chen et al., 2011; Hong, 2011; Qi et al., 2014).

Cacho et al. (2005) analysed the relationship between CDM FCS and poverty alleviation and stated that new demands for carbon emission quotas should be created in response to global warming. The exchangeable carbon-storing products do not have to be transported and have no quality difference, thus opening a new market for small farmers in remote areas and bringing a new opportunity to solve poverty problems. It has been estimated that trades in the FCS market will help between 2.5 and 5 million low-income forest farmers in developing countries to eliminate poverty by 2030 (Milder et al., 2010).

The *Annual Reports of Forest Carbon Market Situations (2011–2021)* demonstrate that FCS trading in the global carbon market is on the rise. According to the latest reports from Ecosystem Marketplace, voluntary carbon markets funds in 2021 were on track to achieve an annual market value record of 1 billion+USD for the first time as the all-time market value hit 6.7 billion USD. The trading volume and turnover of the non-Kyoto FCS project market are far larger than those of the Kyoto FCS project market. Since the voluntary market is non-profit, its healthy development is good news and a positive sign for poverty-stricken areas. In China, the initiation of the national carbon emission right trading market, the *PS involuntary* emission reduction, and several forest management and carbon sequestration trading policies from Chinese local government undoubtedly enhance the feasibility of PAFCS.

2.2.2 FCS Project Design and Poverty Alleviation

FCS projects must be designed to highlight poverty alleviation and explore the potential of PAFCS, and their pro-poor performance must be evaluated. After considering the multiple goals, functions, and benefits of a project, as well as its practical feasibility, some researchers focus on the importance of designing to achieve poverty alleviation. A well-designed FCS project can address climate change by reducing greenhouse gas emissions and improve the living standard of farmers by legitimating their control and management of forest resources (Smith & Scherr, 2003). Developers of FCS projects focus on bringing together funds and buyers, improving projects' market competitiveness, and incorporating poverty alleviation into project design.

Li et al. (2006) investigated FCS projects' triple functions, including adapting to climate change, mitigating the climate change, and promoting sustainable development and believed that one or two of these must be ignored if the benefits of a project are to be maximised. Hence, some issues must be considered at design stage, including how the project is arranged overall, trade-offs among multiple benefits, the synergies between different functions, alleviating rural poverty, and guaranteeing community residents' benefits. Huang et al. (2013) analysed the feasibility of designing an FCS project and stated that poverty alleviation and regional economic growth are the social benefits it confers. Whether a project can be successfully implemented largely depends on the cooperation and support of local communities. Hence, local economic development, resident living standard improvement, and employment increment must be included in the project design.

According to Huang et al. (2015), alleviating global warming is the primary goal of FCS projects, and the tree species and land selection and use of labour in afforestation should meet local conditions. Moreover, the technical requirements of FCS projects must be respected, and the bonus brought by project exploitation must be shared by local farmers or rural collective organisations. Case studies show that if the protection of the basic rights of the poverty-stricken people is properly considered at project design stage, farmers' living standards and local sustainability will eventually improve (Asquith et al., 2002). The design of CDM FCS projects must welcome local residents' participation, expand project scale, and optimise the compensation mechanism and biodiversity protection, which, taken together, will decrease transaction costs and project implementation risks (Smith & Scherr, 2003).

2.2.3 *Implementation and the Poverty Alleviation Effect of FCS Projects*

FCS projects are key to the exploration of PAFCS' potential, which, in turn, is beneficial to poverty alleviation. Carbon sequestration AR projects have extensive poverty alleviation effects. Currently, relevant research focuses on two levels: poverty-stricken areas and poverty-stricken farmers. The research at the first level is based on regional ecology, economy, and social development ability, while at the second level it focuses on sustainable livelihood.

2.2.3.1 Effects on Poverty-Stricken Areas

At the macro level, growing evidence proves that less developed areas have more lands suitable for afforestation, lower opportunity costs of forest land management, and lower labour force costs, all of which contribute to decreasing exploitation costs and increasing investment returns. In China, FCS projects are mainly located in remote poverty-stricken regions (Cheng, 2015). These projects not only bring economic benefits to local regions, but also promote regional sustainable development. And FCS not only provides a new way for poverty-stricken areas to participate in efforts to address climate change but also makes poverty-stricken areas less vulnerable to hill landslip, debris flow, flooding, drought, and other natural disasters caused by aggravated climate change (Xu & Ju, 2009). Carbon neutralisation technology and the carbon sequestration mechanism can be applied to the natural reserves in poverty-stricken areas to increase fiscal revenue, increase resident incomes, and accelerate poverty alleviation (Liu et al., 2011).

Reportedly, FCS projects implemented in Ethiopia moderately improved communities' managing and coordinating abilities, enabling them to better undertake poverty relief works (Bluffstone et al., 2016). Surveys show that some projects state in contracts that afforestation companies must assist local areas in building roads or schools, which objectively improves local traffic conditions and education levels and reinforces market exchanges with the external world (Byron & Arnold, 1999). Implementation of PAFCS projects in Yuexi County, Sichuan Province, not only brought economic income, employment opportunities, and new technology to remote poverty-stricken areas but also offered external resources to break down the resource trap, mobilised internal resources, and attracted policy resources (Ma et al., 2012).

The key to improving the self-developing ability of poverty-stricken areas is encouraging local farmers to participate in the projects, diverting their attention to the development of poverty-stricken communities, and enhancing their initiatives (Ma et al., 2012). Empirical research suggests that the construction of PAFCS models provides poverty-stricken areas with optional means for economic development and offers a role model for the design of participatory poverty alleviation (Ding & Ma, 2013). In poverty-stricken areas with less-developed infrastructure, infrastructure construction is more effective than compensation funds in helping farmers to benefit from projects. Liu et al. (2011) probed into the poverty alleviation projects in rocky desert areas of the Karst region in Guizhou and found the implementation of FCS projects not only contributed to the governance of rocky desertification but also achieved a balance among environment, economy, society, and sustainable development. Wei (2015) reported similar conclusions in rocky desertification areas of Yunnan province. Carbon finance was interrelated with rural finance as poverty alleviation projects provided poverty-stricken areas with carbon finance to develop their regional economies (Ji & Wang, 2012; Zhang et al., 2011; Zou, 2011). Empirical research of reforestation projects in the Pearl River basins of Guangxi suggests that FCS trading innovatively solves financial exclusion issues in Chinese rural areas and effectively guides capital to rural poverty-stricken areas. Financial institutions also support carbon sink trading, thus improving the finance ecology in such areas (Lan et al., 2013).

Nevertheless, FCS projects also come with risks, and researchers have made different suggestions in this regard. Some researchers hold that project exploitation will cause environmental, economic, and social risks due to multiple factors, including natural disasters and human activities (Chia et al., 2013; Galik & Jackson, 2009; Wu & Zhang, 2007; Xiang et al., 2009). Some projects have also created conflicts with indigenous knowledge and traditional cultures (Walker et al., 2014). Hence, the benefits of FCS projects, in some cases, compete with other poverty-alleviating and farmer-benefiting ones.

Lin and Li (2006) believe that the effects of CDM-AR on the regional environment and society are highly uncertain. Such projects may cause diseases and insect pests, depletion of topsoil nutrients, intensified soil sheet erosion, loss of surface water, loss of native tree species, a decrease in biodiversity, and other environmental risks. FCS projects may also induce other risks; for example, local unique natural landscapes may be changed and harmed, and communities and farmers may lose control of forest products and conflict with external investors. Local governments often transfer lands at low prices and provide other preferential policies, which will cause poverty-stricken areas to lose their natural resources and damage their interests in other ways. A comparative analysis by Li et al. (2015) indicates that since site selection is affected by many natural factors and lands must be managed on a large scale, individual poverty rights may be ignored, which will enlarge the gap between the rich and the poor in the region concerned.

2.2.3.2 Effects on Farmers' Livelihoods

At the micro level, the literature regarding the effects of projects on local farmers' livelihoods covers the areas of economic incomes, feasibility, and development opportunities for poverty-stricken populations.

Regarding economic incomes, project sites are often in areas with a concentrated poverty-stricken population. By participating in FCS projects, farmers can gain a certain income from their work and compensation for the loss in grazing (Estrada & Corbera, 2012; Hejnowicz et al., 2015; Pagiola, 2008; Wu et al., 2015). In the meantime, forest farmers can also gain income through sharing, renting, and transferring forest lands. This transition in the use and management of land can help pull poverty-stricken land users out of poverty (Jack et al., 2008; Richards & Stokes, 2004).

In addition to derived incomes from afforestation management and protection and from sales of forest products, another important source of income comes from CER trade (Bulte et al., 2008; Lu et al., 2013). An appropriate selection of fast-growing tree species with high economic values can bring additional income to farmers (Cacho et al., 2005; Jiao & Ai, 2011). Chen (2010) investigated the Pearl River basin reforestation projects in Guangxi and discussed their effects on farmers' livelihoods, finding that farmers' incomes come from five sources: renting lands, wood sales, CER sales, participating afforestation, and afforestation management. Land collective ownership ensures that all farmers can benefit from projects in cooperation with the afforestation entity. Hence, FCS projects are commercial projects whose operation is based on ecological compensation rather than simple projects to alleviate poverty (Chen, 2010).

The non-income effects of projects on farmers, such as feasibility improvement and development opportunity creation, are also extensively covered. Jindal et al. (2008) found that the effect of FCS projects on farmers' living ability in Mozambique surpassed that of direct incomes. First, forest farmers acquire new afforestation and forest management skills directly through training (Grieg-Gran et al., 2005; Parajuli et al., 2015). Farmers' practical skills and industrial experience in seedling selection, culturing, fostering, and forest disease and comprehensive governance of insect pests can be improved by learning while working (Nuthall, 2006; Roberts et al., 2012). Second, farmers can increase their incomes by transferring their acquired abilities to other occupations, enterprises, and economic activities (Felman, 2015; Mchenry, 2009). For instance, forest farmers can apply the management skills acquired from a project to running ecological tourism businesses (Brandth & Haugen, 2011; Konu et al., 2015). Third, farmers' social networks, social capital, and market knowledge expand (Ferraro et al., 2015; Grabowski & Chazdon, 2012). In northwest Sichuan, China, CDM-AR projects on deteriorated lands were supplemented with edible fungus breeding and other projects. The book *Dried Mushroom Bag-Planting Activity Experience* was also distributed to farmers, and experts and professionals were invited to give skills training and guidance. Consequently, the projects successfully secured support from community residents by creating new development opportunities for them (Lv, 2014).

Projects' negative effects on farmers have also been widely covered in the literature. For instance, to avoid carbon leakage, activities including forest planting, wood cutting, and grazing in project areas and peripheral areas are often forbidden (Chen, 2010; Jindal et al., 2008). The income from Chinese medicinal herbs and mushroom digging in mountainous areas has decreased over certain periods (Li et al., 2015). Cost income analysis shows that forest farmers' incomes from FCS projects are, in some areas, lower than opportunity costs. For example, a local forest farmer in Tanzania can acquire 0.02 USD per tree from carbon sequestration afforestation projects, which is far lower than the income from activities such as breeding and planting, thus making farmers less willing to participate in projects (Stringer, 2012).

2.2.4 Challenges in Exploiting the Potential of PAFCS

There are many challenges to exploiting the potential of PAFCS. The first is that poverty-stricken areas can have difficulty directly participating in FCS projects. The FCS enhanced mechanism, which uses forest protection and forest sustainable management, creates an opportunity for local people in less developed areas to directly participate in FCS trading. There have already been successful cases of farmers' participation in forest management carbon sequestration trading in China. However, many uncertainty risks as regards skills, market, policies, and disagreements still exist, as there is globally a controversy about the REDD+mechanism regarding its political, policy-related, and technical aspects, particularly about how basic skills can be monitored, reported, and verified (MRV) (Ma & Xu, 2015; Sheng & Wu, 2012; Xue et al., 2012).

Carbon sequestration AR projects generally face challenges; for example, they involve large amounts of land, a long project period, stricter standards, higher risks, and more investments than ordinary afforestation projects. The financial benefits of CDM projects are only apparent after the project's end, thus those projects have strict standards for land feasibility, carbon absorption, and carbon sequestration behaviours. They also involve complex project preparation and registration procedures and high technicality. Consequently, small and micro enterprises and farmer households have less opportunity to participate in these projects. The majority of CDM projects in developing countries are limited to large enterprises. In China, only big companies can organise and implement CDM projects (Li & Cheng, 2009; Seres, 2013; Thomas et al., 2010; Wang & Cao, 2015).

The second challenge is the participation barrier faced by poverty-stricken populations. Farmer participation is reportedly key to the success of FCS projects (Huang & Chen, 2015), but many factors, including transaction costs, forest land property rights systems, and resource endowment, limit the valid participation of poverty-stricken populations. Smith and Scherr (2003) hold that developing countries should enjoy a sustainable livelihood with CDM projects under the provisions of *the Kyoto Protocol*. However, it is difficult to attain the social and economic benefits that simultaneously focus on poverty relief as projects with high social benefits are often accompanied by high transaction costs. FCS projects in developing countries are allowed to plant only certain tree species, which has resulted in fewer economic benefits from carbon sequestration forests, and these projects are thus not very attractive to low-income farmers (Stringer et al., 2012; Sunderlin et al., 2004).

Reportedly, FCS projects implemented in Africa improved residents' incomes and natural resources, but the instability of land ownership is a barrier to investment, and the transaction costs are high (Jindal et al., 2018). Moreover, one-to-one negotiation between multiple project entities raises trading costs; therefore, to reduce project start, execution, and monitoring costs, project organisers tend to select landowners with larger lands for trading negotiation and exclude those with smaller lands or land-free forest farmers (Benessaiah, 2012; Wu & Zhao, 2013). Project implementation is often accompanied by unequal political gaming, with women, small landowners, the poverty-stricken, and indigenous people usually excluded (Cacho et al., 2005; Zhu et al., 2015) or seeing their legal benefits embezzled, making them less willing to participate. In addition, farmers' income levels have been empirically proven to significantly affect their willingness to engage in FCS (Zhu et al., 2015). High-income farmers, who have a higher risk tolerance, are better placed to take risks and more easily change the traditional forest management model so are thus actively engaged in and benefit from carbon sequestration exploitation (Zhu et al., 2015).

2.2.5 Literature Review Summary

In summary, FCS already plays a critical role in combating poverty, and poverty alleviation has gradually become an important part of FCS projects in developing countries. There is an increasing amount of research on FCS directly associated with poverty alleviation. Researchers in this area have consolidated knowledge of the functions, potentials, basic paths, and challenges of PAFCS and provided a certain theoretical basis and references for further study. However, most studies focus on the sustainability of FCS, establishment, and operation of the carbon market, industrial development, and FCS project development, and there is less research on FCS from the perspective of poverty alleviation. Hence, research on PAFCS is still a relatively new topic. Moreover, the concepts, elements, principles, and essential features of PAFCS are not clearly defined. There is more theoretical than empirical research and more qualitative analysis than quantitative research. No quantitative analysis or theoretical research has been conducted on the pro-poor effects of different projects or the same project implemented in different areas.

At least two basic scientific problems must be solved to further unearth the poverty-relieving potential of FCS and undertake in-depth study of PAFCS. The first is how to create benefits and develop opportunities for poverty-stricken populations. Many studies focus on poverty-stricken areas and locals' livelihood but pay far less or no attention to poverty-stricken populations. Although there is a considerable overlap between poverty-stricken areas and poverty-stricken populations, not all those living in such areas are necessarily poverty-stricken people. FCS exploitation is dominated by the market mechanism and certainly focuses on efficiency rather than equity and hardly at all on poverty alleviation, that is, benefiting poverty-stricken populations and creating development opportunities for the poor. If PAFCS is simply reduced to developing FCS projects, the goals of poverty alleviation will be missed or deviated; however, doing so may lead to a practical problem of costs surpassing benefits, which goes against the original intention of PAFCS.

Furthermore, the conflict between poverty alleviation and project marketisation must be acknowledged. Further efforts are needed to highlight the goal of poverty alleviation during project planning and design, organisation and operation, management, and monitoring and evaluation. Governments must offer support and intervene where necessary to compensate for market failure. Further research is expected to design a pro-poor FCS system that targets poverty-stricken populations, designing poverty alleviation routes, and improving pro-poor effect evaluation. Arguably, pro-poor FCS should be a core issue of PAFCS research.

The second basic problem is how to balance FCS and poverty alleviation. Interactions and conflicts between these two have already been studied, and research shows that the relationship between forest ecosystem supply and poverty alleviation should be balanced with carbon sequestration at its core (Lai et al., 2008; Wang & Zhong, 2009; Xu et al., 2015). Unilateral stress on poverty alleviation will decrease the effectiveness of ecological services, going against the original intention of FCS. Hence, further research is needed to envisage the balance between poverty alleviation goals and FCS supply and to study how to build a win–win practice model, sustainable paths, and a monitoring evaluation index system to address climate change and poverty alleviation.

Other promising research areas include the participation of project community farmers and relevant key influencing factors, elite drive and elite capture, pro-poor evaluation index and method selection, macroscopic stratagem and policy system, and case studies and quantitative analyses of different types (rules) of FCS projects and the same project implemented in different areas. All these studies will contribute to the design of theoretical innovation in poverty alleviation and poverty alleviation tools and the path selection and institutional arrangements of eco-compensation.

References

Antle, J. M., & Stootvogel, J. J. (2009). Payment for ecosystem, services, poverty and sustainability: The case of agricultural soil carbon sequestration. *Payment for Environmental Service in Agricultural Landscapes, 1*, 133–161.

Asquith, N. M., Rios, M. T. V., & Smith, J. (2002). Can forest-protection carbon projects improve rural livelihoods? Analysis of the Noel Kempff Mercado climate action project, Bolivia. *Mitigation and Adaptation Strategies for Global Change, 7*(4), 323–337.

Benessaiah, K. (2012). Carbon and livelihoods in post-Kyoto: assessing voluntary carbon markets. *Ecological Economics, 77*, 1–6.

Bluffstone, R., Mekonnen, A., & Beyene, A. (2016). Community forests, carbon sequestration and REDD+: Evidence from Ethiopia. *Environment & Development Economics, 21*(2), 249–272.

Brandth, B., & Haugen, M. S. (2011). Farm diversification into tourism–Implications for social identity? *Journal of Rural Studies, 27*(1), 35–44.

Bulte, E. H., Lipper, L., Stringer, R., et al. (2008). Payments for ecosystem services and poverty reduction: Concepts, issues, and empirical perspectives. *Environment and Development Economics, 13*(3), 245–254.

Byron, N., & Arnold, M. (1999). What futures for the people of the tropical forests? *World Development, 27*(5), 789–805.

Cacho, O. J., Marshall, G. R., & Milne, M. (2005). Transaction and abatement costs of carbon-sink projects in developing countries. *Environment and Development Economics, 10*(5), 597–614.

Chen, C. C., Mccarl, B., Chang, C. C., et al. (2011). Evaluation the potential economic impacts of Taiwanese biomass energy production. *Biomass & Bioenergy, 35*(5), 1693–1701.

Chen, J. H., & Song, W. M. (2006). Evaluation index system of CDM forestry carbon sequestration projects in China. *Journal of Northeastern Forestry University, 34*(1), 87–88.

Chen, C. Y. (2010). Forestry carbon sequestration and farmers' livelihoods: A case study of the first global forestry carbon sequestration project. *World Forestry Research, 23*(5), 15–19.

Chen, J. I. (2015). Barriers and legal countermeasures to forestry carbon sequestration in China. *Journal of Northwest Agriculture and Forestry University, 5*, 154–160.

Chia, E. L., Somorin, O. A., Sonwa, D. J., et al. (2013). Local vulnerability, forest communities and forest- carbon conservation: Case of southern Cameroon. *International Journal of Biodiversity & Conservation, 5*, 498–507.

Ding, Y., & Ma, P. P. (2013). Research on forest carbon Sink and economic development of minority areas in West Sichuan Province: Cases from Yuexi County, Liangshan Yi Autonomous prefecture, Sichuan Province. *Rural Economy, 5*, 38–41.

Estrada, M., & Corbera, E. (2012). The potential of carbon offsetting projects in the forestry sector for poverty reduction in developing countries. In *Integrating Ecology and Poverty Reduction*. Springer.

Fan, X. J. (2011). Review and prospects of poverty alleviation and development in China. *LaoQu Construction, 21*, 10–12.

Felman, S. (2015). Environmental degradation and poverty in India. *International Review* (Spring): *45*.

Ferraro, P. J., Hanauer, M. M., Miteva, D. A., et al. (2015). Estimating the impacts of conservation on ecosystem services and poverty by integrating modeling and evaluation. *Proceedings of the National Academy of Sciences, 112*(24), 7420–7425.

Galik, C. S., & Jackson, R. B. (2009). Risks to forest carbon offset projects in a changing climate. *Forest Ecology & Management, 257*(11), 2209–2216.

Gong, Y. Z., & Li, N. Y. (2006). Demand analysis and design ideas of forestry carbon sequestration projects in China. *Forestry Economics, 6*, 36–38.

Guo, X. B., & Luo, Z. (2005). On the evolution of the concept of poverty. *Jiangxi Social Science, 11*, 38–43.

Guo, X. B. (2005). On the connotation of the concept of poverty. *Shandong Social Science, 12*, 49–54.

Grabowski, Z. J., Chazdon, R. L. (2012). Beyond carbon: Redefining forests and people in the global ecosystem services market. In *SAPIEN. S. surveys and perspectives integrating environment and society*, (1st ed.)

Grieg-Gran, M., Porras, I., & Wunder, S. (2005). How can market mechanisms for forest environmental services help the poor? Preliminary lessons from Latin America. *World Development, 33*(9), 1511–1527.

Hejnowicz, A. P., Kennedy, H., Huxham, M. R., et al. (2015). Harnessing the climate mitigation, conservation and poverty alleviation potential of sea grasses: Prospects for developing blue carbon initiatives and payment for ecosystem service programmes. *Frontiers in Marine Science, 2*, 32.

He, Y., Chen, X. T., & Su, D. (2016). *Forestry carbon sink knowledge reader*. China Forestry Press.

Hong, M. (2011). A preliminary exploration of forest carbon sequestration industrialization. *Ecological Economy, 1*, 113–115.

Huang, Y. L., Qin, H. Y., & Huang, P. (2013). Feasibility design of forestry carbon sequestration project development. *Resource Development and Market, 8*, 806–808.

Huang, Z. S., & Chen, Q. (2015). Analysis of countermeasures for carbon sequestration forestry development based on carbon sequestration perspective. *Forestry Economics, 11*, 86–89.

Huang, D. (2008). Forest carbon sequestration: an important way to reduce emissions in the post-Kyoto era. *Forestry Economics, 10*, 12–15.

Jack, B. K., Kousky, C., & Sims, K. R. E. (2008). Designing payments for ecosystem services: Lessons from previous experience with incentive-based mechanisms. *Proceedings of the National Academy of Sciences, 105*(28), 9465–9470.

Jindal, R., Swallow, B., & Kerr, J. (2008). Forestry-based carbon sequestration projects in Africa: Potential benefits and challenges. *Natural Resources Forum. Blackwell Publishing Ltd, 32*(2), 116–130.

Ji, X., & Wang, X. L. (2012). Carbon Finance Innovation and "Low Carbon Poverty Alleviation". *Agricultural Economic Issues, 1*, 79–87.

Jiao, S. L., & Ai, Q. S. (2011). Analysis of the economic benefits of carbon sequestration in the retreating forest project in Qianzhong Karst region. *Ecological Economy, 10*, 69–72.

Konu, H. (2015). Developing a forest-based wellbeing tourism product together with customers–An ethnographic approach. *Tourism Management, 49*, 1–16.

Lai, L., Huang, X. J., & Liu, W. L. (2008). Research progress on ecological compensation theory and method. *Journal of Ecology, 28*(6), 2870–2877.

Lan, H., Zhu, Y., & Mu, Q. S. (2013). On the innovative model of resolving rural financial exclusion—an empirical analysis of forestry carbon sequestration trading to guide capital flow back to rural areas. *Economic Theory and Economic Management*, *4*, 43–50.

Li, J. H., Ming, H., & Yu, W. Y. (2015). A comparative analysis of forestry carbon squestration project implementation in Sichuan Province. *Journal of Sichuan Agricultural University, 3*, 332–337.

Li, J. Q., Niu, S. K., & Liu, Y. H. (2017). *Forest ecology* (3rd ed.) Higher Education Press.

Li, N. Y., Gong, Y. Z., & Zhang, S. D. (2006a). Triple function analysis of forestry carbon sequestration projects. *World Forestry Research, 19*(3), 1–5.

Li, N. Y., Gong, Y. Z., & Zhang, S. D. (2006b). The Triple function analysis of forestry carbon sequestration project. *World Forestry Research, 19*(3), 1–5.

Li, N. Y. (2007) *Forest carbon sequestration in China*. China Forestry Press.

Li, N. Y. (2016). China forestry policy and action on climate change. *Chinese Journal of Social Sciences, 4*.

Li, N. Y., & Su, D. (2018). Development trend and prospect of forestry carbon sequestration in China. *China Environment, 9*, 28–31.

Li, X., & Cheng H. Q. (2009). Research on forest carbon sequestration trading based on transaction cost theory. *Forestry Economic Issues, 3*, 269–273.

Liu, S. X. (2013). *Community psychology*. Kaiming Publishing House.

Lin, D. R., & Li, Z. Y. (2006). Policy options for CDM afforestation reforestation carbon sequestration projects in China. *World Forestry Research, 19*(4), 52–56.
Liu, N., Sun, M., Gao, X. D., et al. (2011). Research exploration on low carbon economic poverty alleviation model in China's nature reserves. *China Collective Economy, 1*, 38–39.
Liu, S. Y., Zhang, X. J., et al. (2011). Exploring the ecological path of poverty alleviation and development in stone desertification areas based on CDM forestry carbon sequestration. *Business Times, 23*(1), 142–143.
Lu, J., Zhang, Y., & Li, N. Y. (2013). International experience that can be learned from forestry carbon sequestration trading. *China Population—Resources and Environment, 12*, 22–27.
Lu, H. W., & Huang, C. W. (2016). *China's targeted poverty alleviation development report (2016): Targeted poverty alleviation strategy and policy system*. Social Science Literature Press.
Lv, Z. (2014). *Forest carbon sequestration practices and low carbon development in China*. Peking University Press.
Ma, P. P. (2012). *Research on forest carbon sequestration and development of poverty-stricken minority areas in west Sichuan—a case study of poverty alleviation through carbon sequestration in yuexi, liangshan*. A Master's Thesis from Sichuan Academy of Social Sciences
Ma, T., & Xu, Y. D. (2015). Hot spots and controversies in the development practice of REDD+ mechanism. *World Agriculture, 5*, 60–64.
Mchenry, M. P. (2009). Agricultural bio-char production, renewable energy generation and farm carbon sequestration in western Australia: Certainty, uncertainty and risk. *Agriculture Ecosystems & Environment, 129*(s1–3), 1–7.
Milder, J. C., Scherr, S. J., & Bracer, C. (2010). Trends and future potential of payment for ecosystem services to alleviate rural poverty in developing countries. *Ecology and Society, 15*(2), 4.
Montagnini, F., & Nair, P. K. R. (2004). Carbon sequestration: An underexploited environmental benefit of agroforestry systems. *Agroforestry Systems, 61*(1–3), 281–295.
Nair, P. K. R., Kumar, B. M., & Nair, V. D. (2009). Agroforestry as a strategy for carbon sequestration. *Journal of Plant Nutrition and Soil Science, 1*, 10–23.
Noble, I., & Scholes, R. J. (2001). Sinks and the Kyoto protocol. *Climate Policy, 1*, 5–25.
Nuthall, P. L. (2006). Determining the important management skill competencies: The case of family farm business in New Zealand. *Agricultural Systems, 88*(2), 429–450.
Pagiola, S. (2008). Payments for environmental services in Costa Rica. *Ecological Economics, 65*(4), 712–724.
Parajuli, R., Lamichhane, D., & Joshi, O. (2015). Does Nepal's community forestry programme improve the rural household economy? A cost–benefit analysis of community forestry user groups in Kaski and Syangja districts of Nepal. *Journal of Forest Research, 20*(6), 1–9.
Paratihast, A., Herold, M., et al. (2012). Mobile devices for community-based REDD + monitoring: A case study for central Vietnam. *Sensors, 13*(1), 21–38.
Perez, C., Roncoli, C., Neely, C., et al. (2007). Can carbon sequestration markets benefit low-income producers in semi-arid Africa? Potentials and challenges. *Agricultural Systems, 94*(1), 2–12.
Qi, Y. B., Zhang, Y.,& Jia, Y. (2014). Study on the operation mechanism of pilot forest carbon sequestration trading in China. *Agricultural Economic Issues, 4*, 73–79.
Richards, K. R., & Stokes, C. (2004). A review of forest carbon sequestration cost studies: A dozen years of research. *Climatic Change, 63*(1–2), 1–48.
Roberts, D., Boon, R., Diederichs, N., et al. (2012). Exploring ecosystem-based adaptation in Durban, South Africa: 'Learning-by-doing' at the local government coal face. *Environment and Urbanization, 24*(1), 167–195.
Sheng, J. C., & Wu, Y. (2012). A comparative study of forest emission reduction policies in five developing countries: A policy assessment methodology based on the structural variable "REDD + Mechanism." *Chinese Soft Science, 9*, 175–183.
Smith, J., & Scherr, S. J. (2003). Capturing the value of forest carbon for local livelihoods. *World Development, 31*(12), 2143–2160.
Scurlock, J. M. O., & Hall, D. O. (1998). The global carbon sink: A grassland perspective. *Global Change Biology, 4*(2), 229–233.

Seres, S. (2013). Hats off to the Kyoto protocol and the CDM: A giant success story. *Carbon Management, 4*(1), 23–25.

Stringer, L. C., Dougill, A. J., Thomas, A. D., et al. (2012). Challenges and opportunities in linking carbon sequestration, livelihoods and ecosystem service provision in drylands. *Environmental Science & Policy, 11*(5), 121–135.

Sunderlin, W. D., Angelsen, A., & Wunder, S. (2004). Forests and poverty alleviation. *State of the World's Forests, 4*, 61–73.

Thomas, S., Dargusch, P., Harrison, S., et al. (2010). Why are there so few afforestation and reforestation CDM mechanism projects?". *Land Use Policy, 27*(3), 880–887.

Walker, W., Baccini, A., Schwartzman, S., et al. (2014). Forest carbon in Amazonia: The unrecognized contribution of indigenous territories and protected natural areas. *Carbon Management, 5*, 1–26.

Wang, Q., & Cao, Y. K. (2015). A review of foreign research on incentive mechanisms for forestry carbon sequestration projects. *World Forestry Research, 10*, 10–14.

Wang, J. N., & Zhuang, G. T. (2006). *Ecological compensation mechanism and policy design*. China Environmental Science Press.

Wang, L. A., & Zhong, F. L. (2009). Research progress on the relationship between ecological compensation and poverty alleviation. *Forestry Economic Issues, 29*(3), 201–205.

Wei, X. F. (2015). Exploring the ecological path of poverty alleviation and development in stone desertification areas of Yunnan Province based on CDM forestry carbon sequestration. *Biotechnology World, 4*, 17.

Wu, G. C., & Zhao, B. B. (2013). Market development of forest environmental services in Latin America and the impact on farmers' livelihoods. *China Forestry Economy, 1*, 30–32.

Wu, S. H., & Zhang, X. Q. (2007). Non-persistent risk analysis of CDM forestry carbon sequestration projects. *Forestry Science, 8*, 123–126.

Wu, G. Z., Wang, P., Zhou, N. D., et al. (2015). Analysis of social benefits of forest carbon sequestration industry development in Hunan Province. *Forestry Economics, 11*, 90–93.

Xiang, Z., Wu, X. P., et al. (2009) Analysis of the current situation and prospects of forest carbon sequestration emission reduction projects. *Environmental Pollution and Prevention, 2*, 94–96.

Xian, Z. D., Wang, P. P., & Wu, W. (2016). Poverty criteria and poverty monitoring in rural China. *Statistical Research, 33*(9), 3–12.

Xu, J. Y., Liu, X. X., Feng, L., et al. (2015). Advances in research on ecological compensation trade-off relationships. *Journal of Ecology, 35*(20), 6901–6907.

Xue, M., Wu, S. H., & Cheng, S. Q. (2012). Measurement, reporting and verification system of REDD+ actions in China. *Forestry Science, 3*, 128–131.

Xu, Y. L., & Ju, H. (2009). Climate change and poverty: A case study of China. *World Environment, 4*, 50–53.

Yang, G. T., Zhou, H. J., & Li, Y. X. (2012). Review of the connotation, evolution and development of the concept of poverty. *Journal of Ningxia University (Humanities and Social Sciences Edition), 34*(6), 139–143.

Zeng, W. Z., Liu, S., Yang, F., et al. (2017). A review of research on forest carbon sequestration from the perspective of poverty alleviation. *Agricultural Economic Issues, 2*, 102–109.

Zhang, Y., Qi, Y. B., & Jia, Y. (2011). Research on the mechanism of benign interaction between low-carbon agriculture and carbon finance. *Agricultural Economic Issues, 6*, 96–102.

Zhu, Z., Shen, Y. Q., & Bai, J. D. (2015). Risk attitudes of forest farmers and carbon sink supply decisions in southern collective forest areas. *Chinese Soft Science, 7*, 148–157.

Zou, X. Y. (2011). Study on the interaction between carbon finance and rural finance—based on the concept of localization of carbon finance and rural financial innovation. *Agricultural Technology and Economics, 6*, 70–76.

Zhang, W. K., Fu, X. H., & Zeng, W. Z. (2012). Research on ecological compensation in China's watersheds: a review from an economic perspective. *Journal of Central South University of Forestry and Technology (Social Science Edition), 6*(5), 55–60.

Zhuo, Y. (2004). *Public sector performance management*. Fujian People's Publishing House.

Chapter 3
Theoretical Framework for PAFCS

PAFCS, as a state-of-the-art and cross-disciplinary topic, has been widely researched in the fields of management, economics, and sociology. China's experience in developing the FCS industry and afforestation projects has enabled it to promote the topic while participating in and leading global climate change cooperation. It has become essential to further condense and abstract the most prominent problems in current research and practice, in a scientific manner, to formulate theories and thoughts guiding the development of the FCS industry and related poverty alleviation projects, thus providing effective theoretical tools, means, and methods for in-depth empirical research as well as an impetus to practice. To this end, this chapter attempts to deepen the understanding of fundamental connotations, basic elements, essential characteristics, stakeholders, and constituent factors related to PAFCS; analyse in depth the constituent factors of the FCS system and the incentive mechanism to incorporate FCS project development and poverty alleviation; and explore the effect of FCS projects on alleviating poverty in poverty-stricken communities and populations. By elaborating on the contemporary implications of this discipline and the theoretical guidelines and practice logic of targeted poverty alleviation, this chapter aims to scaffold a theoretical framework for PAFCS from the perspective of targeted poverty alleviation.

3.1 Argumentation and Analysis of PAFCS

PAFCS is a special form of poverty alleviation, implemented via the ecological industry, which offers a means to eradicate poverty and control climate change simultaneously. In this section, this study defines the concept of PAFCS based on a systematic review of the literature, proposes the core characteristics which distinguish it from other FCS projects and poverty alleviation projects, and lays a foundation for the construction of a theoretical framework of PAFCS.

W. Zeng and F. Yang, *Poverty Alleviation Via Forest Carbon Sequestration*, International Research on Poverty Reduction, https://doi.org/10.1007/978-981-99-8191-5_3

3.1.1 Definition of PAFCS

Although the inherent connection between FCS and anti-poverty measures has been discussed for decades, there is still no clear definition of PAFCS; a few scholars' definitions of low-carbon poverty alleviation and poverty alleviation through carbon sequestration come close. Ji and Wang (2012), for example, believe that low-carbon poverty alleviation refers to alleviating poverty while developing a low-carbon economy, and vice versa. Lu (2012) defines low-carbon poverty alleviation as the use of methods and tools for a low-carbon economy to exploit the carbon resources in naturally endowed poverty-stricken areas and develop carbon trade products in exchange for funds in the carbon-trading market.

Meanwhile, the low-carbon economy and poverty alleviation can ultimately be integrated by developing industries, thus providing a rich array of eco-products for society. Chen (2014) points out that low-carbon poverty alleviation, as a systematic project, is an important approach to promote poverty alleviation using the carbon trading market. Such an approach is not confined to developing low-carbon industries but should also foster notions and behaviours supporting low-carbon poverty alleviation. Xu (2015) argues that low-carbon poverty alleviation will minimise or control the emission of carbon dioxide and other greenhouse gases while alleviating poverty, creating a model for ecological and low-carbon sustainable development in a virtuous cycle. Shen (2017) proposes that PAFCS is intended to transfer the newly added carbon sequestration in poverty-stricken areas to enterprises in developed areas to offset corporate quotas on carbon emission reduction. Thus, PAFCS is a new ecological poverty alleviation model that works alongside improvements to the carbon trading market mechanism.

This study holds that PAFCS, following a market mechanism and with the support of government and society, ought to be based on the development of forest land and other resources in underdeveloped areas. PAFCS should be operated to create benefits and development opportunities for the poverty-stricken populations. It relies on FCS projects and focuses on the participation of poverty-stricken populations and the construction of a pro-poor mechanism. Ultimately, a novel poverty alleviation model will surface to secure a *win–win* situation that both reduces the negative impact of climate change and alleviates poverty when facilitating FCS industrial development and the sustainable operation of projects. Its fundamental connotations include the following four aspects (Zeng et al., 2016).

3.1.1.1 PAFCS is a Means of FCS Industry Development Oriented to Poverty Alleviation

PAFCS is an integration of the FCS industry and poverty alleviation. However, it is essential to combine administrative and market approaches and acknowledge the roles of the government and the market. Based on market-oriented operations and balancing fairness and efficiency, poverty alleviation potential in developing FCS projects will be explored to achieve the goals of creating benefits and development opportunities for poverty-stricken people. PAFCS can easily become a paradox if it deviates from the objectives of poverty alleviation. Therefore, the development of the FCS industry is merely a vehicle, and poverty alleviation is the goal. Likewise, the form serves the content, and the content relies on the form; PAFCS inevitably emphasises that FCS industry development begins with poverty alleviation and arrives at the breadth and depth of pro-poor projects. In other words, poverty alleviation is not only one of the important goals set by FCS project development but also a method to develop the poverty alleviation-oriented FCS industry.

3.1.1.2 PAFCS Focuses on the Participation of Poverty-Stricken Populations

In practice, the participation of poverty-stricken populations in FCS development is the most effective way for such populations to gain economic benefits, transform their understanding of development, and enhance a project's viability. Lack of effective participation by poverty-stricken populations will significantly undermine the poverty alleviation function and pro-poor performance. Although poverty-stricken areas and populations heavily overlap, if PAFCS is reduced to implementing FCS projects in poverty-stricken areas while overlooking effective participation of the poverty-stricken population there, FCS projects will inevitably become disconnected from the goal of poverty alleviation, resulting in elite capture and thus *helping the rich but not the poor.* Worse still, target farmers would be excluded from sharing the economic benefits, and PAFCS could deviate from its objectives or result in project replacement. As such, the key to realising PAFCS is that poverty-stricken populations are fully engaged in FCS initiatives and their rights of knowing, participating, choosing, and supervising are respected. Moreover, obstacles that hinder their participation ought to be removed, and their ability and opportunities to participate, as well as the degree to which they can do so, should be improved. Efforts should be made to mobilise their enthusiasm, proactivity, and creativity within their participation to help them share the benefits of participating in the development of the FCS industry.

3.1.1.3 The Essence of PAFCS is the Construction of a Pro-Poor Mechanism

Poverty reduction is an important reason that FCS projects are popular in developing countries, especially in poverty-stricken areas. However, FCS projects are dominated by a market mechanism, which focuses their attention on investment efficiency. Therefore, the construction of a pro-poor mechanism not only governs the direction of PAFCS industry development but also decides whether poverty-stricken people can benefit from FCS project development. As such, PAFCS should not only comply with the basic FCS market transaction rules, adhering to industrial operation and anti-poverty objectives, but also address the contradiction between poverty alleviation goals and economic operations. It is essential to remove those elements restricting the synergy between the FCS industry and anti-poverty goals, establish a net benefit and guarantee mechanism for the poverty-stricken people, and direct FCS industry development to create benefits and development opportunities for poverty-stricken people. Meanwhile, it should be ensured that poverty-stricken populations reap the benefits brought by FCS project development without harming the interests of the non-poor and enjoy, as far as possible, the welfare generated by the external poverty alleviation resources introduced as a means of exploring the poverty alleviation potential of FCS.

3.1.1.4 PAFCS Depends on the Sustainable Development of the FCS Industry

As a derivative of the carbon dioxide emission trading system, PAFCS is rooted in international market-based ecological compensation. Typically, it aims to realise a *win–win* goal of both addressing climate change and alleviating poverty through economic incentives. From the ecological service market perspective, the value of FCS is far greater than that of pure carbon absorption. FCS market transactions have realised only partial compensation for the values of forests in multiple ecological services, such as water conservation, soil conservation, carbon sequestration, and biodiversity conservation and protection. Prominent industrial development prospects and poverty reduction potential exist. However, the benefit and development of poverty-stricken populations are subject to the effective market demand for forest carbon trading products. The basic premise is that the marginal benefits of FCS afforestation or reforestation are greater than the marginal costs. If the FCS industry fails to bring benefits to poverty-stricken households, efforts to alleviate poverty via FCS are futile. Therefore, PAFCS must fully consider special land qualification requirements, afforestation land clearing and land preparation, carbon sequestration measurement and monitoring during FCS afforestation, and the practical difficulties of realising the value of carbon sequestration. Moreover, to achieve a *win–win* outcome and accomplish poverty reduction while promoting the prosperity of the FCS market and sustainable project operation, an incentive mechanism is needed to incorporate the sustainable development of carbon sequestration into anti-poverty operations.

3.1.2 PAFCS Characteristics

As a complicated FCS industry and a new method of alleviating poverty, PAFCS needs to be investigated from different perspectives, including its objective and subjective effects and relationship with policies. Therefore, four kinds of characteristics are discussed and verified below.

3.1.2.1 Accurately Identifying Targets

With the goal of *truly alleviating the poor, and alleviating the truly poor,* prerequisites to implementing PAFCS are that the land is suitable for afforestation, the forestry productions are competitive in the market, and the land meets the eligibility of carbon sequestration afforestation or sustainable management. The primary poverty alleviation target is communities in poverty-stricken areas (especially those that are ecologically fragile or climate-sensitive), and the further target is poverty-stricken populations who have the right to use natural resources such as forest land or the ability to work and who are willing to participate in PAFCS projects. Practical measures should be taken to help these populations by adjusting measures to local living conditions, creating benefits and developmental opportunities for them, and enhancing their capability to eradicate poverty. Fundamentally, the target of a PAFCS project is the poorest people in poverty-stricken areas, such as the poverty-stricken households identified and filed in previous targeted poverty alleviation programmes.

3.1.2.2 Diversity of Entities

Aiming to sustain the growth of the local economy and empower the poverty-stricken people with the means to prosper, PAFCS is a developmental and open project involving diverse entities, such as governments, enterprises, society, and communities, which aims to meet poverty alleviation targets. Based on the *Kyoto* or *non-Kyoto* primary rules of the carbon trading market, PAFCS is a process and result of active participation and mutual assistance under market and policy incentives working around FCS quotas. These entities in multi-faceted poverty alleviation include the country or region, on the one hand, and administrative agencies, legal persons, NGOs, communities, and individuals, on the other hand. Hence, they are diverse, complex, dispersive, and dynamic.

3.1.2.3 Effects on Multidimensional Levels

The pro-poor effects of PAFCS are reflected on the macro level in its impact on the capability of poverty-stricken areas to develop sustainably. On the meso scale, they are reflected in strengthening the collective economy for poverty-stricken communities, enhancing organisational management and growth capabilities, and promoting the synergy and interaction between the FCS industry and specific industries with local characteristics. On the micro level, they are reflected in the income, capability, and growth opportunities open to the poverty-stricken people. The effects can be divided into positive and negative and into short and long terms. Poverty-stricken areas and populations in ecologically fragile or climate-sensitive regions take time to realise the non-economic benefits of the pro-poor effects. In this sense, PAFCS effects are characterised by diversity, spatial differences, temporal dynamics, and lagging.

3.1.2.4 A Policy-Based Approach

On the one hand, PAFCS is a method of forest ecological compensation dominated by a market mechanism. It is self-evident that the invisible hand of the market can't be solely relied on to guide the work of poverty alleviation, attract social attention, and achieve pro-poor goals. The project may tend to *help the rich but not the poor*, an outcome also known as *benefits spillover* and *elite capture*. On the other hand, to realise the *help offered by* poverty alleviation, the social welfare element of PAFCS is emphasised, as it is driven by non-market incentives as opposed to simply aiming to develop FCS projects. Thus, it is critical to consider this social welfare element and acknowledge the government's role in macro control. The role of the government includes ensuring policy guidance, resource integration, and market failure compensation by combining administrative and market approaches. A further critical point is that the poverty-stricken people must be given the right to know, participate, make decisions, manage, supervise, and earn income to ensure they are the main participants in PAFCS and promote the *win–win* cooperation of all relevant stakeholders. To sum up, PAFCS adopts a government-led approach under market conditions and is strongly policy-based.

3.2 Analysis of PAFCS Stakeholders

Ultimately, the behaviours of all participants in their specific institutional environments culminate in the achievement of PAFCS goals. Participants have complex, intertwined, and interactive stakes in FCS development, play different roles, and exert different influences. What directly influences and determines the objective performance of PAFCS is FCS stakeholders' understanding, approval, sense of mission, and responsibilities for poverty alleviation. Therefore, to ensure the reform and transformation of PAFCS, stakeholders must be correctly defined, and the interests and appeals of participating entities must be clarified.

3.2.1 Definition of PAFCS Stakeholders

Stakeholder theory was first proposed by Dodd (1932). In 1984, the American economist R. Edward Freeman, in his book *Strategic Management: A Stakeholder Approach*, clearly posited the theory of stakeholder management. A stakeholder is any individual or group who can affect or is affected by the achievement of an organisation's objectives. Based on stakeholder theory, this study defines stakeholders as entities, organisations, or individuals relating to PAFCS. From the perspective of FCS market players, the main stakeholders in PAFCS include those demanding and supplying FCS, third-party independent certification agencies, and others that play important roles in guiding, promoting, and supervising the development of the FCS

Table 3.1 Definition of PAFCS stakeholders

Stakeholders	Specific stakeholders covered in this category
Governments	Governments at all levels, and their development & reform commissions, forestry, poverty alleviation, science & technology departments in charge
Enterprises	Certification Emission Reduction (CER) purchasers including enterprises in need of carbon reduction, CER developers and suppliers including afforestation entities
Communities	Communities where the projects are implemented
Farmers	Farmers involved in and having influence on the development of FCS
Experts	Experts providing professional services for project planning, application, registration, monitoring, evaluation, and decision-making consultation, and technical guidance and training for afforestation and forestry management
Media	Printed media, TV, Internet, etc.
NGOa	Carbon funds of all levels and types, intermediary of carbon capital management, environmental protection associations, social groups
Financial institutions	Financial and insurance institutions providing low carbon financing, insurance service, etc.
Public	Natural persons in support of eliminating carbon footprint, caring and supporting the course of coping with climate change and poverty alleviation

market, such as governments, NGOs, and intermediary groups (Wang & Gao, 2011). From the perspective of a framework for ecological service compensation, the main stakeholders in PAFCS projects include ecological service providers (e.g., farmers, afforestation entities), beneficiaries of ecological services (e.g., biocarbon funds), and other intermediary groups (e.g., international and domestic NGOs, governmental agencies) that directly or indirectly facilitate transactions between beneficiaries and suppliers of ecological services (Chen, 2010).

In summary, taking a PAFCS AR project as an example, the main stakeholders include governments and administrative authorities, afforestation entities, communities, and farmers in the areas where the projects are implemented. Other stakeholders include institutions and personnel that provide consulting, organisation, management, funding, legal, and technical backup and services for project planning, application, approval, registration, implementation, and monitoring, such as NGOs, professional associations, research institutes, financial and insurance institutions, social media, and benefactors who pledge to practise low-carbon environmental protection and poverty alleviation. The stakeholders are defined in more detail in Table 3.1.

3.2.2 Classification of PAFCS Stakeholders

At present, the most used method to classify stakeholders is the Mitchell Score-based Approach (Jia & Chen, 2002). Based on stakeholder theory, this study adopts this approach with three selected indicators, namely urgency, authority, and legitimacy,

Table 3.2 Classification of PAFCS stakeholders

Stakeholder	Urgency	Authority	Legitimacy	Type
Government	High	High	High	Confirmed
Enterprise	High	High	High	Confirmed
Community	High	Low (progressive increase)	High	Confirmed
Farmer	High	Low (progressive increase)	High	Confirmed
Expert	Medium	High	Medium	Anticipated
Media	Low (progressive increase)	Medium	High	Anticipated
NGOs	Medium (progressive increase)	Low (progressive increase)	Medium (progressive increase)	Anticipated
Financial institution	Low (progressive increase)	Medium	Low (progressive increase)	Potential
Public	Low (progressive)	Low (progressive increase)	Medium (progressive increase)	Potential

to score the main stakeholders in PAFCS. Then, the stakeholders are classified by the scores into three types: confirmed, anticipated, and potential, as shown in Table 3.2.

Notably, governments, enterprises, and poverty-stricken communities and farmers are the core stakeholders of those listed above. In other words, they are most closely connected to PAFCS development and have the greatest initiative, significance, and influence. On the second level, experts, media, and NGOs are closely related to the core stakeholders, being themselves indirect stakeholders. Financial and insurance institutions and the public are the potential and relatively marginal stakeholders, which currently have limited participation channels and degrees of participation, but given time and changes in the carbon trading system and innovations in mechanisms, they will inevitably pay more attention to, care more about, and extend greater support to PAFCS.

3.2.3 *Interests of Core Stakeholders*

The process of realising PAFCS involves the participation and coordination of multiple entities, dominated by rational individuals, within the constraints of international climate change agreements and supported by domestic policies concerning emission reduction and poverty alleviation. Overall, all entities share the same interests: in the process of realising PAFCS, they share an interest in the development of the FCS industry, especially the benefits derived from participating and the social

Table 3.3 Interests of core stakeholders of PAFCS

Stakeholder	Roles	Interests	Focus
Governments	Guide Regulator Supervisor Beneficiary	Social and cultural interests, Eco-environmental interests, Economic interests	Win–win in coping with climate change and tackling poverty, obtaining foreign capital and technology, increasing employment opportunities, promoting regional socioeconomic development, and implementing targeted poverty alleviation
Enterprises	Leader Investor Executor Beneficiary	Investment returns	Maximisation of carbon emission rights and profits, sustainable development of FCS projects, establishment of a good public image, and improved social reputation
Community/ Farmers	Participant Beneficiary Evaluator	Sharing the benefits of participation	More economic income, employment opportunities, and new technologies; respect for and protection of local traditional culture and natural landscapes

public welfare it brings about. However, from a local perspective, different entities tend to have different interests and focuses. The fundamental interest of participating individuals is not completely consistent with, and can even deviate from, poverty alleviation goals. For example, FCS purchasers wish to maximise the utility of ecological services such as carbon absorption and biodiversity protection, while FCS providers, for example afforestation entities, expect to maximise profits, as shown in more detail in Table 3.3.

3.2.3.1 Governments

PAFCS development is a market-oriented model in which stakeholders will not automatically follow and coordinate pro-poor activities to achieve the goal of poverty alleviation. Therefore, governments will inevitably intervene as public interest representatives.

In the first place, the central government, especially the CNDC, needs to intervene, being the government agency responsible for tackling climate change and focusing on national, long-term interests. The government is the policymaker that enacts important laws and regulations for China's carbon emission trading system, FCS rules for voluntary emission reduction to encourage poverty alleviation, and details of how trading is managed. The central government regulates market operations; decides on the establishment, approval, and filing of PAFCS projects; and works towards a *win–win* outcome of reducing both climate change and poverty. It also guides the policies designed to promote the FCS industry to create benefits and development opportunities for poverty-stricken people.

The provincial government is in second place, especially provincial or municipal development and reform commissions and departments responsible for forestry, poverty alleviation, science and technology, etc. These are the decision makers in promoting PAFCS planning and preferential policies. They are also the supervisors, promoting and supervising the achievement of poverty alleviation goals through administrative and industrial management.

The third level involves local governments, mainly those of the cities or states, counties, and townships or towns where the projects are developed and their related administrative departments. The local governments substantively exercise territorial jurisdiction over PAFCS and development. They are the supervisors, regulators, monitors, and beneficiaries. Led by policy frameworks concerning ecological construction and poverty alleviation through industrial development and the guidance of relevant departments, local governments integrate poverty alleviation resources, promote synergy between the development of FCS projects and planning of targeted poverty alleviation, and balance the interests of various stakeholders during FCS project operation. Moreover, local governments safeguard the legitimate rights and interests of community residents as well as the benefits that farmers can gain from project development. These efforts will jointly lead to regional socioeconomic development, targeted poverty alleviation, ethnic unity, and other political achievements.

3.2.3.2 Enterprises

Being artificial and open, the global FCS market involves multiple levels, such as international, regional, national, and local, as well as multiple types, such as the *Kyoto rules* and *non-Kyoto rules* market, project and quota market, and mandatory compliance and voluntary emission reduction market. In carbon sequestration AR projects, on the one hand, enterprises in compliance with emission reduction obligations are the main investors in FCS projects. They obtain carbon emission rights in key invested fields while maximising the utility of ecological services centred on carbon absorption. Simultaneously, to honour their social responsibilities, curate their public image, enhance their social reputation, and engage in charity, these enterprises often emphasise the poverty alleviation function of FCS and meet the poverty alleviation goals through contracts and other methods in their capacities as PAFCS investors and promoters.

On the other hand, enterprises leading the implementation of carbon sequestration AR, namely the project owners, play an important role as links and bridges in PAFCS development. Undoubtedly, they are the key leading force. They strive to maximise their profits from the development of FCS projects and, by strengthening poverty alleviation goals, secure more funds or preferential policies from governments, FCS buyers, and even society. Meanwhile, they foster the enthusiasm of the community and its farmers to participate in project implementation, work to ensure a better understanding and greater cooperation among community residents, enhance investment return, and achieve sustainable operation of PAFCS projects. These enterprises are PAFCS investors, executors, and leaders.

3.2.3.3 Community Farmers

Community farmers are not only the most direct and important participants in the development of carbon sequestration AR projects but also the beneficiaries as they share development benefits, and the bearers of the negative impacts and potential risks of project development. More importantly, they are also judges of the multi-faceted performance of FCS projects. On the one hand, the successful development and long-term operation of PAFCS projects rely on farmers' participation, approval, and cooperation. Through collective bargaining and action, they wish priority to be given to employment opportunities, increasing the economic benefits gained from natural resources such as suitable forest land, learning new forestry technologies, and gaining experience in management and environmental protection. They anticipate that their interests will be accepted and respected; local production, living, and ecological conditions will be improved; and negative impacts on local traditional livelihoods, social cultures, and ecological environment will be reduced.

On the other hand, the judgement of project implementation performance by community farmers, especially poverty-stricken populations who rely more on traditional agricultural livelihoods, not only directly correlates with their attitudes and behaviours for continuous participation, but also plays a key role in securing farmers' long-term support and cooperation, reducing resistance, avoiding confrontation, and consolidating the achievements of early AR and its long-term operation. Moreover, their judgement is critical to discovering the potential of FCS to reduce poverty and achieve the *win–win* goal of addressing climate change and achieving poverty reduction.

3.3 Systematic PAFCS Analysis

As an emerging industrial and economic activity and a novel poverty alleviation model, PAFCS is essentially a subsystem of the socioeconomic system. Therefore, it is of positive theoretical value and practical significance to use systems theory to explore the systemic elements of PAFCS and reveal the mechanism combining project development and poverty alleviation.

3.3.1 *PAFCS Constituents*

Systems theory points out that a system is a dynamic and organic whole with specific functions achieved through the mutual dependence, interaction, and friction of two or more elements. There are flows of matter, energy, and information among the different segments of the system. PAFCS is not a one-way, static, and closed system, but a dynamic, open, and composite system driven by the carbon trading market, government promotion, and social assistance. It mainly comprises entities in PAFCS, pro-poor resources and methods, and the closely related incentive system, participation system, and support system which together constitute the PAFCS operating system. These components continue to interact to stabilise and improve the system and thereby promote scarce resources such as capital, technology, talents, and information to converge on poverty-stricken areas to meet the objective of *alleviating the truly poor and truly alleviating the poor.* The system's operation depends on the effective coupling of various subsystems, the interaction and combination of various components in the system, and the ability of the external environment to support the PAFCS model. It is a dynamic process of continuous cultivation, improvement, and optimisation, as shown in Fig. 3.1.

3.3.1.1 Main Entities in PAFCS

The main entities in PAFCS are the institutions and individuals who directly participate in the development of FCS projects or provide backup and services for the

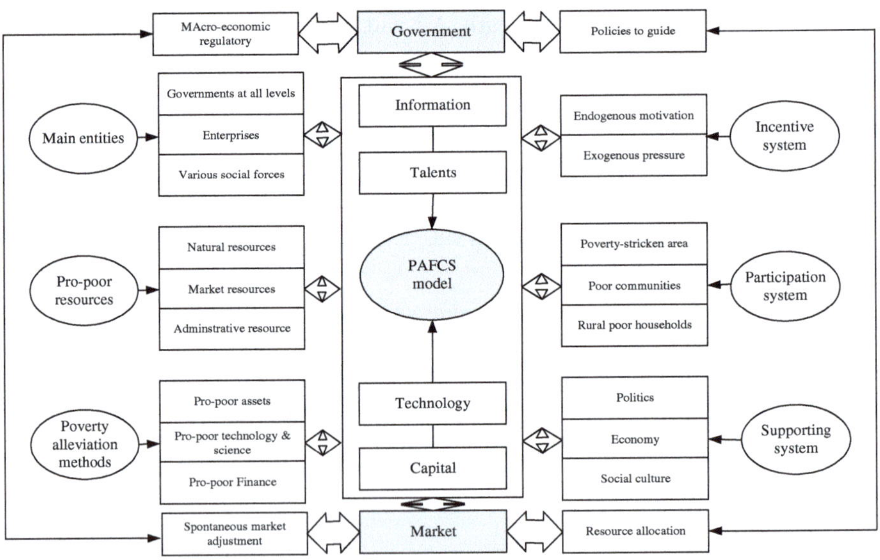

Fig. 3.1 PAFCS system constituents

process to generate actual impact and contribute to exploiting FCS potential. They include governments and institutions at all levels, private carbon sequestration businesses, research institutes, media, financial, and insurance institutions, NGOs, and the public. They are constantly evolving and updating *variables*. In essence, the entities are closely connected to but distinct from PAFCS stakeholders, and they are homogeneous, staged, diversified, and dynamic. Currently, in the pilot and initial stages the FCS industry still faces many challenges. Consequently, governments should offer more policy guidance, institutional constraints, supervision, and management. Moreover, the enthusiasm and creativity of FCS project owners (e.g., afforestation companies, major professional households, and community elites) should be mobilised to reduce poverty.

3.3.1.2 Pro-poor Resources

Pro-poor resources are the material guarantee to operate the PAFCS model. Depending on their origins, they can be divided into internal and external resources in poverty-stricken areas. Internal resources include natural resources which are valuable for carbon sequestration development, for example suitable forest land or forests, appropriate climate and soil, and abundant labour resources, and local material and non-material cultural resources accumulated over long-term historical development. External resources are not limited to capital, but also include market resources driven by the carbon trading market with FCS development at the core, such as funds, technology, operation, management, and information. External resources also include administrative resources, for example guidelines, policies, development strategies, management systems, FCS emission reduction standards that highlight the function of poverty alleviation, and preferential and regulative policies functioning as guidance, incentives, and guarantees. These resources are formulated by governments to offset the impact of market failure, explore PAFCS potential, and achieve the core goals of social poverty alleviation. Other external resources are those injected to attract social forces to participate in poverty alleviation.

3.3.1.3 Poverty Alleviation Methods

PAFCS is an open system without a universal standard model. Poverty alleviation methods are often affected by macro factors, such as international and domestic standards for FCS emission reduction, and micro factors, such as the guiding and support policies set by governments at all levels, operation models for FCS development projects, and a mechanism linking the benefits of PAFCS. In essence, PAFCS integrates market orientation and poverty alleviation to improve social welfare. It has been proven to be a fair, efficient poverty alleviation strategy that encourages development and participation. PAFCS, obviously intended for public welfare, has unique advantages and functions. Given that poverty alleviation resources are limited, PAFCS integrates multiple poverty alleviation methods, such as using technology, industry, and financial resources. Together, these methods are used with a focus on achieving the *win–win* goal, forming an interdependent yet complementary relationship that is mutually promoting and competing.

3.3.1.4 Incentive System

The incentive system involves various forces that promote the establishment, operation, and development of the PAFCS model. Generally, it can be divided into intrinsic and extrinsic incentives, depending on the reason it is formed, while incentives can be market-driven and government-driven, depending on their origin; direct and indirect, depending on their different functions; and dominant or auxiliary, according to their intensity. Relevant poverty alleviation entities are the *juncture* of various driving forces. Each entity drives the operation of the PAFCS model through a system of resources and methods. Various joint forces synergise and, in return, have an impact on the main entities, thus generating new forces to promote the sustainable operation of the PAFCS model. The following section further analyses the causes of the formation of the PAFCS incentive system shown in Fig. 3.2.

Intrinsic Incentive for PAFCS

The intrinsic incentive for PAFCS is the inherent motivation of various poverty alleviation entities to demonstrate behaviours when proactively practising the philosophy and principles of PAFCS. It covers three aspects: the driver of economic interests, the driver of emission reduction obligations, and the establishment of a public image.

Economic Interests as a Driver

FCS buyers, intermediaries, and suppliers are all essentially organisations that pursue economic interests. Even the government has various direct or indirect interests at stake in the development of PAFCS. Therefore, economic benefits are not only the

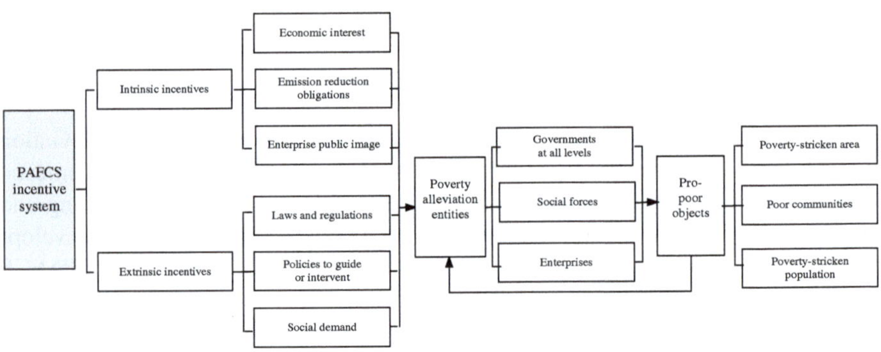

Fig. 3.2 PAFCS incentive system

bond that connects the participating entities, but also the fundamental internal driving force that promotes the operation of the PAFCS model.

Emission Reduction Obligations as a Driver

Poverty reduction is not only an important function of FCS under *Kyoto rules* and *non-Kyoto rules but* also a critical guarantee for fulfilling emission reduction commitments, winning widespread support from community farmers, and realising the smooth progress and long-term operation of FCS projects.

Creating a Public Image

As a major way to achieve a *win–win outcome*, FCS is currently pivotal in promoting global emission reduction, sustainable development, inclusive growth, and low-carbon transformation. Given the increasing concern over climate change, both governments and private sector organisations undertake their share of social responsibilities. Thus, enhancing their social image and social reputation is another intrinsic motivation for various entities to participate in PAFCS.

Extrinsic Incentive for PAFCS

The extrinsic incentive for PAFCS refers to various external forces that originate from outside poverty alleviation entities and have an impact on their behaviours. This incentive category has three main aspects: legal constraints, government policy intervention, and social demand.

Legal Constraints

While fulfilling global emission reduction commitments and obligations, China must formulate rules for FCS voluntary emission reductions that can meet both the requirements of international carbon trading rules and those of targeted poverty alleviation after 2015. Laws and regulations on the pricing, registration, measurement, verification, and monitoring of carbon market transactions are modified to regulate and restrain the behaviours of PAFCS entities.

Government Policy Intervention

Governments ought to make full use of the support and intervention function of policy instruments including poverty alleviation, fiscal policy, taxation, land, finance, insurance, and other policy means to create a benign policy environment for the development of the FCS industry and projects. Moreover, they should continually campaign for, support, and incentivise projects and behaviours that highlight PAFCS. Governments should also regulate and restrict behaviours that are not conducive to the reasonable exploitation of natural resources or the fair participation of community farmers in poverty-stricken areas.

Social Demand

From a demand perspective, with the progression of climate change and international negotiations, the market demand for FCS is ever-growing, and there is huge potential for industrial development and business opportunities. The ever-growing demands from global government organisations and the public for market-oriented FCS ecological services are powerful external forces that drive PAFCS model operation.

3.3.1.5 Participation System

The PAFCS participation system consists of three dimensions: poverty-stricken areas (macroscopic), communities (mesoscopic), and populations (microscopic). In essence, the participation of poverty-stricken people is a key component in exploring the potential of PAFCS, but participants in PAFCS projects certainly should not and cannot be limited to poverty-stricken populations. Therefore, the participation of communities and their farmers, as well as poverty-stricken populations, is a vital factor in determining both the sustainable operation of the project and the realisation of the poverty alleviation goal. The participation system is undoubtedly the key to the operation of the PAFCS model. As far as the extensive pilot carbon sequestration AR projects in China are concerned, community farmers may participate in four ways: project planning, project construction, monitoring and evaluation, and benefits distribution.

Project Planning

Community farmers participate in project plot selection, boundary delineation, and environmental assessments such as biodiversity evaluation and reach consensus with afforestation entities on alternative lands for project development, tree species selection, participation methods, benefit distribution plans, and contract signing. These types of participation effectively reflect the interests of the majority of local people and enhance the rationality of project planning and the feasibility of implementation.

Project Construction

Community farmers can obtain economic or non-economic benefits by buying in or transferring their stakes in land suitable for forestry through technical training and by providing services such as afforestation land preparation, planting, and tending. Subsequently, communities will supervise the follow-up management and protection and urge farmers to fulfil contractual obligations to avoid or reduce negative impacts on sustainable FCS project development.

Monitoring and Evaluation

At different stages of project implementation, especially in the process of CER carried out by qualified third-party monitoring and evaluation agencies, community farmers can join in the evaluation of the effects of PAFCS. This process serves as an important safeguard that protects community farmers' legitimate rights and interests, reinforces positive effects while reducing negative impacts, corrects deviation from poverty alleviation actions, and improves PAFCS's efficiency.

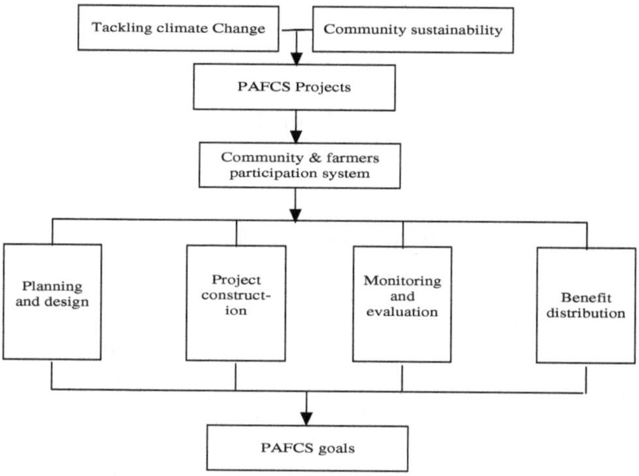

Fig. 3.3 PAFCS participation system

Benefits Distribution

The fundamental goal of PAFCS is to enable poverty-stricken people to fairly and reasonably share project development benefits without infringing on non-poverty-stricken people and to maximise the benefits resulting from market intervention by governments and additional input of administrative resources to alleviate poverty. It is worth noting that farmers can directly participate in PAFCS by buying or transferring their stakes in low-efficiency land suitable for forestry, technical training, services, or follow-up management and protection. They can also indirectly participate in PAFCS by completing the preliminary baseline survey and participating in mid- and late-stage monitoring and evaluation. As Fig. 3.3 demonstrates, farmers' participation is not only a voluntary economic decision but also a management choice under the leadership of the local government and the project owner.

3.3.1.6 Environment System

Subject to various regional, national, and international factors, robust PAFCS operation relies on the support of the external environment. The environmental system entails unifying the external political, economic, and cultural environments in which PAFCS operates, including macroscopic and microscopic, external, and internal aspects. The political environment—especially the strategic orientation, path selection, and China's related laws and regulations in response to climate change—is the institutional foundation on which the effective operation of the PAFCS model rests, while the economic environment, particularly building the carbon sequestration system and developing the FCS industry, is its economic foundation. The cultural environment refers to the value orientation, concepts, and cultural atmosphere advocated by society in practising ecological civilisation and jointly promoting green development, low-carbon development, and inclusive growth. This is the social foundation for the operation of the PAFCS model.

3.3.2 The Operating Mechanism of PAFCS

Based on systems theory, the operating mechanism of PAFCS refers to the functional structure, relationship among various elements that maintain the overall normal operation of PAFCS, and process, operation, and mechanism whereby these elements have an impact and exert their functions, as shown in Fig. 3.4.

Regarding the structural functions of the PAFCS operating mechanism, several basic functions serve as interactive, combined, organic, and coordinated units. There are five types of functions in PAFCS. The first is the integration function, which connects the elements of PAFCS as an organic whole which, with relative stability, is capable of automatic and adaptive adjustments. The second is the directional function, which ensures the PAFCS system operates in the direction of the established goals and achieves the essential objective of creating benefits and development opportunities for the poverty-stricken people. The third is the motor function, which provides a continuous impetus for the operation of PAFCS and promotes the coordination of poverty alleviation entities, integration of poverty alleviation resources, and linkage of poverty alleviation methods. The motivation function also maintains the healthy and sustainable operation of the system. The fourth is the regulating function. When there is a deviation during system operation, the regulating function continuously corrects it and restores the system to normal status through effective data transmission, information feedback, and management activities. The fifth is the development function. While ensuring the normal operation of a PAFCS system, this

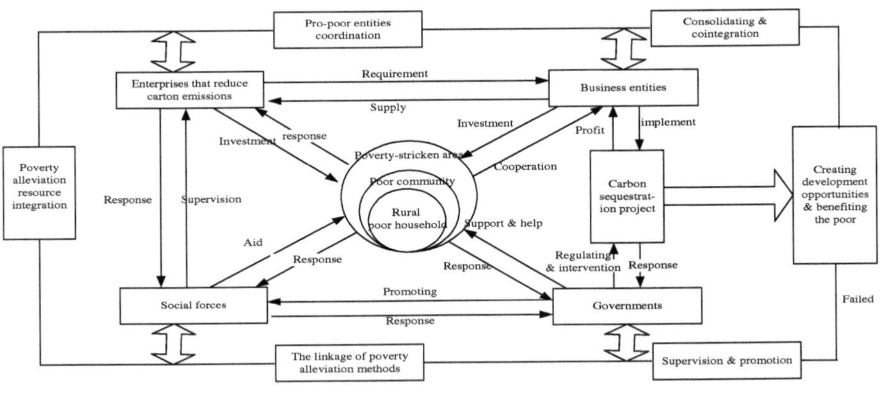

Fig. 3.4 PAFCS operating mechanism

function promotes its development and evolution through incentives, competition, and coordination.

In terms of operation mode, the operating mechanism of the PAFCS system is driven by a dual compound that combines government administrative intervention and a market mechanism. Observing the logic of poverty alleviation, the function of this mechanism is mainly manifested in four aspects: promoting diversity of poverty alleviation entities, market-based allocation of poverty alleviation resources, innovative poverty alleviation methods for poverty-stricken populations, and discontinuing the vicious cycle of poverty and ecological degradation.

3.3.2.1 Promoting Diversity of Poverty Alleviation Entities

The main entities in PAFCS are the leading forces that promote the PAFCS model operation. Therefore, an effective cooperation platform is established for diverse PAFCS entities to honour social responsibilities, participate in alleviating poverty and benefiting farmers, and gradually transform into major poverty alleviation entities. More importantly, under China's new round of reform on collective forests from 2008, which is fundamentally completed, forestland contractual management rights are granted to households. Against this background, PAFCS projects will inevitably and increasingly adopt a basic operation mode of *management entities* + *farmers and management entities* + *collective economic organisation in villages* + *farmers,* laying a foundation in property rights to turn local village collective economic organisations and community elites into major anti-poverty entities.

3.3.2.2 Market-Based Allocation of Poverty Alleviation Resources

In operating the PAFCS system, the market plays a determining role in allocating resources and combining production factors, promoting the conversion of abundant forest land, labour, and other resources into capital in poverty-stricken areas, and increasing the supply of scarce poverty alleviation factors such as funds, technology, and management, Therefore, in the institutional arrangements for PAFCS, governments no longer play a leading role. Poverty alleviation funds from fiscal revenues are no longer the most important external variable. Poverty alleviation resources are not limited to natural resources, civil affairs department support, microfinance, and other poverty alleviation funds; rather, they are market-oriented, focusing on poverty-stricken farmers at the core, bound by interests, and underpinned by industry development. Based on a market mechanism and coordinated with government guidance and market adjustment, a dual market (i.e., international and domestic) and the resources available in the development of the modern forestry economy are not only used, but are fully integrated, prioritising the coordinated allocation of poverty alleviation resources and reducing poverty reduction costs, which eventually has a positive impact on market-based anti-poverty initiatives.

3.3.2.3 Innovative Poverty Alleviation Methods for Poverty-Stricken Populations

Unlike traditional poverty alleviation methods, which are oriented towards the product market and focus on the development of agricultural products, PAFCS meets the needs of the global ecosystem service market through FCS projects focused on developing tradable *carbon sequestration quotas*. Community farmers, including poverty-stricken populations, are expected to acquire or transfer their stakes in forest land and secure income from labour services, sales of forest trees and forest by-products, and, more importantly, the revenues from CER sales. Undoubtedly, PAFCS creates a novel poverty alleviation method that provides an effective solution to outstanding problems, such as long forestry production cycles and short-term difficulties in securing income for poverty-stricken populations. Ultimately, PAFCS integrates multiple means to alleviate poverty: science and education, industry solutions, assets, income, and community poverty alleviation. It builds a solid platform for poverty alleviation through science and technology, represented by training and demonstrating state-of-the-art afforestation, management, and protection technology.

3.3.2.4 Discontinuing the Vicious Circle of Poverty and Ecological Degradation

Climate change exacerbates poverty both directly and indirectly. Direct impact comes particularly from *extreme-climate-event* damage to farmers' livelihoods, lives, properties, and production infrastructure. Indirect impact results from *climate-change-caused* damage to the production, life, and eco-environment of poverty-stricken areas. In China, poverty-stricken and ecologically fragile areas are heavily coupled because poverty-stricken populations are usually distributed in ecologically fragile areas. Therefore, ecological improvement is fundamental to eliminating poverty.

PAFCS positively combines poverty alleviation with ecological and environmental protection. Developing FCS projects not only provides channels for poverty-stricken areas to participate in addressing global climate change, but also improves the regional microclimate, regulates the hydrological cycle, and makes areas less vulnerable to landslides, mudslides, floods, droughts, and extreme weather events caused by climate change. PAFCS thereby reduces the risk of a return to poverty due to disasters and brings new opportunities for inclusive growth that combines poverty reduction and ecological protection, thereby generating an effective approach to avoid the *poverty trap*, namely the cycle of *poverty-stricken population—resource exploitation—environment degradation—accelerated exploitation—environment deterioration—worsened poverty*.

3.4 Analysis of PAFCS from a Targeted Perspective

PAFCS is an emerging concept based on the reality that carbon sequestration AR project locations immensely overlap in remote and poverty-stricken areas. As emphasised above, the starting point of PAFCS is poverty eradication, which is highly consistent with the strategic thinking of targeted poverty alleviation after 2015. Therefore, it is necessary to further analyse the practice and research framework of PAFCS from the perspective of targeted poverty alleviation theory to facilitate the transformation of poverty alleviation goals—net benefit and development opportunities for the poverty-stricken people—into research and practice guidance in achieving a *win–win* outcome addressing climate change and poverty alleviation.

3.4.1 Contemporary Implications of Targeted Poverty Alleviation

President Xi Jinping advanced the important concept that *seeking truth from facts, adapting to local conditions, accurately classifying guidance, and targeting poverty alleviation* for the first time in public during an inspection tour of Shibadong Village, Huayuan County, Hunan Province, in November 2013. Since then, academia has carried out in-depth and extensive research on targeted poverty alleviation; hence, the concept has been enriched and improved. Initially, it referred to the definition given in the *Implementation Plan for Establishing a Targeted Poverty Alleviation Work Mechanism* issued by China State Council's Poverty Alleviation Office. The definition is embodied in optimising various poverty alleviation resources through targeted identification, targeted assistance, targeted management, and targeted assessment of poverty-stricken households and villages to alleviate poverty therein, gradually building a long-term scientific mechanism and laying a solid foundation for poverty alleviation.

Through further research and practice, scholars have generally come to believe that targeted poverty alleviation will be an extensive counterpart to poverty alleviation in general that covers scientific methods of overall planning, coordination, and classification in new stages and new situations to transform *large-scale alleviation* into targeted *drip-with-precision alleviation.* It is a novel and integrated approach to solving poverty issues based on the environment and conditions of poverty-stricken households in different poverty-stricken areas and to implementing a comprehensive new poverty alleviation method through targeted identification, targeted assistance, targeted management, and targeted assessments that follow scientific and effective procedures (Wang, 2014).

At its core, targeted poverty alleviation addresses the prominent issues in previous poverty alleviation attempts in which poverty-stricken villages and populations were not precisely targeted (Zuo et al., 2015). It focuses on the issues of *whom to support, how to support, who will help, and how to exit* (Huang et al., 2016), so poverty alleviation resources and measures can be better directed to *alleviate the truly poor and truly alleviate the poor.* The targeted poverty alleviation method mobilises both wisdom and ambition, empowers the poverty-stricken populations to use their initiative, and enhances their abilities to develop themselves (Huang & Qin, 2015). The goal is to reduce the number of people living in poverty-stricken areas, eradicate absolute poverty, and build a prosperous society leading to full-scale development. This connotation of *targeted* poverty alleviation is expressed as targeting a certain *recipient–resource–entity* with specific *goals, methods, and processes* at *micro, meso, and macro* levels (Zhuang et al., 2016).

3.4.1.1 Connotations of Recipient–Resource–Entity in Targeted Poverty Alleviation

The connotations of *recipient–resource–entity in* targeted poverty alleviation need to be theoretically clarified in terms of how to precisely identify the targets, extent, and causes of poverty; select, allocate, and use pro-poor resources; define poverty alleviation entities, configure them, and set responsibilities and assignments to achieve targeted poverty alleviation.

Targets, Extent, and Root Causes

Targeted poverty alleviation focuses on precisely identifying the target, extent, and root causes of poverty. To do so, it is necessary to overcome bias as regards scale exclusion, regional exclusion, and subjectivity and objectivity exclusion in the identification process, and to prevent *elite capture* (Deng, 2014). Moreover, scientific, reasonable, and easy-to-operate identification standards and procedures should be established to ensure farmers are the main participants and that their enthusiasm is mobilised and their participation is increased. Furthermore, attention must be dedicated to ensuring ethical standards in target identification and full respect for the identified targets' poverty situation. *Critical farmers* peripheral to the identification standards should also be carefully screened (Ge & Xing, 2015; Tang et al., 2015).

Resources for Targeted Poverty Alleviation

To avoid misalignment of supply and demand, pro-poor resources must be precisely selected, allocated, and used according to the root causes of poverty. Likewise, to prevent both excessive allocations, which result in waste, and insufficient allocation, resources must be effectively allocated to the poverty-stricken people according to the extent of their poverty (Ge & Xing, 2015). Moreover, the use of pro-poor resources must be accurate to prevent the rent-seeking of poverty alleviation entities and the embezzling of poverty alleviation resources for other purposes, making it difficult to utilise the resources for poverty alleviation purposes.

Entities in Targeted Poverty Alleviation

It is essential to precisely define, configure, and set responsibilities for poverty alleviation entities. First, each poverty alleviation entity must act within its assigned responsibilities, neither going beyond these nor ignoring them. Second, to achieve synergy these entities must be precisely matched and coordinated to interact internally and externally with the poverty-stricken people. In addition, the responsibilities—particularly administrative and financial—of poverty alleviation entities must be assigned by obeying the rule of equality in poverty alleviation works.

3.4.1.2 Goal, Method, and process in Targeted Poverty Alleviation

The goal is the direction, the method is the support, and the process is the guarantee. Targeted poverty alleviation must be carried out according to its objectives; otherwise, target deviating may occur. Furthermore, alleviation methods and technical means must be advanced, scientific, and precise. Issues of *mistargeting, resource wasting, and entity misplacing* in traditional poverty alleviation can be solved only by using a precision approach, which provides technical support for targeted poverty alleviation. The poverty alleviation process *and* workflow must be accurate. Every step of the work should be accurately implemented and effectively connected with other steps to ensure that targeted poverty alleviation proceeds towards its predetermined goals.

Goals of Targeted Poverty Alleviation

The goals of targeted poverty alleviation cover three levels. The first, in recent years, was the short-term goal of eradicating absolute poverty by 2020. Based on eradicating absolute poverty, the second level is the medium-term goal, which is to continually improve people's livelihoods in poverty-stricken districts, consolidate the achievements of previous poverty alleviation initiatives, prevent a return to poverty, and eliminate poverty vulnerabilities. After achieving the medium-term goal, the third level is the long-term goal, which is to assist poverty-stricken populations to achieve long-term sustainability and stable poverty alleviation. The relationship among the short-, medium-, and long-term goals is progressive, and the advancement of the goals in the latter phase must be based on the realisation of the goals in the previous phase. Stagnation in achieving the goal of a previous phase will affect the realisation of the goal in the current phase. In this sense, the short-term goal of targeted poverty alleviation—namely that all the poverty-stricken people would be lifted out of absolute poverty by 2020—was not only a critical task for building a moderately prosperous society in all respects but also a necessary premise to continually improve livelihoods, eliminate poverty vulnerabilities, and stably alleviate poverty. Hence, its important role and significance as a milestone in achieving the goals of targeted poverty alleviation and developing socialism are self-evident.

Means of Targeted Poverty Alleviation

The means employed to implement targeted poverty alleviation, for example the technical and management means, must be advanced, scientific, and precise. Science and technology are the primary productive forces. Targeted poverty alleviation must rely on the support of advanced technology to reduce identification costs, boost production development, and reduce transaction costs. First, the use of advanced technology not only improves accuracy in identifying targets and offering a scientific underpinning for the accurate supply of poverty alleviation resources but also reduces identification costs and improves overall benefits. Second, using scientific production technology helps targets solve key technical problems that hinder production and development, thereby increasing production, quality, and income. Finally, *digital technologies* and new technology systems, that is, t*echnology-economic paradigm* changes, will comprehensively enlighten the poverty-stricken people in terms of technology, concepts, and behaviours and help reduce market transaction costs, change their thinking, broaden their horizons, and enhance their human capital and viability to eradicate poverty.

Process of Targeted Poverty Alleviation

The importance of precision in the entire process of poverty alleviation should be emphasised in terms of ideas, content, and standards. A linear poverty alleviation workflow and a complete process and evaluation mechanism are pivotal. Specifically, target identification, targeted assistance, targeted management, and targeted assessment must be established throughout the entire poverty alleviation process.

Targeted Identification

The ideal identification of poverty alleviation targets maximises accuracy while minimising identification costs (Zuo et al., 2015). As Fig. 3.5 to the right demonstrates, certain costs are inevitably incurred in identifying targets. Under normal circumstances, these costs are proportional to accuracy: greater accuracy requires higher costs, as demonstrated by Zones II and IV. Using creative technical approaches, the cost in Zone I can be reduced. However, unscientific and unreasonable identification methods may result in low accuracy at high costs, as in Zone III. Accurately identifying poverty-stricken villages is relatively easy, but identifying poverty-stricken populations is relatively difficult (Ge & Xing, 2015). First, the status of poverty-stricken populations is multidimensional, so it cannot be accurately described from a single dimension such as income or consumption. Second, in the case of asymmetric information, potential targets may hide their true situations to secure poverty alleviation resources, thereby complicating the process. Therefore, innovative identification methods and procedures are needed to seek a balance between comprehensiveness and representativeness in terms of identification dimensions. Simultaneously, poverty-stricken populations should be identified through democratic, scientific, and transparent identification procedures.

Targeted Assistance

Targeted assistance means implementing differentiated policies according to the root causes and extent of poverty. Mainly aimed at addressing the phenomenon of *one-size-fits-all* and *big-and-all-encompassing*, targeted assistance is designed to respect and follow the *particularity-of-contradiction* law. Because China is so vast and situations vary from place to place, the root causes and extent of poverty differ tremendously. Thus, poverty alleviation should follow basic principles while being tailored

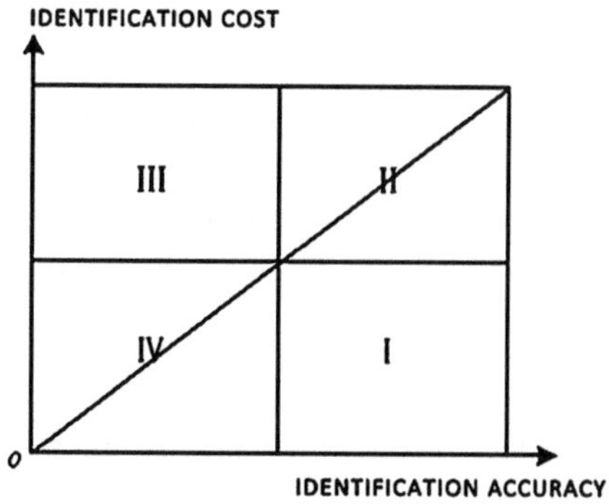

Fig. 3.5 The relationship between the accuracy and the cost of poverty target identification

to local geographic conditions, time conditions, different people, and specific situations. Differentiated assistance measures and approaches and degrees of assistance must be adopted in different poverty-stricken areas according to different root causes and poverty levels.

Targeted Management

To ensure practice effectiveness, targeted management entails comprehensive, information-based, and dynamic supervision of both poverty alleviation progress and poverty alleviation entities' assistance. First, target management must be used throughout the process. Second, information technology must be employed for overall and dynamic supervision and management of the root causes and extent of poverty, assistance measures, poverty alleviation entities' assistance methods, and target populations' progress towards poverty alleviation, and all changes should be updated. Meanwhile, targets and assistance must be adjusted to disqualify or qualify targeted populations, ensuring the aid is directed to those truly in need.

Targeted Assessment

Targeted assessment evaluates the effect of targeted poverty alleviation, and the target is mainly local governments. Central and local governments have separate responsibilities for rural poverty alleviation after 2015. The former is mainly responsible for regional development and area identification, while the latter is mainly in charge of targeted poverty alleviation. Targeted assessment is designed to urge local governments in poverty-stricken areas to prioritise their work (Wang & Wu, 2015). By monitoring the information system on the population in poverty, higher level authorities can clearly and accurately determine the rigour and accuracy of the work performed by lower level departments in the identification of poverty-stricken households and populations and make timely checks on the use of funds and project implementation. To mobilise the enthusiasm of lower level departments and their staff and maintain the necessary driving force towards poverty alleviation, higher level departments can also make timely assessments of the effectiveness of poverty alleviation in various places, allocate resources, and promote and appoint personnel according to the effectiveness of poverty alleviation (Ge & Xing, 2015).

3.4.1.3 Micro, Meso, and Macro Levels of Targeted Poverty Alleviation

Targeted poverty alleviation can be analysed at different levels: poverty-stricken farmers, households, and villages at the micro level; poverty-stricken counties at the meso level; and poverty-stricken areas at the macro level. Although different levels focus on different targets and priorities, they share the same objective: targeted poverty eradication.

Targeted Poverty Alleviation at the Micro Level

Targeted poverty alleviation at the micro level focuses mainly on poverty-stricken farmers, households, and villages. At this level, it requires accurate identification of poverty-stricken targets, analysis and judgement of poverty's root causes, formulation of assistance measures, a combination of assistance resources and factors, implementation of assistance, and evaluation of assistance effectiveness in the process of system design and subsequent practice. Targeted poverty alleviation at the micro level should focus on cultivating poverty-stricken populations' capability, empowering the assisting entities, and mobilising the targets' initiative and participation enthusiasm to achieve the goals of sustainable and stable poverty alleviation.

Targeted Poverty Alleviation at the Meso Level

At the meso level, targeted poverty alleviation focuses mainly on key counties because, in theory, assistance to poverty-stricken farmers, households, and villages is interrelated and nested in a certain external environment. If individual development is not supported by the external environment, it will inevitably become *water without a source or trees without roots* and so will not last long. The external environment most closely connected with poverty-stricken farmers, households, and villages is the administrative county where they are located.

Notably, targeted poverty alleviation at the meso level emphasises overall county development, especially the leading role of industrial development. Economic development is the foundation of poverty alleviation, so it is necessary to look for economic development solutions. Using various resources in poverty-stricken areas to develop industries with local characteristics is an important means. Without industrial support, the results of poverty alleviation are often unstable, and poverty is likely to reoccur. During industrial development, precision is required in intra-organisational technical cooperation, production factor cooperation, and inter-organisational market exchange, that is, maximising economic benefits or minimising development costs. A county's marketing position, resource endowment, and comparative advantages should be considered in its coordinated industrial development, based on local conditions and sound planning. Using technology to build an entire industry chain as the starting point for the breakthrough, poverty-stricken counties should aim to promote the scientific and sustainable development of the industrial economy according to market demands and local industry characteristics. Through this process, poverty-stricken people can benefit from industrial development.

Targeted Poverty Alleviation at the Macro Level

Targeted poverty alleviation at the macro level primarily focuses on the extremely poor located in geographically connected regions. The goal was overall regional poverty eradication by 2020. Generally, although poverty is manifested in individuals, it is mostly related to regional characteristics. Whether from a global perspective or that of China's actual situation, poverty exists in entire regions or areas and is seen as not only an economic problem, but also a complex social and political problem. Macroscopically, targeted poverty alleviation focuses on solving regional development and environmental problems, including natural and social environments. Regarding the natural environment, green development should be endorsed to alleviate poverty and protect the environment. Regarding the social environment, regional infrastructure must be upgraded, the quality of regional public services should be improved, the regional market environment should be optimised, and connectivity between intra- and extra-regional markets should be enhanced. Simultaneously, the institutional environment for developing poverty-stricken areas should be fostered and improved to remove the regional causes of poverty.

3.4.2 Theory Development for Targeted PAFCS

As the practice of targeted poverty alleviation progresses in China, the ideology and concepts of targeted poverty alleviation—whereby *the vital, critical, and determining factor of successful poverty alleviation is targeted*—have become popular after 2015. This idea has become the guideline, basic principle, basic procedure, and practical path providing a theoretical reference for further research on PAFCS and offers theoretical guidance for promoting the implementation of targeted identification, targeted assistance, targeted management, and targeted assessment of PAFCS in practice.

3.4.2.1 Target Identification

Target identification is the prerequisite for achieving the goal of *alleviating the truly poor and truly alleviating the poor* for PAFCS. In essence, it addresses the primary issue of *whom to support*. In a broader sense, the target of PAFCS includes interconnected areas of poverty, counties, communities, and ecologically fragile, climate-sensitive, and poverty-stricken populations who possess certain advantages and endowments in forest carbon trading resources. Yet, from the perspective of targeted poverty alleviation, the narrowly defined target refers to the poverty-stricken people who can participate in PAFCS transactions to change their poverty status and improve their ability to lift themselves out of poverty.

In the past, a *blood-donation* type of poverty alleviation was implemented for those households whose annual per capita net income was less than 2,300 RMB

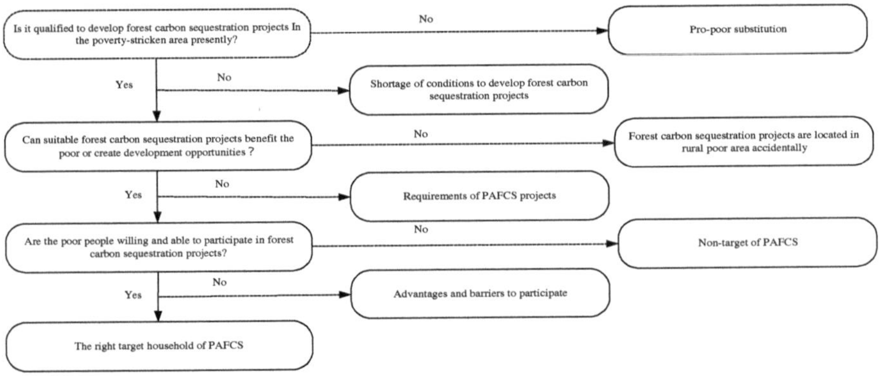

Fig. 3.6 Targeted poverty alleviation routine of PAFCS

(about 333 USD) at a constant price of 2010. Unlike this crude method, targeted poverty alleviation requires the consideration of *two assurances (food and clothes) and three guarantees (basic education, basic health care, and personal housing security)* (Zhang & Huang, 2019) to identify the poverty-stricken population sustainably. Further, populations with PAFCS alleviation potential (Deng et al., 2017) should be identified and their wishes and value judgements respected, as shown in Fig. 3.6. Regarding carbon sequestration AR, in strict academic terms the target populations are, on the one hand, the poverty-stricken people who have certain human capital, natural capital, and social capital and are willing to participate in FCS projects; on the other hand, in practice, PAFCS participants and beneficiaries should not and cannot be limited to poverty-stricken populations. This idea inspired the development of poverty alleviation theory.

Therefore, the potential direction of theoretical exploration lies in fully respecting the wishes and value judgements of farmers, including poverty-stricken people, and examining farmers' willingness and the key factors influencing participation in the development of PAFCS. Other issues include discussing how to improve the accuracy of targeting, enabling poverty-stricken people to enjoy equal opportunities for participation, and putting forward supporting policies that can enhance the enthusiasm and initiative of them to continue participating in FCS projects. These potential theoretical research areas are at the top of the research agenda in this field.

3.4.2.2 Targeted Assistance

Targeted assistance is the essence of *alleviating the truly poor and truly alleviating the poor*, primarily addressing the issue of *who will help* and *how to help* in PAFCS. As mentioned above, the main entities of PAFCS are a *variable* that is constantly developing and renewing around the goal of the *win–win* outcome of FCS project development and poverty reduction. *Elite capture* of pro-poor resources should be avoided. Thus, the village community elite, including the chiefs and party secretaries of poverty-stricken villages, major professional AR households, and heads of the co-ops in the project implementation areas, should function as active organisers and role models and guide the poverty-stricken people to participate in and share the benefits of participation, turning them into participating and assisting PAFCS project entities.

Further, PAFCS targets should be poverty-stricken populations in poverty-stricken areas. Moreover, based on accurately identifying the target population, differentiated assistance measures, namely measures tailored to local conditions, time conditions, and individuals, should be taken to meet individual needs for assistance in different places. Poverty-stricken people should be given equal opportunities and rights to participate; and their anti-participation barriers, such as low education, poor afforestation skills, market inaccessibility, high dependence on traditional livelihoods, and weak organisational strength, should be continuously removed or reduced. The pertinence and effectiveness of assistance measures should be continuously improved to achieve *win–win* goals while facilitating sustainable project operations.

Moreover, the effect of PAFCS is quite diversified, time dynamic, and time-lagged. Community farmers are not only the main beneficiaries of poverty alleviation but also the main victims of ecological construction projects. They make great contributions, but they pay costs relating to production and livelihood accordingly. In this sense, they should receive ecological compensation, including carbon sequestration market compensation. It is vital to implement policies tailored to individuals and local conditions so that the costs paid by local people will not exceed the benefits received.

In this regard, attention should be given to three issues. The first is how to *develop industry* in poverty-stricken areas and create good conditions for sustainable PAFCS development while providing targeted assistance to poverty-stricken households to *help people.* Bottleneck barriers to participation should be removed to encourage poverty-stricken people to participate more effectively at each stage and enable them to equally, fully, and genuinely voice their concerns and interests and fairly and reasonably attain participation benefits. The second is how to fully account for the negative impact of PAFCS projects on community farmers and actively promote the effective integration of ecological compensation and poverty alleviation policies. The third is how to fully respect traditional norms in agricultural and animal husbandry and ethnic–cultural customs in poverty-stricken areas while following international standards of emission reduction, project transaction rules, and the contract terms thereof. In this process, functions of both formal and informal systems should be fully used to accurately implement policies and guide cultural practices.

3.4.2.3 Targeted Management

Referring to the implementation of a series of management initiatives throughout the PAFCS process, targeted management is the key to *alleviate the truly poor and truly alleviate the poor.* Management tools include institutional arrangements, policy regulation, planning, organisation, adjustment, incentives, and performance assessment, all of which essentially address the issues of *who will manage, what to manage,* and *how to manage.*

First, PAFCS managers include governments on five levels: central, provincial, city (prefecture), county, and township, particularly development and reform commissions, forestry departments, authorities for poverty alleviation, science and education, and finance above the county level. The diversity of management entities makes works complicated and obligations uncertain, which can easily lead to practical problems such as *ignoring responsibilities* and *kicking the can.* The managing entities under the framework of targeted poverty alleviation should promote effective connection and coordination among many administrative departments and integrate managing entities issuing policies such as FCS market transactions, poverty alleviation, and forestry ecological construction. For instance, on the macro level, apart from the central and provincial leadership teams for climate change, energy conservation, and emission reduction, a leading PAFCS organisation capable of interdepartmental coordination should be established.

Second, as an emerging poverty alleviation method through industry development, FCS projects function as a bridge connecting the main PAFCS entities, poverty alleviation resources, methods, and poverty-stricken populations. FCS is a key link in a systematic PAFCS project. PAFCS project management directly affects not only the comprehensive benefits of the project—especially the economic and ecological benefits—but also the contents, means, opportunities for, and extent of poverty-stricken populations' participation, which, in turn, determines the scope and extent of benefits for poverty alleviation targets. Therefore, the focus of management includes targeted management for location selection at an early stage, planning demonstration, review, and filing. Targeted management also concerns the integration of poverty alleviation resources, confirmation of responsibilities, refinement in assistance details, methods, the pro-poor mechanism in the mid-term construction and operation, and mid-term evaluation of poverty alleviation performance and industry supervision throughout project development.

Finally, targeted management requires transparent and dynamic supervision based on an information system on the participation of the poverty-stricken population and the assistance provided by the helping entities and its effectiveness. However, as mentioned before, market-oriented PAFCS is intertwined with multiple entities' different interests. The cost of obtaining all the information needed to achieve Pareto optimality is extremely high, often resulting in moral hazard, adverse selection, and other issues on the part of participating entities acting as *homo economicus.* This dilemma requires a focus on *information efficiency, market failure,* and *incentive compatibility,* fully acknowledging governments' guiding, promoting, and supervisory roles. Supporting policies on PAFCS incentives, guarantees, and norms as well as assessment and evaluation methods should be adjusted to match local conditions. Moreover, the pro-poor mechanism centred on the fair participation of poverty-stricken farmers and protection of their interests should be continuously improved to minimise *elite capture.*

While satisfying the multiple entities' pursuit of their interests, poverty alleviation should be achieved following a market approach of industry development to create benefits and development opportunities for the poverty-stricken people. In this regard, it is necessary to empirically study three basic issues. The first is how to define the relationship between farmers' livelihood capital and direct benefits from participating, given that participating farmers have diverse profiles, and how to differentiate policies for different groups and optimise PAFCS approaches. The second is how to enable local communities where projects are implemented to mobilise, manage, coordinate, and achieve a *win–win* outcome where FCS project development and transaction costs are reduced, and pieces of afforestation lands are connected and managed collectively. The third is how to maximise the leading role of rural elites and promote the active participation of community farmers, including the poverty-stricken people, to operate sustainably and share development achievements. The reality is that FCS project development depends on the lead and example of the elite. Thus, an in-depth analysis is needed of PAFCS's pro-poor nature and errors in poverty alleviation practice, and further discussion is required for effective approaches to achieve *win–win goals* and establish and improve pro-poor mechanisms.

3.4.2.4 Targeted Assessment

Focused on issues including *who will assess* and *how to assess*, targeted assessment of PAFCS is the guarantee to *truly alleviating the poor and alleviating the true poor*. Targeted assessment under the framework of targeted poverty alleviation is performed by third-party evaluation agencies independent of stakeholders, for example, consulting firms in this industry, private-sector evaluators, and professional evaluation organisations. In their capacity as *observers*, they professionally, objectively, and fairly evaluate whether a poverty alleviation approach is effective, whether poverty reduction meets the anticipated goal, particularly whether the benefits to the poverty-stricken people are greater than the costs they incur, and whether the development opportunities are guaranteed. The evaluation conclusion not only serves as a basis for decision-making by relevant governmental departments to make timely adjustments to the PAFCS strategy and optimise policies, but also urges the poverty alleviation entities to improve their work methods and approaches in a timely fashion to correct errors in their practice to increase PAFCS transparency, the compliance rate of the poverty alleviation entities, and their performance. All these point to the characteristics of poverty alleviation, including diversity, spatial heterogeneity, time-dynamicity, and time-lag.

In this regard, two basic issues are worthy of further discussion. On a macro level, the first is how to establish a comprehensive system of measurable, assessable, controllable, and inspectable indicators to dynamically monitor and quantitatively evaluate the comprehensive poverty alleviation performance. Second, on the microscopic level, the farmers' perception of the poverty alleviation effect should be evaluated, highlighting the humanistic care for poverty-stricken groups and enhancing their sense of acquisition, recognition, and satisfaction.

3.4.3 Rationale for the Workflow in PAFCS

Targeted poverty alleviation practice aims at internalising the core concepts of *seeking truth from facts, adapting to local conditions, classifying guidance, and targeting poverty alleviation.* Together with the previous stakeholder analysis, PAFCS entities can be divided into five categories: governments at all levels, business entities, community elites, poverty-stricken populations, and third-party assessment agencies. They fulfil their respective responsibilities and continually practise target identification, targeted assistance, targeted management, and targeted assessment, including project planning and design, organisational construction, and continuous operation, monitoring, and evaluation. Their working mechanism can be summarised as follows.

Governments at all levels function as managers, regulators, and supervisors. Using supporting policies, they strengthen system design, integrate poverty alleviation resources, compensate for market failures, formulate targets, meet the needs of poverty-stricken farmers, and encourage afforestation entities to involve as many poverty-stricken farmers as possible in afforestation. Moreover, governments should dynamically and comprehensively manage and trace all project development aspects, promoting the reform and transformation of PAFCS from a *one-wheel-driven* focus on poverty-stricken areas to a *two-wheels-driven* focus on both the region as a whole and individuals within it.

Regarding the business entities in the projects, they, as the project development leaders, executors, and investors, should fully consider the actual conditions and interests of poverty-stricken communities and poverty-stricken population, balance the relationship between FCS increase and poverty alleviation, and proactively execute PAFCS practice while adhering to relevant policies and respecting poverty-stricken populations' wishes. They should also provide feedback to governments about their successes and difficulties, enabling governments to adjust policies. Furthermore, village chiefs, village party secretaries, first secretaries, major AR households, heads of professional co-ops, respected elders, clan chiefs, and other community elites are expected to act as role models who organise and guide the poverty-stricken people to participate and share the benefits of participation. They are also expected to provide feedback to governments about the effects of poverty reduction and existing problems, enabling governments to adjust policies.

A third-party assessment agency refers to a third-party evaluation organisation, which is different from the first party, which comprises all levels of governments, business entities, and community elites, and the second party, which comprises poverty-stricken populations. The agency is usually an independent or commissioned third-party, industry, or private consulting agency or a professional evaluation organisation, which is independent of stakeholders. It acts in the capacity of an *observer* who objectively and impartially evaluates the target identification, poverty reduction approaches, and effects of pro-poor measures. As shown in Fig. 3.7, governments can use its assessment as a reference to make timely revisions to their strategies and policies, and it can drive business entities and community elites to improve their

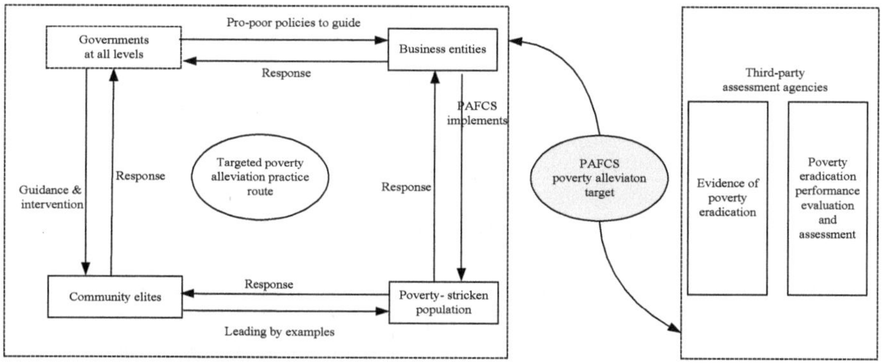

Fig. 3.7 Rationale for the workflow in PAFCS

work methods and correct deviations in poverty alleviation practices, thereby creating better benefits and development opportunities for the poverty-stricken people.

3.4.4 PAFCS Analytical Framework from a Targeted Perspective

In summary, PAFCS aims to create a *win–win situation* in response to climate change and to ensure poverty alleviation, relies on FCS project development, focuses on creating participation opportunities and protecting the rights of poverty-stricken populations, and culminates in *alleviating the truly poor and truly alleviating the poor*. It is underpinned by the sustainable development of the FCS industry and institutional innovation. PAFCS is an effective whole incorporating target identification, targeted assistance, targeted management, and targeted assessment throughout the process of FCS projects. How to fully respect the wishes and value judgements of farmers, including the poverty-stricken people, enhance farmers' willingness to participate, remove barriers to their participation, mitigate cultural conflicts while respecting ethnic traditional customs, mobilise communities and their elite organisations to play an exemplary role while avoiding *elite capture,* and highlight the comprehensive performance and third-party evaluation of farmers' perceptions of the effects of poverty alleviation? All these questions are vital in ensuring the effectiveness of *alleviating the truly poor, truly alleviating the poor.* The theoretical analysis framework of this research is shown in Fig. 3.8.

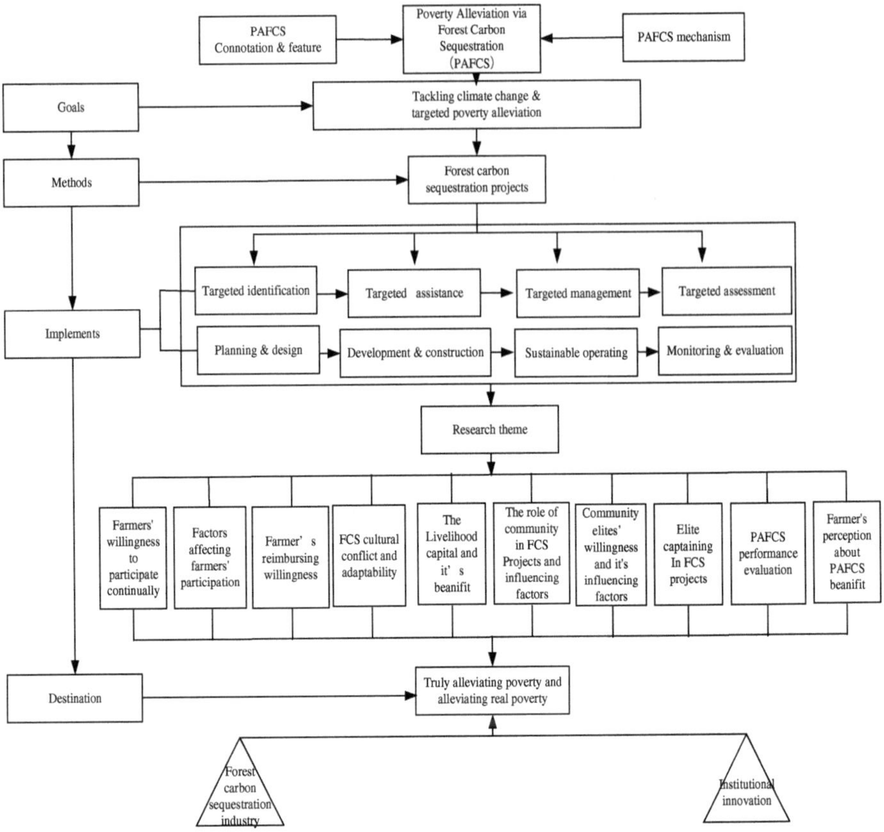

Fig. 3.8 PAFCS Research schema from the perspective of targeted poverty alleviation

3.5 Summary

Countries around the world are coping with climate change, and China is dedicated to targeted poverty alleviation. Against this background, this chapter presents a theoretical analysis framework for research on PAFCS from a targeted perspective, based on the results of existing research and theoretical analysis. The main research conclusions and implications are listed below.

First, PAFCS is China's emerging poverty alleviation model, which has arisen from its participation in and leading of global actions in addressing climate change. This model features development, participation, and *non-blood-donation-type* poverty alleviation. Poverty reduction is the goal of many ecological compensation projects, including FCS projects. PAFCS has an identified target for poverty alleviation with diverse participatory entities, multidimensional effects, and strong policy support. The intrinsic incentive includes obligations for emission reduction, economic benefits, and enhancement of corporate public image; the extrinsic incentive includes legal and regulatory constraints on emission reduction, governmental policy promotion, and social demands.

Second, the core of PAFCS lies in the effective participation of poverty-stricken populations. The beneficiaries of PAFCS should not and cannot be limited to poverty-stricken populations, but the participation of them is undoubtedly vital in gaining benefits, changing understandings of development, and improving development capabilities. PAFCS encompasses empowerment and equal opportunities. FCS projects provide poverty-stricken populations with fair participation opportunities and ensure that they gain reasonable participation benefits so they can enjoy the welfare derived from external poverty alleviation resources to the greatest extent possible.

Third, as a systematic project, PAFCS is not only a poverty alleviation-oriented industrial and economic activity but also a realistic segment in the innovation of ecological poverty alleviation and green poverty alleviation. It is also a dynamic, open, and composite system driven by the emerging carbon trading market, government policies, and social assistance. It is composed of PAFCS entities, pro-poor resources, poverty alleviation methods, and an incentive, participation, and support system in close connection therewith.

Fourth, the PAFCS effect is comprehensive. On a macro scale, PAFCS promotes socioeconomic development in poverty-stricken areas and enhances the organisation and management capabilities of poverty-stricken communities. Microscopically, PAFCS increases the economic incomes of farmers and the poverty-stricken people, enhances their capabilities, and creates development opportunities for them. It generates both economic and non-economic benefits, short- and long-term effects, and positive and negative consequences. Hence, the effects of PAFCS have the characteristics of diversity, spatial heterogeneity, temporal dynamics, and time-lagging.

The above conclusions bear policy implications. First, a *win–win* solution for climate change and poverty alleviation is a prerequisite for PAFCS. FCS projects undoubtedly have poverty reduction potential that must not be neglected. However, whether such potential can be transformed into an incentive depends on the basic philosophy of PAFCS, pro-FCS policies, and actions highlighting its poverty alleviation function. Poverty alleviation must be achieved in the process of promoting the FCS market and industry development, particularly the sustainable PAFCS project operation. Specifically, the concepts of inclusive growth, green poverty reduction, and a *win–win* response to climate change and poverty alleviation should be incorporated into FCS industry development strategies and top-level institutional design to formulate scientific and systematic implementation plans to improve the effectiveness and sustainability of PAFCS.

Second, the effective participation of poverty-stricken populations is essential. PAFCS stresses that community elites play an active role in driving the participation of poverty-stricken populations and avoiding *elite capture*. More importantly, participation by poverty-stricken populations should be improved in terms of their capabilities, opportunities, and depth of involvement, ensuring the needs of poverty-stricken populations are prioritised and treated in a differentiated manner, promoting the reform and transformation of PAFCS from a *one-wheel-driven* focus on poverty-stricken areas to a *two-wheels-driven* focus on both the region as a whole and households and individuals.

Third, one of the important strengthening strategies is the integration of poverty alleviation resources. As a systematic project, PAFCS cannot rely solely on industrial and carbon sequestration policies in response to climate change; it requires appropriate governmental intervention that integrates policies for poverty alleviation and ecological construction. In this way, PAFCS can guide and promote synergy among project entities, integration of poverty alleviation resources, and unity of poverty alleviation methods, thus giving renewed impetus and vitality to creating benefits and development opportunities for the poverty-stricken people and improving the efficiency and practical effects of PAFCS in poverty eradication.

Fourth, assessment of the pro-poor effect is an important guarantee of poverty alleviation. Evaluation of PAFCS performance should focus on the objective assessment of direct net benefits and opportunities created for poverty-stricken populations at the macro level and evaluation of farmers' and the poor's subjective perceptions at the micro level. Attention should be paid to the negative impacts of carbon leakage on livelihood activities, such as logging and grazing, and the cultural adaptability of local people.

References

Chen, C. Y. (2010). Forestry carbon sequestration and farmers' livelihoods—take the world's first forestry carbon sequestration project as an example. *World Forestry Research, 23*(5), 15–19.

Chen X Y (2014) Analysis of the theoretical connotation and operating mechanism of the low-carbon poverty alleviation model. *Ecological Economy (Academic Edition), 1*, 102–104

Deng, W. J. (2014). Difficulties, countermeasures and path selection of precise poverty alleviation. *Rural Economy, 6*, 78–81.

Deng, X. H., Zeng, L., & Xiao, H. L. (2017). Analysis of the concept, composition and operation mechanism of tourism precision poverty alleviation. *Jiangsu Agricultural Science, 2*, 265–269.

Dodd, M. E. (1932). For whom are corporate managers trustees? *Harvard Law Review, 45*, 1145–1163.

Freeman, R. E. (1984). *Strategic management: a stakeholder approach* (pp. 1–21). Pitman, Boston.

GaoG. (2015). Effectiveness, problems and thinking turn of poverty alleviation and development in ethnic areas.*Guizhou Ethnic Studies, 5*, 131–135.

Ge, Z. J., & Xing, C. J. (2015). Targeted poverty alleviation: Connotation, practical dilemma and explanation of reasons based on a survey of two villages in Yinchuan, Ningxia. *Guizhou Social Science, 5*, 157–163.

Huang, C. W., Ye, T., & Lai, L. (2016). Innovation of poverty alleviation model-precision poverty alleviation: theoretical research and Guizhou practice. *Guizhou Social Science, 10*, 4–11.

Huang, C. W., & Qin, Z. M. (2015). Research on targeted poverty alleviation and the construction of the national poverty alleviation governance system. *Journal of Yan'an Executive Academy of China, 8*(1), 131–136

Jia, S. H., & Chen, H. H. (2002). Comment on the definition method of stakeholders. *Foreign Economic & Management, 24*(5), 13–18.

Ji, X., & Wang, X. L. (2012). Carbon financial innovation and poverty via low carbon. *Agricultural Economical Problem, 1*, 79–87.

Lu, H. W. (2012). A preliminary study of low-carbon poverty alleviation paths and policies in contiguous destitute areas. *Journal of Guangxi University (Philosophy and Social Sciences Edition), 3*, 23–27.

Shen, M. Y. (2017). Research on related issues of carbon sequestration and poverty alleviation in Tibetan areas of northwest Sichuan. *Sichuan Forestry Investigation and Design, 2*, 1–8.

Tang, L. X., Luo, J. Y., & Li, X. Y. (2015). Policies and practical dilemmas in the implementation of targeted poverty alleviation mechanisms. *Guizhou Social Science, 5*, 151–156.

Wang, S. T. (2014). Targeted poverty alleviation: Changing flood irrigation to drip irrigation. *Sichuan Party Construction (Rural Edition), 4*, 14–15.

Wang, S. G., & Guo, Z. H. (2015). Research on targeted poverty alleviation in China. *Guizhou Social Science, 5*, 147–150.

Wang, X. Z., & Gao, J. Z. (2011). Analyzing the development of the forest carbon sequestration market from the perspective of market entities. *Forestry Investigation and Planning, 36*(1), 117–119.

Xu, C. H. (2015). *Research on the development of low-carbon economy in western China* (p. 332). People's Publishing House.

Zeng, W. Z., Zhang, J. Y., & Yang, F. (2016). Poverty alleviation by forest carbon sequestration: Theoretical discussion and realistic thinking. *Agricultural Economy, 5*, 17–22.

Zhang, Y., & Huang, Y. L. (2019). Will forest carbon sequestration projects help reduce poverty? *Forestry Economic Issues, 1*, 71–76.

Zhuang, T. H., Yang, F., & Zeng, W. Z. (2016). The connotation of targeted poverty alleviation and its dialectical relationship with targeted poverty alleviation. *Inner Mongolia Social Sciences (Chinese Edition), 37*(3), 6–12.

Zhang, Z. B. (2016). An analysis of comrade Xi Jinping's thoughts on poverty alleviation and development. *Policy, 1*, 46–48.

Zuo, T., Yang, Y. X., & Zhong, L. (2015). Targeted poverty alleviation: technical targeting, theoretical analysis, and practical challenges. *Guizhou Social Science, 8*, 156–162.

Chapter 4
Evolution of PAFCS Policies

With increasing recognition of the hazards posed by climate change to human survival and development, global cooperation to address climate change is growing, and consensus has been achieved on the need to reduce carbon emissions and increase carbon sequestration. FCS is one important measure already employed by all countries to address climate change. PAFCS is an innovative practice conducted by developing countries, including China. This chapter summarises the policy of PAFCS development in China and demonstrates how it has been implemented over time. The summary of policy evolution indicates that PAFCS is an innovative creation based on the continual practice, application, and extension of cooperative climate change governance. Its longitudinal development can be divided into three stages: the establishment of frameworks for carbon sequestration cooperative governance, the implementation of pilot projects, and targeted poverty alleviation via FCS.

4.1 Emission Reduction and Carbon Sequestration in Response to Climate Change and Framework for International Cooperation and Governance (1990–2007)

Climate change, a global concern in a global society, poses severe challenges to human survival and development. Global society has reached a consensus on the functions of forests in carbon sequestration and believes that forestry measures are important ways to address climate change. China has also chosen forestry development as a strategy to respond to climate change.

W. Zeng and F. Yang, *Poverty Alleviation Via Forest Carbon Sequestration*, International Research on Poverty Reduction, https://doi.org/10.1007/978-981-99-8191-5_4

4.1.1 Formation of International Cooperative Governance Framework (1990–2000)

The UN Convention in December 1990 decided to establish an Inter-governmental Negotiating Committee (INC) to draught a protocol for the global governance of climate change. The first INC conference was held in February 1991. After five conferences, the FCCC was signed in May 1992 in Rio de Janeiro, Brazil. A manifestation of the world's expectation and passion for sustainable development and determination to combat climate change, it took effect on 21 March 1994. The FCCC was the first international convention to control emissions of carbon dioxide and other greenhouse gases and to tackle the adverse economic and social effects of global warming. It was also the first basic framework of international cooperation addressing climate change. By April 2022, there was a total of 197 parties to the convention (UNCC 2022).[1] The most important outcomes of the FCCC are threefold: its objectives, its basic principles, and the commitments made by all parties. First, the second article of the FCCC presents the following definite objective:

> To achieve stabilisation of greenhouse gas concentrations in the atmosphere at a level that would prevent dangerous anthropogenic interference with the climate system. Such a level should be achieved within a timeframe that is sufficient to allow for the natural adaption of ecosystems to climate change, to ensure that food production is not threatened, and to ensure that economic development proceeds sustainably.

This objective is twofold: first, to stabilise atmospheric greenhouse gas concentrations to address climate change; and second, to prevent the risks caused by climate change, including ecosystem adaptability, food production safety, and sustainable economic development.

Second, the FCCC established the basic principles for international governance, including equity, common but differentiated responsibilities, respective capabilities, sustainable development, cooperation, respecting national specific conditions, and cost-effectiveness. These principles consider all aspects of the international governance of climate change and guarantee that such governance would develop in the desired direction in the following 20 years.

In addition to objectives and principles, the FCCC set out the commitments of all parties and states in terms of taking active carbon sequestration and emission reduction measures to decrease emissions in 2000 to the level of 1990. The parties were divided into Annex I and non-Annex I countries, with the former including the Russian Federation and other industrialised countries in the Baltic Sea that were to take much historical responsibility for climate change, and the latter being developing countries. Given that the priority of developing countries is to achieve socioeconomic development and poverty elimination, no greenhouse gas emission reduction quota was set for them. Instead, developed countries were required to take the lead in emission reduction and offer funds and technical support to developing countries.

[1] From the UNCC website https://unfccc.int, accessed 28th April 2022.

After obtaining funds and technical support from developed countries, developing countries were to take relevant measures to address climate change.

From 1995 to 2018, after 24 conferences among the parties to the FCCC, *the Kyoto Protocol* in 1997, *Copenhagen Accord* in 2009, *Durban Platform* in 2011, and *Paris Agreement* in 2016 were signed, all of which have continually shaped the frameworks of the international governance of climate change.

After the FCCC was passed, 149 countries or regions signed *the Kyoto Protocol* in Kyoto, Japan, in December 1997, which aimed to reduce global greenhouse gas emissions. *The Kyoto Protocol* set the first quantitative indicators under legal restrictions, setting up a collective goal that all developed countries must reduce greenhouse gas emissions between 2008 and 2012 to a level 5.2% below that in 1990 and assigning each party a specific emission reduction task. Developing countries, including China and India, could voluntarily set their emission reduction goals.

Another important achievement of *the Kyoto Protocol* was that three flexible mechanisms, namely Joint Implementation (JI), Emissions Trading (ET), and CDM, were presented for parties to undertake emission reduction obligations via economic means, based on the principle of cost-effectiveness established by the FCCC. For instance, developed countries could jointly carry out projects to reduce greenhouse gas emissions or increase carbon sequestration and could also share emission reduction units. Developed countries were encouraged to transfer emission reduction cancellation to developing countries by launching greenhouse gas emission reduction projects there.

These three flexible mechanisms achieved the objectives and proved the effectiveness of economic means in emission reduction. Additionally, CDM conducted in developing countries promoted the development of renewable energy sources and made developing countries more conscious of climate change and more confident about emission reduction than before. The effects of CDM laid a solid foundation for subsequent developments.

4.1.2 Development of the CDM Mechanism (2001–2003)

The Kyoto Protocol acknowledged that FCS contributes to relieving global warming and requested that all parties strengthen their sustainable forest management and vegetation recovery and protection. Moreover, developed countries were to provide developing countries with funds and technology and use the carbon sequestration quota generated from the AR projects in developing countries to offset their emissions. In many international negotiations of the FCCC since 1997, global society has paid considerable attention to the role of forests in clustering and absorbing carbon dioxide. The *Bonn Political Agreement* in 2001 and the *Marrakesh Accord* in 2001 both incorporated AR activities into the CDM of *the Kyoto Protocol* and encouraged all countries to use AR to offset their industrial carbon dioxide emissions. In principle, both these documents agreed to set AR as the first commitment phase of CDM, meaning that developed countries could undertake CDM FCS projects conducted in developing countries to offset their emissions to some extent.

When the ninth Conference of Parties (COP9) of the FCCC was organised in December 2003, the international community had already agreed to include AR activities in carbon sequestration CDM projects and formulated new operation rules to initiate CDM projects. CDM projects helped Annex I countries meet their commitments to greenhouse gas emission reduction at low cost, while developing countries could, by introducing advanced techniques, formulate sustainable development strategies, offering them a win–win choice.

CDM projects in China started in 2002 when the Dutch and Chinese Governments signed a cooperation contract for the Huitengxile Wind Farm Project in Inner Mongolia Autonomous Region. Through CDM projects, China has actively participated in international carbon market trading. With the rapid development of CDM projects since 2002, China replaced India in 2006 as the second-largest country in terms of CDM projects. By 2014, a total of 5058 CDM projects had been approved by CDRC, and 3802 CDM projects were successfully registered at the UN CDM-EB. The global accumulative CER was 1.468 billion tons of carbon dioxide, of which China ranked first and accounted for over 60%.[2] The period 2006–2013 witnessed particularly rapid development of CDM projects in China, with 3,740 projects registered and documented. Since 2013, due to the European Union's restriction on CDM projects, the number of such projects in China has seen a sudden decline. According to UNFCCC data, the number of registered CDM projects in China in the last five years has dropped to nearly 0.[3]

4.1.3 Chinese Forestry as a Means of Addressing Climate Change (2004–2007)

China was among the first parties to sign the UNFCCC and *the Kyoto Protocol* and has issued multiple policies and taken several measures to address climate change.

First, China has actively explored carbon sequestration pilot work. In 2004, the first carbon sequestration pilot began, consisting mainly of carbon sequestration evaluation and implementing CDM projects. With support from the World Bank in 2006, the first CDM project in the world, the Pearl River Basin Reforestation Project, was implemented in Guangxi as a role model for subsequent CDM worldwide. Later, Sichuan and Yunnan also used international funds to initiate carbon sequestration pilot works. In 2006, China State Forestry Administration released *Guiding Suggestions on Afforestation and Reforestation Carbon Sequestration Projects under the*

[2] From the Carbon Trade Market website: http://www.tanjiaoyi.com/article-5811-1.html, released on 20th December 2014.

[3] From the *Sighted Economist* https://baijiahao.baidu.com/s?id=1707681005793117148&wfr=spider&for=pc, accessed 30th April 2022.

Clean Development Mechanism, aiming to promote AR projects and relevant works under the CDM.

Second, China prioritised the formulation of national strategies to address climate change. *The Chinese Government Working Report* (2007) announced the strengthening of water conservancy, forestry, and grassland construction, the enhancement of rocky desertification governance and climate changeability construction, and the acceleration of ecological restoration. China State Council formally issued China's *National Climate Change Programme* in 2007, which was China's first comprehensive policy on climate change and the first national programme in this field in any developing country. The *National Climate Change Programme* fully elaborated on China's main measures before 2010 for climate change and named forestry as one of the six key fields of climate change alleviation and one of the four key fields of climate change adaptation. Moreover, policy measures were proposed to increase greenhouse gas absorption sink in forests, maintain and enhance the overall functions of forest ecosystems, and build an eco-environment. Furthermore, the special position and role of forestry in climate change were highlighted.

Third, China actively promoted national capability construction to address global climate change. In early 1990, China State Council established a National Climate Change Countermeasure Coordinating Team, which was responsible for putting forward measures to address climate change in China. Since 2003, the coordination team has successively set up a climate office, a carbon sequestration office, an energy office, and other forestry management offices and taken the lead in planning climate change works. In 2007, the National Leading Group to Address Climate Change was founded, which was responsible for formulating major national strategies, policies, and measures to address climate change. Thereafter, the Climate Change Department at CDRC and the National Energy Committee were successively founded.

Noticeably, forestry development has always been a strategic choice for China to address climate change. FCS plays an irreplaceable role in surfacing an ecological culture and creating a living environment in which humans can live in harmony with nature.

4.2 Promotion and Applications of Pilot Projects and Exploration of Multiple Benefits of FCS (2008–2011)

Driven by global climate negotiations and the emergence of the FCCC and *the Kyoto Protocol*, COP13 in 2007 signed *the Bali Roadmap*, which laid down that emissions must be reduced, due to forest destruction and forest deterioration, and that carbon sequestration should be increased through forest protection, forest sustainable management, and forest area expansion. To date, many countries, including China, have regarded these practices as key optional means to address climate change. A high value has been placed on forests' carbon cost-effectiveness for climate change. Meanwhile, the non-carbon benefits of FCS, including ecological, social, and economic benefits, have been explored, affirmed, and popularised. For instance, funds from FCS projects can be used for local biodiversity protection to guarantee community sustainability and improve farmers' livelihood.

4.2.1 Conducting FCS Pilot Projects

At the UN Summit Conference of Climate Change in September 2009, President Hu Jintao announced to the world that China would try to increase FCS and expand forest areas by 40 million ha and forest reserves by 1.3 billion m^3 from 2005 to 2020 (i.e., twofold). *The Chinese Government Working Report* in 2010 defined the concept of FCS for the first time in history. Specifically, afforestation is a measure to develop FCS, achieve the double-increase objective in forestry, and fulfil China's commitments.

The Notification on Carbon Sequestration Afforestation Pilot Works (henceforth, the Notification) issued by China State Forestry Administration in 2010 signalled the start of carbon sequestration afforestation pilot works. A set of strict standards was proposed for carbon sequestration afforestation. In particular, *the notification* stressed that only forests following certain standards can be considered carbon sequestration forests. Compared with ordinary afforestation, carbon sequestration afforestation—an activity implemented in strict accordance with climate change methodology—highlights the carbon sequestration function, multiple benefits of forests, econometric monitoring, and corresponding technical requirements.

China State Forestry Administration issued the *Carbon Sequestration Afforestation Technique Provision (Tentative)* (henceforth, *the Provision*) in 2011, which provides strict carbon sequestration afforestation methods and addresses technical issues. *The Provision* explicates that carbon sequestration afforestation is to be carried out preferentially in areas covering important ecological zones and vulnerable eco-environments. Site selection must meet the following four conditions. First, the sites must be hills or lands, sand lands, or other types of afforestation lands that have been barren at least since 1 January 2000 (or 1 January 2005 in certain situations). Second, the ownership of the sites must be definite, authenticated, and approved by a government at the county level or above. Third, trees are expected to play an effective role in carbon sequestration. Fourth, multiple benefits are to be generated, including local biodiversity protection, prevention, and control of land deterioration, local economy, and social development. Based on the above four conditions, the geographic zone from Heihe River to Tengchong was chosen as a suitable district for carbon sequestration afforestation.

In addition to the above four conditions, tree species must be capable of carbon sequestrating, which means that eucalyptus and certain other tree species may not be prioritised. Additionally, tree species that are sustainable and resistant to severe conditions must be the first choice. The proportions of broad-leaved and conifer species must be determined according to local conditions, while mixed forests must be created and monoculture of tree species avoided. As concerns planting, *The Provision* provides for soil tillage, planting density, fertilisation, fostering, and management, among other issues.

4.2.2 Ecological Benefits of FCS

The ecological benefits of FCS include biodiversity protection in addition to addressing climate change. The UN COP13 in 2007 signed the *Bali Roadmap* and formally proposed the FCS mechanism, aiming to decrease emissions due to deforestation and forest degradation by utilising the carbon market and to increase carbon sequestration through forest protection, forest sustainable management, and enlarged forest areas (henceforth, the REDD + mechanism). The REDD + mechanism provides developing countries with auxiliary measures to ensure biodiversity safety through REDD + activities. To ensure fund support for the REDD + mechanism, the Durban Climate Conference initiated the Green Climate Fund in 2011, which was mainly used to support REDD + in developing countries.

In response to the REDD + mechanism, China has accelerated afforestation to improve the FCS function. The Afforestation Department of China State Forestry Administration released *The Notification on Enhancing Forestry and Carbon Sequestration Management to Cope with Climate Change,* which encourages afforestation and sustainable forest management to improve forest quality and strictly forbids deforestation and the illegal acquisition of forestland. The China State Forestry Administration launched *The Notification on Enhancing Carbon Sequestration Afforestation Management* in 2009, which requires afforestation management to be reinforced in three ways: a filing process for existing carbon sequestration projects, a registration process for new projects, and a regulated management system. In 2011, the China State Forestry Administration launched *The 12th Five-year Plan for Forestry Development*, which set out to strengthen forest management and improve the coverage rate, stock volume, and carbon sequestration capacity of forests. At the end of 2011, the China State Forestry Administration released *The 12th Five-Year Plan Action Points for the Forestry to Cope with Climate Change*, which required further promotion of afforestation and planting, expansion of forest areas, strengthening of forest management and forest quality, and prevention and control of forest disasters. It also required protection for forestland, wetland, desert ecosystems, and biodiversity, continual increase of forest carbon stocks, and improvement of forestry capabilities to mitigate climate change.

4.2.3 Social Benefits of FCS

FCS projects have critical social benefits. Globally, they give rise to international communication and cooperation and promote the formation of a new pattern and new experience of global cooperation for climate change. Moreover, such projects help developing countries acquire funds and technical support from developed countries, thus promoting their sustainable development.

In China, the launch of FCS projects helps improve local governance and skills while creating an excellent living environment for local people. In 2009, the China State Forestry Administration released the *Forestry Action Plan to Address Climate Change* determined 5 basic principles, 3 goals, and 22 main actions to guide forestry departments in their works. The five basic principles are combining forestry development goals with the national climate change strategy, combining expanded forest areas with increased forest quality, combining carbon sequestration increment with emission control, combining government dominance with social participation, and combining mitigation with adaptation. In the same year, China State Council issued the *Suggestions on Promoting Agriculture Stable Development and Farmer Persistent Income Increment 2009,* which proposed creating modern mountain forests and developing mountainous forest special products, ecological tourism, and carbon sequestration forestry. It also emphasised the multiple benefits of FCS.

In 2012, China State Forestry Administration issued *The 12th Five-year Action Plan of Forestry to address Climate Change,* which encouraged the public and community populations to participate in forest afforestation and enhanced locals' consciousness of natural protection while increasing carbon sequestration. It also set out to improve community living environments and develop local infrastructure and cultural services. In the meantime, the forest administration of forestry departments at different levels was to be further enhanced, as were forest legislation and law enforcement. The reform of forest ownership was to be accelerated to ensure a fair benefit-sharing distribution mechanism and carbon sequestration rights. It was also noted that training and publication on climate change were essential in the community and that the participation of forestry departments and forest populations should be improved. Society, enterprises, groups, and individuals were to be encouraged to participate in planting, afforestation, forest protection, and FCS protection.

4.2.4 Economic Benefits of FCS

The economic benefits of FCS are increased job opportunities, significant revenue growth, and alleviation of poverty. To encourage farmers to obtain incomes through forest management and fully leverage the economic benefits of FCS, China has actively promoted the establishment of a carbon trading market.

In 2010, China set up its first national green mutual fund, the CGCF, for carbon sink enhancement, emission reduction, and climate change governance. The CGCF has built a non-profit platform for enterprises, organisations, and the public to store carbon credits through forestry measures, meet their social responsibilities, and improve farmers' incomes and the environment. It has thus enabled more social participation in FCS projects to address climate change.

In 2011, the CDRC launched *The Notification on Carbon Emission Right Trading Pilot Works* and selected Beijing, Tianjin, Shanghai, Chongqing, Hubei, Guangdong, and Shenzhen as pilot cities for carbon emission right trading. *The Notification of the 12th Five-Year Plan for Energy Conservation and Emission Reduction Comprehensive Working Schemes,* issued by the CDRC in 2011, underlined the need to meet emission reduction objectives to promote energy conservation and emission reduction works through the joint efforts of governments, enterprises, and society. The document also proposed to start pioneering works in emission trading and a voluntary emission reduction mechanism.

The Measures on Greenhouse Gas Voluntary Emission Reduction Trading Management was released by the CDRC in 2012 and proposed to build a carbon emission trading market and promote the role of the market mechanism in economic development. It laid a foundation to build a trading system to promote the economic benefits of FCS projects. After the pilot works, a unified carbon emission trading market in China was formally launched on 6 July 2021, and China Ministry of Ecology and Environment issued a document entitled *Carbon Emission Right Trading Management Methods (Tentative)* at the same time.

Locally, the National People's Congress, the Government, and the Ministry of Housing and Urban–Rural Development of Beijing have launched many policies, provisions, and notifications on carbon emission reduction, including a pilot for carbon emission right trading, trading affiliations, management methods of trading emission right and cancellation, and emission total quantity control. Tianjin has issued nearly 10 policies or notifications since 2013 which cover the issues of carbon trading pilot works, carbon market trading management, quota distribution scheme, and greenhouse gas emission control works. These policies and notifications guarantee the stable operation of the carbon market to achieve the energy conservation and emission reduction goals in Tianjin. Guangdong has released over 10 policies, notifications, or interim measures since 2012 which cover not only carbon trading pilot works, carbon market trading management, and quota distribution schemes, but also voluntary emission reduction quantity, interrogation, examination, and check reporting.

As regards trading systems, China's first trading system for farmer forest management, Lin'an Farmer Forest Management Carbon Sequestration Trading System, was set up in October 2014, with farmers selling carbon sinks from the first day of operation.[4] The innovativeness of this system is that farmers can directly manage forests and acquire FCS incomes. It offers a solution to the problem that forest farmers have no short-term incomes from long-term forestry production and is a valuable reference for developing China's FCS trading market.

[4] From the China Carbon Emissions Trading Network website, http://www.tanpaifang.com/tanguwen/2014/1016/39229.html, accessed 16 October 2014.

4.3 Popularisation of Targeted Poverty Alleviation Focusing on FCS and Initiation of Ecological Industry Development for Poverty Alleviation (2012–Present)

The Chinese Communist Party, the ruling party of China, announced a strategic decision in November 2012 to promote ecological civilisation, aiming to construct an ecological and prosperous country. The UN COP21 passed the *Paris Agreement* in 2015 and agreed to strengthen global responses to the threat of climate change by promoting sustainable development and poverty alleviation endeavours. FCS, which features low organising costs, stable poverty alleviation effects, and persistent poverty relieving potential, is increasingly favoured by policymakers. It has gradually become an important part of poverty alleviation via eco-compensation and an up-and-coming way of alleviating poverty through eco-industrial development.

4.3.1 Promotion of Ecological Culture

In November 2012, the 18th National Congress of the Communist Party drew up a strategy to promote ecological civilisation through a 10-aspect blueprint. The first, second, and third sections of the 18th National Congress Report elaborate on the great achievements, important position, and key objectives of ecological civilisation in China. The whole eighth section of the report profoundly explicates all aspects of ecological civilisation.

The Suggestions on Speedy Promotion of Ecological Civilisation (henceforth, *the Suggestions)*, launched by the Chinese Government in May 2015, underlined the need to establish green development, recycled development, and low-carbon development as the basic means to achieve ecological civilisation. *The Suggestions* fully reflect the Chinese Government's determination to achieve ecological civilisation and present its overall requirements, objective prospects, key tasks, and institutional system. Thus, *The Suggestions* is a key document in the promotion of ecological civilisation in China. It is based on the philosophy that *green water and green mountains are golden, lucid water and lush mountains are invaluable assets and* on the principles of innovation, coordination, greenness, openness, and sharing. Moreover, it calls for ecological civilisation to be integrated into governmental actions in the fields of political, economic, cultural, and social development. In October 2015, strengthening ecological civilisation was written into China's national five-year plan for the first time.

Since then, the Chinese Government, key officials, and various departments have emphasised repeatedly that forestry is a major part of eco-development and the primary task of ecological civilisation. President Xi Jinping has undertaken many discussions of forestry and pointed out that forestry construction is a fundamental part of socioeconomic sustainability. Ecology will prosper only with the prosperity of forestry, and civilisation will prosper only when ecology does the same. Practical, effective methods and measures must be taken to ensure better and faster forestry development, and efforts must be made to meet the requirements of ecological civilisation.

The Forestry Development 13th Five-Year Plan, issued by China State Forestry Administration in 2016, presents the objectives, development patterns, strategic tasks, key engineering projects, and institutional systems for the following five years of forestry development in China. A green and prosperous society also needs forest development and supply-side reforms. For instance, the development of forestry relies on environmental protection, high-quality eco-products, and ecological well-being. Forestry aims to create a happy homeland for all of us with a blue sky, green lands, and clear waters. However, mountainous, forest, and sand areas are usually concentrated in poverty-stricken regions, and it is difficult to develop traditional industries there. Forestry has an entrance threshold, long industrial chains, and huge employment potential which can provide long-lasting incomes for farmers. Consequently, efforts must be made to develop the ecological industry and eco-compensation. There must be a focus on ecological protection while the goal of targeted poverty alleviation in ecologically fragile areas is also achieved.

From 2016 to 2017, China State Forestry Administration launched the *China Ecological Culture Development Outlines* (2016–2020), the General Office of China State Council released *Suggestions on Collective Forestry Property Rights Improvement Regime,* and the Office of China State Forestry Administration issued *The Notification on Pilot Works of Forest Characteristic Town Construction.* These documents aimed to promote the development of eco-tourism, forest leisure, and healthcare businesses, thus not only providing new livelihood opportunities, but also offering new policy guidance for employment.

4.3.2 Ecological Poverty Alleviation

In 2015, the UN Forum on Forests discussed global forest policies for the following 15 years and achieved consensus on the key roles of forests in eliminating poverty and addressing climate change.

In 2016, China State Forestry Administration issued *The Notification on Strengthening Ecological Protection in Poverty-Stricken Areas and Promoting Targeted Poverty Alleviation through Industrial Development.* This document demands that forestry departments at all levels promote poverty alleviation via forestry, strengthen ecological protection and industrial development in poverty-stricken areas, and promote targeted poverty alleviation. Key works include the employment of ecological rangers, returning farmland to forest, and developing woody oil plants. Four specific plans were proposed to benefit from forestry. First, in the following five years, forestry investment in poverty-stricken areas was to be 15% higher than the provincial average. Second, forest tourism, forest park infrastructure, wetland parks, and nature reserves were to be developed alongside planting, aquaculture, and handicraft industries related to tourism. Third, forest fruits and other forest-related industries were to be developed; and fourth, afforestation in poverty-stricken and ecologically fragile areas was to be rolled out. All local governments responded positively. For instance, in 2016, the Department of Forestry of Anhui Province issued *Suggestions on Promoting Targeted Poverty Alleviation* via *Forestry*.

The CDRC, China State Forestry Administration, China Ministry of Finance, China Ministry of Water Resources, China Ministry of Agriculture, and the Poverty Alleviation Office of China State Council jointly issued *Plans on Ecological Poverty Alleviation Works* (henceforth, *the Plans*) in January 2018. *The Plans* aimed to combine ecological protection with targeted poverty alleviation. Ecological poverty alleviation is a reasonable choice given the current ecological crises and the status of poverty-stricken populations: most poverty-stricken areas in China are facing land desertification, rocky desertification, and other ecological problems. As a result, development in these ecologically friable and poverty-stricken areas is unsustainable. Ecological protection must be stressed alongside economic development, as otherwise the friable ecosystem will deteriorate. Ecological poverty alleviation aims at maintaining clear waters and lush mountains while developing the economy.

The Plans presented definite objectives: that the poverty-stricken populations' incomes would be significantly raised by 2020, and their production and living conditions would be considerably improved through their participation in ecological protection, ecological restoration, and eco-industrial development. The ecosystems in poverty-stricken areas were to be effectively improved, with the supply of eco-products and enhanced sustainability, and the eco-compensation level was to match the level of socioeconomic development. The Plans also proposed to establish up to 12,000 professional cooperatives including 10,000 afforestation cooperatives (teams) and 2,000 grassland and husbandry cooperatives, involving the participation of 100,000 poverty-stricken people. Moreover, 400,000 ecological management and protection positions (300,000 ecological rangers and 100,000 grassland protectors) were to be created. The incomes of 15 million poverty-stricken people were to be raised through rapid eco-industrial development.

4.3.3 Poverty Alleviation Via FCS

PAFCS is an innovative ecological poverty alleviation mechanism that integrates eco-development, climate change governance, and anti-poverty actions, effectively converting ecological advantages into economic ones.

The Plans set out to explore carbon trading compensation methods and promote reforms in the trading system of CDM and greenhouse gas voluntary emission reduction, and develop a national carbon emission right trading market. In 2014, China State Forestry Administration issued *Guidance Suggestions on Promoting FCS Trading* to improve CDM FCS project trading and FCS voluntary trading and explore the feasibility of starting FCS trading. All projects were encouraged to employ FCS methodology according to local conditions and to introduce the necessary technical framework for FCS voluntary trading.

In 2018, *the Guidance on Three-Year Anti-poverty Progamme* was proposed by the China State Council to employ ecological rangers and grassland managers from poverty-stricken populations and enhance the infrastructure of forest protection in poverty-stricken areas. It also suggested the trusteeship of natural forests and collective non-profit forests and a mode of *cooperatives + management + poverty-stricken farmers to* engage poverty-stricken populations in forest management and protection. Low-yield and low-efficiency forests in poverty-stricken areas were also to be exploited more efficiently. The collective ownership of forestry was to be reformed so that poverty-stricken populations could offer their woodland management rights as stakes in the afforestation cooperatives, thus increasing their property incomes. Poverty-stricken counties, villages, and farmers could thus benefit from ecological protection under the mechanism of horizontal ecological protection compensation. In the meantime, the major emission enterprises were to be encouraged to purchase forestry carbon sinks from poverty-stricken areas.

Subsequently, China Ministry of Ecology and Environment office issued *The Ministry of Ecology and Environment's Three-Year Action Plan on Targeted Poverty Alleviation (2018–2020)* and *Guidance Suggestions on Ecological Protection and Targeted Poverty Alleviation.* These documents included poverty-stricken counties within the scope of transfer payment in key eco-functional areas. The environmental impact assessment of poverty-stricken areas was to be prioritised at no cost. FCS projects from poverty-stricken areas were to be prioritised for compensation trades in the national carbon emission trading market, and poverty-stricken areas were also given priority regarding developing organic products, cadre assignment, and pilot project selection.

Since the national unified carbon trading market was formally started in 2017, The CDRC and eight other departments issued the *Action Plan on Building a Market-based, Multi-faceted Ecological Protection Compensation Mechanism* at the end of 2018. This plan proposed to prioritise FCS in the carbon trading market and to guide enterprises and other concerned entities to buy emission credits generated by FCS projects in poverty-stricken areas.

The Chinese Central Government issued its *Rural Revitalisation Strategic Plans (2018–2022)* in 2018. The third paragraph of Chapter 21 of this document clearly states that establishing trading regulations of water usage rights, pollution rights, and carbon emission rights is an efficient way to involve forest, grassland, wetland, and other ecological restoration projects in carbon trading.

Locally, all local governments should actively participate in carbon sequestration trading and explore the pro-poor methods of FCS. *The Forest Eco-Construction Plans (2018–2027)* released by Henan Province in 2018 aimed to build an FCS monitoring system. Xin County of Henan Province offers a rare case of an FCS trading project being launched, namely the Xin Shawo Carbon Sequestration Afforestation Project, which covered 3.62 million m^2 forest lands in 2017, with an annual carbon emission reduction of over 6,600 tons, and an annual income over 200 thousand RMB (about 29 thousand USD), benefiting over 200 poverty-stricken farmers.

Since 2018, the Shaoguan Development and Reform Bureau has successfully developed forestry projects in 64 mature provincial poverty-stricken villages and one ethnic village, with a total of 307.8 thousand tons of carbon credits successfully traded in the Guangdong carbon market and a turnover of 5.02 million RMB (about 0.73 million USD). It was predicted that 65 poverty-stricken villages would secure 17 million RMB (about 2.5 million USD), thus 260 thousand RMB (about 37.7 thousand USD) per village.

In 2018, Guizhou Province issued the *Guizhou Ecological Poverty Alleviation Executive Plan (2017–2020)* to carry out FCS projects in 14 poverty-stricken counties, 20 poverty-stricken villages or towns, and 2,760 poverty-stricken villages. Guizhou calculated the carbon credit contained, or the carbon dioxide absorbed, by each tree and sold it publicly on an online platform. The funds earned thereby were credited to farmers' bank accounts. This novel online mode of poverty alleviation has been successful in Guizhou.

In 2019, Yong'an City of Fujian Province issued the *Yong'an Plan on Implementing Bamboo Forest Management of Carbon Sequestration Exploitation, which* initiated bamboo forest projects to leverage bamboo FCS trading to promote the scientific, ecological, and sustainable management of bamboo forests and enhance the competitiveness of the local bamboo industry. Meanwhile, village incomes have increased, which contributes to the rural revitalisation strategy.

Joining global efforts to combat global warming, China has put forward the strategy of ecological civilisation, targeted poverty alleviation, and rural revitalisation. With policy support from the state government and pilot projects carried out across the state, China has made remarkable achievements in terms of eradicating poverty. Meanwhile, the experience gained from numerous cases is of significance and practical value for China to promote targeted poverty alleviation with the help of FCS.

Reference

China State Council. (2007). Chinese Government Working Report 2007. http://www.gov.cn/2007lh/content_553165_3.htm. Retrieved 25 Jun 2022.

Chapter 5
Investigating the Status Quo of PAFCS in Southwest Ethnic Areas

Ethnic regions in Southwest China, rich in natural resources such as rivers, forests, and grasslands but lagging in social development, bring together frontier mountainous areas, ethnic inhabitants, and ecologically fragile and poverty-stricken areas. They are also the regions with the largest number of poverty-stricken ethnic minorities in China. In the past decade, they have set examples of balancing ecological production and human life, poverty alleviation, and ecological protection, and have been pioneers in actively addressing climate change and promoting the practice of PAFCS. This chapter analyses the overall status and difficulties of PAFCS in Southwest China by taking the Novartis Sichuan FCS Project as an example. It summarises the project's effective poverty alleviation means, successful experiences, and innovative models to support the policy-making process.

5.1 Distribution of Ethnic Regions in Southwest China

Ethnic regions usually refer to areas inhabited by ethnic minorities and are characterised by autonomous provinces, autonomous prefectures, autonomous counties, ethnic townships, and ethnic villages. The ethnic region in Southwest China is in four provinces and one city, namely Sichuan, Chongqing, Guizhou, Yunnan, and Tibet (Jiang, 2003). This study adopts the categorisation proposed by Zhuang (2011) and Zheng (2014), which excludes Tibet and covers 14 ethnic autonomous prefectures, 48 ethnic autonomous counties, and non-autonomous counties in Yunnan and Guizhou Provinces (which enjoy the same ethnic policies as autonomous places in terms of economic and social development). Specifically, it includes 14 ethnic autonomous prefectures (3 in Sichuan Province, 3 in Guizhou Province, and 8 in Yunnan Province), 48 autonomous counties or similar counties (4 in Chongqing City, 4 in Sichuan Province, 10 in Guizhou Province, and 29 in Yunnan Province), and some ethnic townships and ethnic villages in some non-autonomous prefectures

W. Zeng and F. Yang, *Poverty Alleviation Via Forest Carbon Sequestration*, International Research on Poverty Reduction, https://doi.org/10.1007/978-981-99-8191-5_5

(counties). The ethnic population stands at about 294.8 million and accounts for 59.14% of the total population of the ethnic areas of Southwest China (see Table. 5.1 for details).

5.2 The Status Quo of Poverty in Ethnic Regions in Southwest China

There are a large number of national-level poverty-stricken counties in Southwest China, mainly contiguous and concentrated in ethnic areas. Since the implementation of the strategy of targeted poverty alleviation in 2013, these areas have experienced significant economic and social development; poverty levels have fallen, and people's living standards have greatly improved. Below, the researchers introduce and analyse the economic development level, regional poverty, and population poverty in Southwest China.

5.2.1 Overall Economic Development of Southwest China

The overall economic development of Southwest China is lower than the national average, but the growth rate is gradually accelerating. By the end of 2021, the resident population of the three provinces (Sichuan, Yunnan, and Guizhou) and one city (Chongqing) in Southwest China (henceforth, these three provinces and one city in Southwest China are referred to as the Southwest China) reached 201.26 million, accounting for 14.4% of China's total population (see Fig. 5.1 and Table 5.2 for details). The urbanisation rate in Southwest China rose from 45.39% at the end of 2013 to 57.57% in 2021, showing an increase of 12.28%, but remained 6.32% lower than the national average of 63.89% at the end of 2021. Meanwhile, the total social fixed asset investment in Southwest China in 2021 was 998.47 billion RMB, with an average annual growth rate of 9.56%, and its share in the national investment increased to 18.34% in 2021 from 10.78% in 2013. It is evident that the economic growth rate of the Southwest region is gradually accelerating, and the gap between it and the national average is narrowing.

The overall industrialisation level in Southwest China is relatively low. As of 2021, the three types of industries in Southwest China were in the ratio of 11.04:37.12:51.84. Compared with 2013, the proportion of primary industry had increased by 1.51%, that of secondary industry had decreased by 9.08%, and that of tertiary industry had increased by 10.55%. The proportion of tertiary industry exceeded that of the secondary industry for the first time in 2015, showing that industrialisation in Southwest China is steadily advancing towards the middle stage. At the same time, the national structure of the three types of industries in 2021 was 7.3:39.4:53.3, a ratio which belongs to the late industrialisation stage. Therefore,

Table 5.1 Regional and population distribution of Southwest ethnic regions

Province	Autonomous prefecture	Autonomous country	Major minorities	Overall population (10,000)	Minority population (10,000)	The proportion of ethnic minority population (%)
Chongqing	—	Shizhu, Pengshui, Youyang, Xiushan	Tujia, Miao, Yao, Dong, Zhuang, Bai, Hui, Manchu, Buyi, Yi, Li, Gelao, Hani, Korean	272.00	192.45	70.75
Sichuan	Liangshan, Ganzi, Aba	Mabian, Ebian, Beichuan	Tibetan, Qiang, Naxi, Hui, Tujia, Mongolian, Yi, Pumi, Lisu	746.68	449.19	60.16
Guizhou	Qian southeast, Qiannan, Qian southwest	Weining, Guanling, Zhenning, Ziyun, Daozhen, Buchuan, Yinjiang, Songtao, Yanhe, Yuping	Gelao, Miao, Tujia, Yaoshui, Mongolian, Dong, Hui, Yi, Buyi, Manchu, Maonan	1739.41	1046.33	60.15
Yunnan	Chuxiong, Honghe, Wenshan, Nujiang, Xishuangbanna, Dali, Dehong, Diqing	Shilin, Luquan, Xundian, Yuanjiang, Xinping, Eshan, Gengma, Shuangjiang, Canyuan, Ninglang, Yulong, Lancang, Ximeng, Menglian, Ning'er, Jinggu, Zhenyuan, Jiangcheng, Mojiang, Jingdong	Yi, Miao, Dai, Bai, Hani, Yao, Zhuang, Lahu, Naxi, Hui, Pumi, Dulong, Nu, Tibet, Lisu, Jingpo	2227.29	1260.15	56.58
Total	14	37	—	4985.38	2948.12	59.14

Source China Statistical Yearbook 2016 and Outline of the implementation of regional ethnic autonomy (China State Council, 1952)

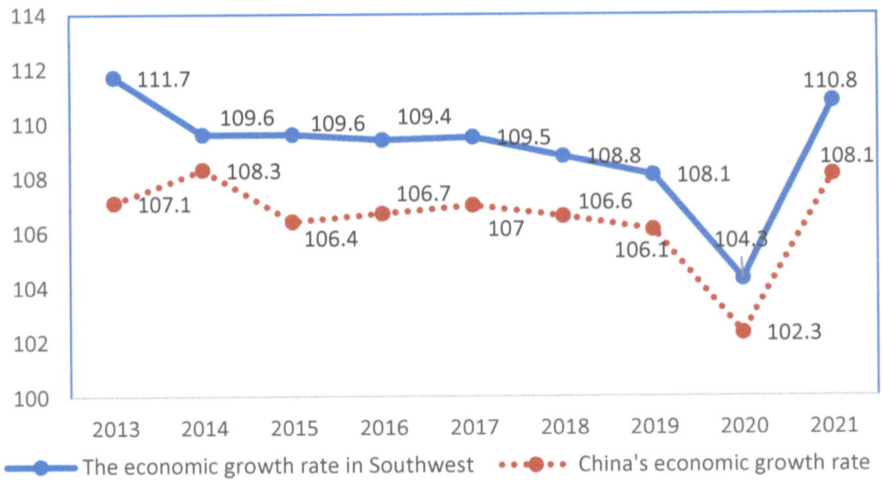

Fig. 5.1 The economic growth rate of Southwest China, 2013–2021 (previous year = 100)

in terms of industrial structure, a gap remains between the industrialisation level of Southwest China and China as a whole.

5.2.2 *Poverty in Ethnic Counties in Southwest China*

Southwest China has been an important region in China's poverty eradication efforts in the past decade. Although absolute poverty had been eliminated in Southwest China by 2020, some ethnic counties are at risk of returning to poverty due to domestic and international challenges such as the COVID-19 epidemic. Due to the special natural geographical environment and economic, social, historical, and many other factors, the poverty-stricken ethnic counties of Southwest China present unique poverty characteristics, which mainly manifest as follows.

First, there is a large overlap between ethnic counties and national poverty alleviation key counties, and the poverty distribution is geographically concentrated and contiguous. Statistics show that, on the one hand, among the 592 national key counties identified by China State Council Poverty Alleviation Office in 2012, 173 are in Southwest China, accounting for 29.22% of all national key counties; moreover, 148 of these are ethnic counties, accounting for 85.55% of the total number in Southwest China. On the other hand, among the 680 counties in 14 highly poverty-stricken areas announced by the state in 2012, 222 are in Southwest China, accounting for 32.65% of the total, including 111 ethnic counties, accounting for 50.45% (see Table 5.3 for details). The poverty in the ethnic areas of Southwest China is serious.

Second, economic development is uneven and disparate across counties. The economic output value of ethnic counties in Southwest China is significantly lower

Table 5.2 Basic socioeconomic situation of Southwest China (2013–2021)

	2013	2014	2015	2016	2017	2018	2019	2020	2021
Resident population (mil)	192.66	193.53	194.93	196.36	197.58	198.72	199.81	201.5	201.26
Share of Chinese population (%)	14.16	14.15	14.18	14.2	14.21	14.24	14.28	14.4	14.4
Urbanisation rate(%)	45.39	46.91	48.49	50.25	51.9	52.4	53.78	53.78	57.57
China's urbanisation rate (%)	53.73	54.77	56.1	57.35	58.52	59.58	60.6	56.51	63.89
GDP (bn.)	5909.5	6488	6989.2	7724	8264.6	9372.8	11021.5	11595	12847.8
Share of China's GDP (%)	9.93	10.07	10.14	10.39	10.44	10.2	11.18	11.44	11.44
Proportion of primary industries (%)	12.5	12.3	12.5	12.4	11.8	11.19	10.6	11.63	11.04
Proportion of secondary industries (%)	46.2	44.4	42.1	41	40.2	38.82	37.11	36.3	37.12
Proportion of tertiary industries (%)	41.3	43.3	45.4	46.6	48	49.99	52.29	52.07	51.85
Fixed asset investment (bn.)	4810.3	5612.8	6432.5	7418.3	8387.9	8506	8916.4	9487.6	9984.7
Share of China's fixed asset investment (%)	10.78	10.96	11.45	12.23	13.08	13.38	16.17	18.29	18.34
Fixed asset investment of farmers (bn.)	133.2	147.3	140.5	143	144	144.6	147.8	149.7	153.7
Share of China's fixed asset investment of farmers (%)	12.63	13.69	13.5	14.35	15.07	15.26	15.87	16.28	16.79

(continued)

Table 5.2 (continued)

	2013	2014	2015	2016	2017	2018	2019	2020	2021
Total revenue of local budget (bn.)	729.5	804.7	882.2	8990	933.1	989.7	1006.1	1025.6	1130.6
Share of China's local budget (%)	10.57	10.61	10.63	10.31	10.2	10.11	11.27	12.4	12.37
Total retail sales (bn.)	2153.2	2567.4	2868.8	3230.5	3612.5	3834.5	4194	4522.1	5194.3
Share of China's total retail sales (%)	9.05	9.44	9.53	9.72	9.86	10.07	10.19	11.54	11.79
Growth rate (%)	13.98	12.83	11.63	12.5	11.8	6.15	9.38	7.85	7.85
Growth rate across China (%)	13.1	12	10.7	1..4	10.2	9	8	9	12.5

Source China Statistical Yearbook (2014–2022)

Table 5.3 Distribution of national minority poverty-stricken counties in Southwest China

Areas	Number of counties in concentrated contiguous poverty with special difficulties	Number of national poverty alleviation counties	Number of ethnic autonomous counties	Number of poverty-stricken ethnic counties	Percentage of poverty-stricken ethnic counties
China	680	592	637	267	45.10
Southwest China	222	173	179	112	64.74
Yunnan	85	73	78	51	69.86
Guizhou	65	50	46	36	72.00
Sichuan	60	36	50	20	55.56
Chongqing	12	14	5	5	35.71

Source List of 592 National Key Counties for Poverty Alleviation and Development in the New Era (State Council Poverty Alleviation Office, 2021), '*Note* on the Announcement of the List of Sub-counties in the National Contiguous Poverty Areas with Special Difficulty' (China State Council Poverty Alleviation Office, 2012a, 2012b)

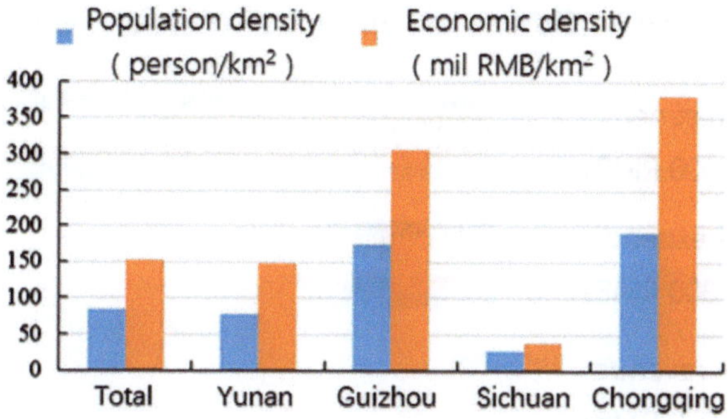

Fig. 5.2 Economic density and population density in Southwest China

than the national average, and the gap in economic development between counties in the area is obvious. In terms of population density, the highest is the ethnic counties in Chongqing, with about 190 people/km^2, followed by the ethnic counties in Guizhou, and the lowest is the ethnic counties in Sichuan, with about 28 people/km^2. In terms of economic density, ethnic counties in Sichuan are the lowest, at only 390.8 thousand RMB (about 56.6 thousand USD)/km^2, while the value for ethnic counties in Chongqing is 3810.8 thousand RMB (about 552.3 thousand USD)/km^2, 9.75 times more than those in Sichuan (see Fig. 5.2 for details). Moreover, in terms of GDP per capita, fixed asset investment, fiscal revenue, and expenditure, inter-county differences in Southwest China are also obvious, showing uneven development (see Fig. 5.3 for details). For example, the lowest GDP per capita in 2021 is in Meigu County, Sichuan Province, with only 16,989 RMB (about 2462.2 USD)/person, while Shangri-La, Yunnan Province, which is also a large agricultural ethnic county, reaches 95,294 RMB (about 13,811 USD)/person, about 5.6 times more than Meigu County.

5.2.3 Population Poverty in Southwest China

Since the implementation of the strategy of targeted poverty alleviation in 2013, the incidence of rural poverty in Southwest China has dropped significantly, and the relative gap between the income and consumption levels of rural residents and the national average has been narrowing. However, internal gaps still exist. Specifically, they manifest in the following three aspects.

First, the incidence of poverty has dropped significantly, and the rate of poverty reduction shows differentiated characteristics. Although all provinces and cities completed the task of eliminating absolute poverty in 2020, the process of poverty reduction differs across provinces and cities. According to statistics, at the end of 2013 in Southwest China, the incidence of poverty in Guizhou Province was as high as 21.3%, while Chongqing had the lowest rate, namely 6%. Specifically, from 2013 to 2017, Chongqing ranked first in terms of poverty reduction rate: from 1.39 million

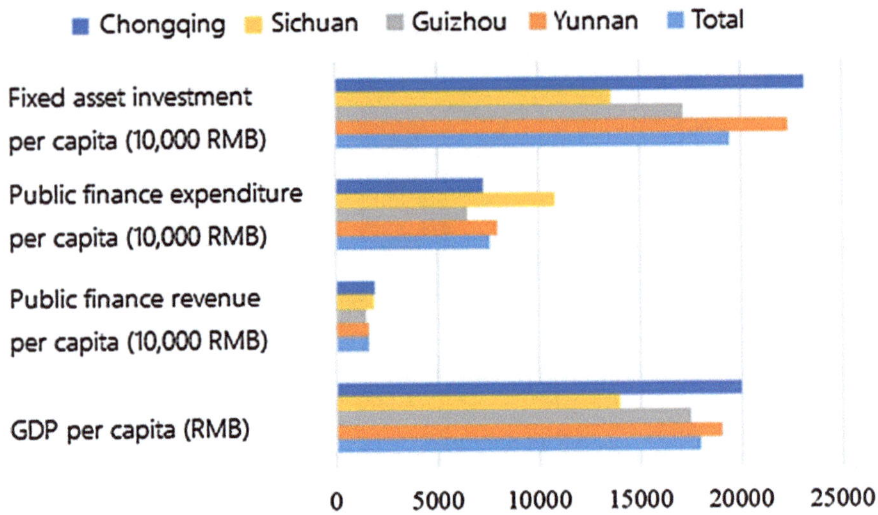

Fig. 5.3 Economic development of ethnic counties in Southwest China. *Source* Compiled from China county statistical yearbook 2021

poverty-stricken people at the end of 2013, there were 225 thousand at the end of 2017, showing a poverty reduction rate of 83.81%. However, Guizhou Province, which is relatively backward due to its history and special geographical conditions, helped over 7 million people out of poverty in 2019 (see Table. 5.4 and Fig. 5.4 for details).

Second, the per capita disposable income has been growing steadily, but there remains a significant gap between it and the national average. As of 2021, the annual per capita disposable income of residents in Southwest China reached 28,065 RMB (about 4100 USD), which was 23.31% lower than the national average, and the income gap is tending to widen. The annual per capita disposable income of rural

Table 5.4 Poverty population/poverty incidence in Southwest China (2013–2019) (unit: 10,000/%)

Area Year	Yunnan		Guizhou		Sichuan		Chongqing	
	Pop	Poverty incidence (%)	Pop	Poverty incidence (%)	Pop	Poverty incidence (%)	Pop	Poverty incidence (%)
2013	661	21.3	745	17.8	602	8.6	139	6.0
2014	574	17.8	623	15.5	509	7.3	119	5.3
2015	471	14.7	507	12.7	400	5.7	88	3.9
2016	363	11.6	402	10.1	306	4.4	45	2.0
2017	248	7.8	280.3	6.6	171	2.7	22.5	1.1
2018	179	4.8	132.3	4.3	67	1.1	13.88	0.7
2019	42	1.32	30.85	1.02	20.3	0.3	2.44	0.1

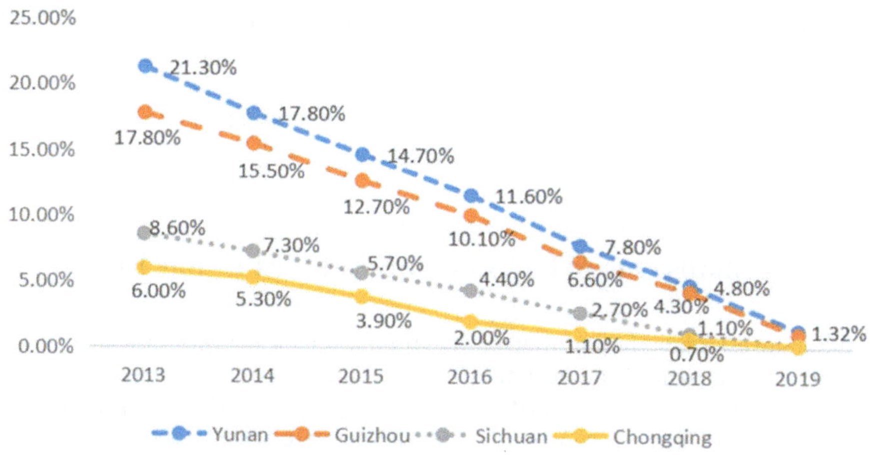

Fig. 5.4 Poverty incidences in Southwest China, 2013–2019. *Source* Compiled from data released to the public by each province

residents was 15,753 RMB (about 2300 USD), which is 16.78% lower than the national average, and 0.86 percentage points lower than that in 2013. Specifically, for all residents' per capita disposable income, Chongqing took first place, while Guizhou rural residents' annual per capita disposable income was only 12,856 RMB (about 1900 USD), which is 32.09% lower than the national average (Table 5.5).

Third, the level of consumer spending is growing faster than the national average, but the internal gap is obvious. As of 2021, the annual per capita consumption expenditure of residents in Southwest China reached 21,127 RMB (about 3100 USD), which was 12.33% lower than the national average, while the gap has been narrowing in recent years. The annual per capita consumption expenditure of rural residents was RMB 14,153 (about 2100 USD), 11.07% lower than the national average. Specifically, Chongqing ranked first in per capita consumption expenditure of residents; in terms of per capita consumption expenditure of rural residents, however, Sichuan

Table 5.5 Rural residents' income and expenditure in 2021 in Southwest China

	Per capita disposable income	Among them: Rural residents	Per capita consumption expenditure	Among them: Rural residents
China	35,128	18,931	24,100	15,916
Southwest China	28,065	15,753	21,127	14,153
Yunnan	25,666	14,197	18,851	12,386
Guizhou	23,996	12,856	17,957	12,557
Sichuan	29,080	17,575	22,530	16,444
Chongqing	33,803	18,100	24,598	16,096

Source Compiled from the 2022 statistical yearbook of each province

surpassed Chongqing, reaching 16,444 RMB (about 2400 USD), which was higher than the national average. It is noteworthy that the statistics show that Yunnan had the lowest annual per capita consumption expenditure of rural residents, at only RMB 12,386 (about 1800 USD), which shows that the consumption demand in rural Yunnan is relatively low (Table 5.5).

5.3 The Status Quo of Pafcs in the Southwest Ethnic Region

The Southwest ethnic region was the initial pilot region for FCS in China, and at the end of 2017, fourteen influential FCS projects had been carried out. This section gathers pre-2017 statistics to analyse the overall characteristics of these projects, point out the problems encountered, explicate the core stakeholders' involvement in PAFCS through the lens of game theory, and, lastly, introduce the successful experience of the Novartis Sichuan FCS Project.

5.3.1 *An Overview of the Pilot FCS Projects*

In 2007, the world's first CDM AR registered project with Gold Certification of CCB Standard, the Northwest Sichuan FCS Project, was launched in Sichuan Province. In 2007, the world's first FCS project to meet the CCB Gold Standard, the Tengchong FCS Project, was launched in Tengchong City, Yunnan Province. According to incomplete statistics, by the end of 2017, a total of 110 FCS projects had been registered and implemented by CDRC, among which 14, accounting for 12.73%, had been carried out in Southwest China. The following three features are typical of these projects (Fig. 5.5).

First, the pilots are mainly AR FCS projects. The 110 pilot projects registered in CDRC include 99 CCER projects, 5 CDM standard projects, and 6 VCS international CER projects.[1] Based on the classification of FCS by CCER, this study classifies the projects into two categories: forest management and AR. The projects involve 148 counties (cities and districts) in 23 provinces. Of these, seventy-five AR projects have the largest share, accounting for 68.18% of the overall FCS projects in China and involving 118 counties (cities and districts) in 21 provinces, while forest management projects account for 31.82%. FCS projects at this stage mainly involve AR.

Second, the southwest is one of the essential regions for AR carbon sink in China. Seventy-five AR FCS projects are implemented in 118 counties (cities and districts) in 21 provinces. The largest number is in Central China, with 18 projects implemented in 32 counties, followed by South China and Southwest China, with 12 projects in 25

[1] 8 The CGCF project developed by the CGCF Foundation was piloted in 2014 in Lin'an District, Hangzhou, Zhejiang Province. CGCF projects are not included in this study.

Fig. 5.5 Spatial distribution of national FCS projects

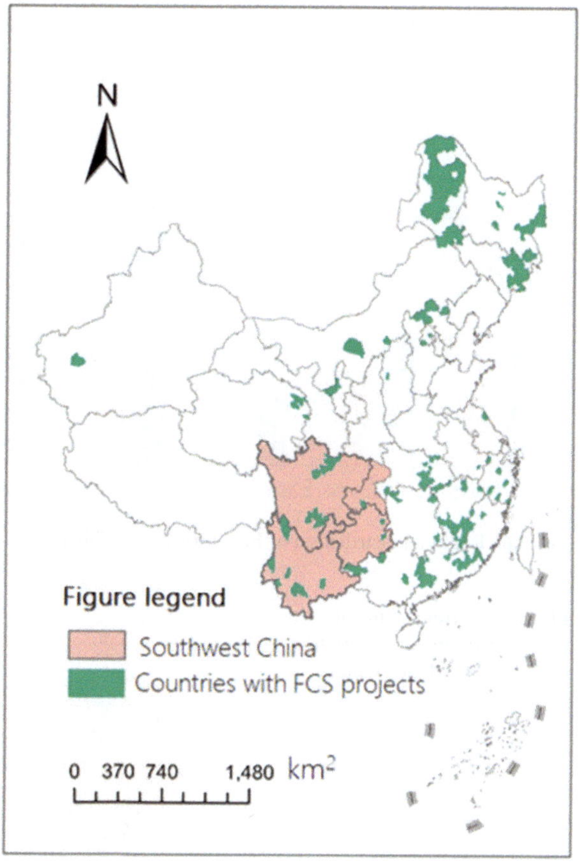

counties and 11 projects in 20 counties, respectively (see Fig. 5.6 for details). Thus, the centre, south, and southwest are the main areas for AR FCS projects in China.

Third, AR FCS projects are mainly located in poverty-stricken ethnic areas in Southwest China. Eleven AR FCS projects have been carried out in Southwest China (see Fig. 5.7), involving a total of 20 counties (cities and districts), of which 8 are national poverty-stricken counties and 13 are ethnic counties. The total project area is 49,113 ha, with an estimated total emission reduction of 11.43 million tons. Within these projects, 46,405 ha were implemented in ethnic areas, accounting for 94.49%, while the estimated total emission reduction was 10.49 million tons, accounting for 91.76%.

In summary, the pilot FCS projects in Southwest China are mainly AR and mainly in poverty-stricken ethnic areas, especially in Yunnan and Sichuan (see Table 5.6), demonstrating that PAFCS projects have targeted poverty-stricken areas. However, from the perspective of economics and management, there are two main reasons to

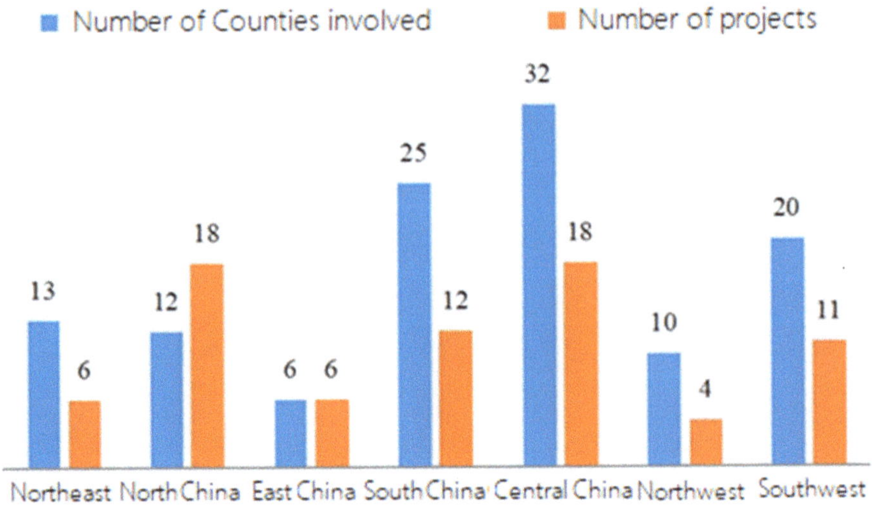

Fig. 5.6 Regional distribution of FCS projects in China

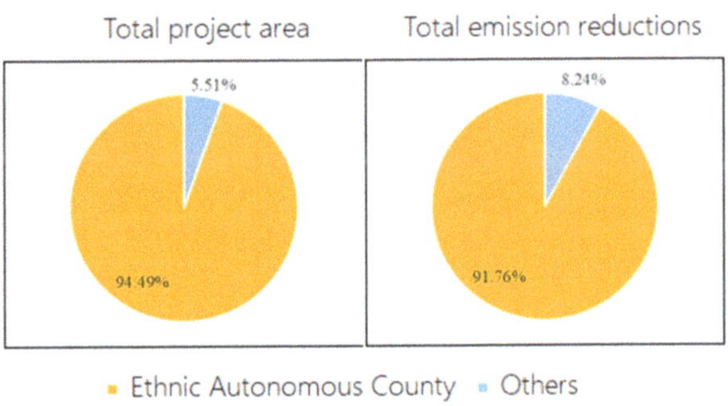

Fig. 5.7 Current status of FCS project implementation in southwest ethnic counties

target poverty-stricken areas. First, the degraded barren mountains and wastelands and mine reclamation lands which meet the *land eligibility* requirements for carbon sink afforestation are mostly located in remote and economically underdeveloped ethnic areas with poor infrastructure and inconvenient transportation. Second, and subjectively, implementation in poverty-stricken ethnic areas will produce better socioeconomic performance and help to win more financial and preferential policy support from the government and society.

5.3.2 *Game Analysis of Core PAFCS Stakeholders*

In recent years, many domestic scholars have conducted a great deal of fruitful research on FCS projects. Among them, Lv (2014) brings together the most typical

Table 5.6 Summary of FCS projects implemented in Southwest China

Area	Project name	Size (ha)	Norm	Type	Counties involved
Yunnan	Yunnan Yunjing Forestry Development Co. FCS Project	7213.6	CCER	Afforestation/ Reforestation	Jinggu Dai and Yi Autonomous County
	Yunnan Pu'er Kemao Forestry & Chemical Co. FCS Project	3189.9	CCER	Afforestation/ Reforestation	Ning'er Hani and Yi Autonomous County
	Disney FCS Project for Revegetation of Degraded Land	1869.97	CCER	Afforestation/ Reforestation	Fengqing County, Zhenkang County, Jianshui County
	FCS Project for Restoration of Penguin Vegetation in Shangri-La	156.1	CCER	Afforestation/ Reforestation	Shangri-La City
	Yunnan Jinggu Evergreen FCS Project	7543	CCER	Afforestation/ Reforestation	Jinggu Dai and Yi Autonomous County
	Tengchong Forest Management Carbon Sink Demonstration Project	3054.93	CCER	Forest Management	Tengchong City
	Kuming forest management improvement projects in Luquan and Xundiann	6879.2	VCS	Forest Management	Luquan Yi and Miao Autonomous County, Xundian Hui and Yi Autonomous Counties
	Tengchong small-scale reforestation landscape restoration project	467.7	CCB	Afforestation/ Reforestation	Tengchong City
Guizhou	Jiangkou Carbon Sink Afforestation Project	678.7	CCER	Afforestation/ Reforestation	Jiangkou county
	Jianhe Carbon Sink Afforestation Project	21,518.7	CCER	Afforestation/ Reforestation	Jianhe county
	Guizhou Carbon Sink Afforestation Demonstration Project in Zazo Forestry	398.13	CGCF	Afforestation/ Reforestation	Huishui county

(continued)

Table 5.6 (continued)

Area	Project name	Size (ha)	Norm	Type	Counties involved
Sichuan	Audi Panda Habitat Multiple Benefit Forest Restoration FCS Project	335	CCER	Afforestation/ Reforestation	Mianning County in Liangshan Yi Autonomous Prefecture
	Northwest Sichuan Degraded Land AR Project in China	2251.8	CDM	Afforestation/ Reforestation	Li, Mao, Beichuan, Qingchuan County, Pingwu County in Aba Tibetan and Qiang Autonomous Prefecture
	Novartis Sichuan Southwest Forestry Carbon Sequestration, Community and Biodiversity Project	4196.8	CDM	Afforestation/ Reforestation	Ganluo, Yuexi, Meigu, Zhaoge and Leibo County in Liangshan Yi Autonomous Prefecture
	Xinjing Reforestation Project	159.2	VCS	Afforestation/ Reforestation	Xingying County in Ya'an City
Chongqing	Nanchuan District Carbon Sink Ecological Engineering Project	6535.9	CCER	Forest Management	Nanchuan district

and representative cases and discusses the current situation, obstacles, experiences, countermeasures, and other practical issues, such as low-carbon development in China. This section briefly introduces the typical Kyoto rules and non-Kyoto rules AR FCS projects implemented in Southwest China and applies game theory to the core stakeholders, to lay a certain foundation for further empirical research.

5.3.2.1 Typical Organisational Model

The Yunnan Xishuangbanna Bamboo Afforestation Project was launched in November 2010 and is the first project implemented under the PS framework, the first voluntary emission reduction standard in China. This project is operated by Yunan Mengxiang Bamboo Industry Co. and funded by Yunan Investment Group and Xishuangbanna Development and Investment Co. in eight townships and 16 administrative villages in Jinghong, Mengla, and Menghai Counties. The project is the first bamboo afforestation FCS project using the PS and bamboo afforestation methodology on degraded land, creating about 3,582.34 ha of dragon bamboo forest. The land used for the project is both state-owned and collectively owned and is leased by the project owner to obtain the land use right (50 years). The farmers only receive the land lease fee, and all bamboo products and the carbon sink income generated are owned by Yunnan Mengxiang Bamboo Industry Co. The project organisation model does not involve the local government; rather, the project owner and farmers are directly involved. This study refers to the organisation model of establishing links between the project owner and the farmers as the autonomous type (Fig. 5.8).

The Tengchong FCS project was officially launched in 2005 with 467.70 ha of new forestation, which is expected to absorb 150,000 tons of carbon dioxide over the 30-year project crediting period, benefiting 2,100 villagers from the community, mainly Bai, Dai, and Hani. It is the first FCS project in the world to meet the CCB Gold Standard and the first project in China funded by enterprises and developed by NGOs in cooperation with the government, while previous FCS projects have been entirely run by the government or enterprises. The project covers five administrative villages in Qu Shi Township, Jietou Township, Monkey Bridge Township, and Sujiang Forestry, Tengchong County, Yunnan Province. Individuals and collective landowners sign land use contracts with Sujiang Forestry to obtain 10% of the carbon sink revenue and all the timber revenue. All costs required for project development are borne by enterprises, and project development, operation, and management are carried out by the project owner. The local government is responsible for providing technical support. In this project, the local government, as an independent third party, promotes the project development by providing guidance and support to Sujiang Forestry and townships. The organisational model is shown in Fig. 5.9. This study refers to this type of organisational model as consultative.

The Northwest Sichuan FCS project was officially launched in 2007 and was UN CDM-EB registered and approved by the CDRC in 2009. It is the first CDM AR registered project in the world to have the CCB Standard Gold Certification and was funded by a 3 M company with Dadu River Afforestation Bureau as the project owner. Approximately 2251.80 ha of land were planned to be artificially afforested in twenty-one townships in five counties in northwestern Sichuan Province, namely Li, Mao, Beichuan, Qingchuan, and Pingwu, and more than 3,200 farming households in the project communities will benefit from the project, including Tibetan, Qiang, Hui, and other ethnic minorities, accounting for over 42% of households. Farmers provide forest land without forest rights disputes, do not bear the cost of afforestation funds, and gain a certain income by participating in land preparation, planting, and other activities. Farmers are expected to receive 30% of the carbon sink

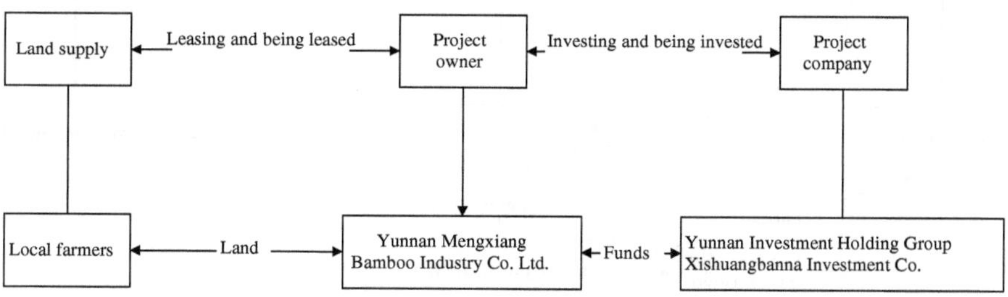

Fig. 5.8 Organisational model of the Yunnan Xishuangbanna bamboo afforestation project

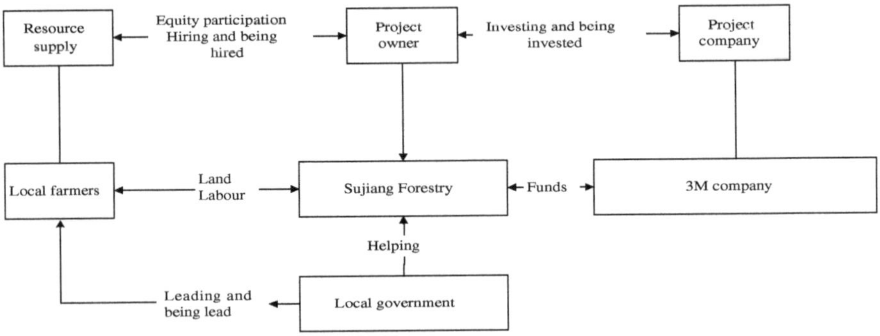

Fig. 5.9 Organisational model of the Tengchong FCS project

revenue and enjoy all the forest trees and forest by-products. The local government cooperates with the Dadu River Afforestation Bureau and signs land use contracts with farmers. The township forestry stations carry out specific afforestation activities, are responsible for post-afforestation nurturing, replanting, custodianship, and forest management, and enjoy a 20% share of carbon sink proceeds. This study calls this type of organisational model a compulsory type (see Fig. 5.10).

The Novartis Sichuan FCS Project is the first CDM-AR afforestation project in China where a domestic company (Sichuan Dadu River Afforestation Bureau, as the supplier and project owner) and a foreign company (Novartis Group, Switzerland, as the purchaser) directly cooperate to pay future carbon sink trading funds in advance for afforestation. It is the first project in the world to be certified by the Climate, Community, and Biodiversity Alliance (CCBA) and the first afforestation and carbon reduction project in the world. The project is expected to plant approximately 4,196.80 ha of forests in five counties of Liangshan Yi Autonomous Prefecture, namely Zhaojue, Yuexi, Ganluo, Meigu, and Leibo; and three nature reserves, namely Yuexi Shengozhuang, Ganluo Ma'anshan, and Leibo Mamize, with a total budgeted investment of 100 million RMB, of which Novartis will invest 57%. It is

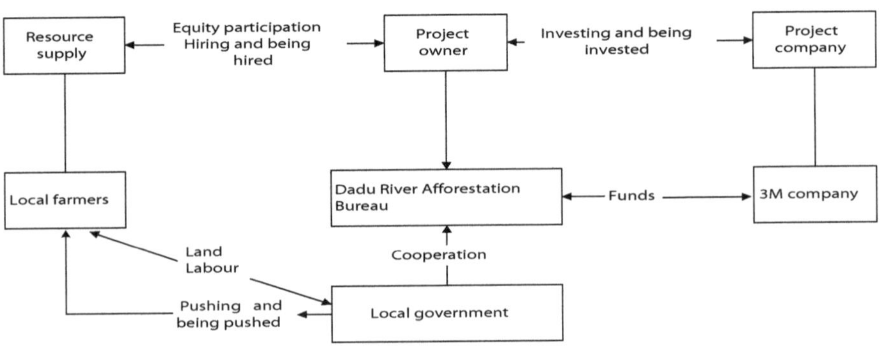

Fig. 5.10 Organisational model of the Northwest Sichuan FCS project

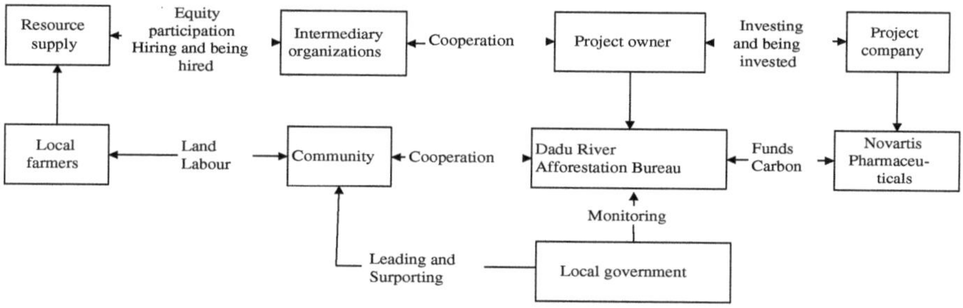

Fig. 5.11 Organisational model of the Novartis Sichuan FCS project

expected to absorb 1.05 million tons of carbon dioxide during the 30-year project-crediting period, and more than 18,000 villagers will gain training, employment, and income, 97% of whom are from ethnic minorities.

Project planning began in 2009, and it was launched in 2011, passed the validation of the international third-party certification body (DOE) in 2012, and was approved by the CDRC and registered with the UN CDM-EB and awarded CCBA Gold Certification in 2013. The Dadu River Afforestation Bureau signs contracts with individual and collective landowners to use their lands, and they participate in FCS projects in the form of shareholding. The timber revenue generated from the projects goes to the farmers, and the carbon sink revenue goes to Novartis Pharmaceutical Co. Local farmers gain a secure income by participating in land preparation, digging, weeding, fertilising, and afforestation. As a third party, the local government, together with the Nature Conservancy, the Mountain and Water Nature Conservation Centre, and other organisations, set up a project steering committee to guide and coordinate overall implementation. Thus, the organisational model of the project is shown in Fig. 5.11. This study refers to this type of organisational model as community-based.

5.3.2.2 Model Construction

Based on the identification of stakeholders and the organisational models, this study explores the game decisions of farmers, project owners, local governments, and communities based on game theory. According to the type and number of stakeholders, the game model involves three categories: without government and community intervention, with government and without community intervention, and with government and community intervention. The game tree is shown in Figs. 5.12, 5.13 and 5.14.

Without government and community involvement. In the context of no government and no community involvement, the only stakeholders are the project owner and local farmers, and the relationship of supply and demand of land and labour is directly established between them. In this model, the game between project owners and farmers is relatively simple and of a typical rational type, influenced by information asymmetry and trust degree. The benefits of individual farmers are little affected

Fig. 5.12 Stakeholder gaming without government and community involvement

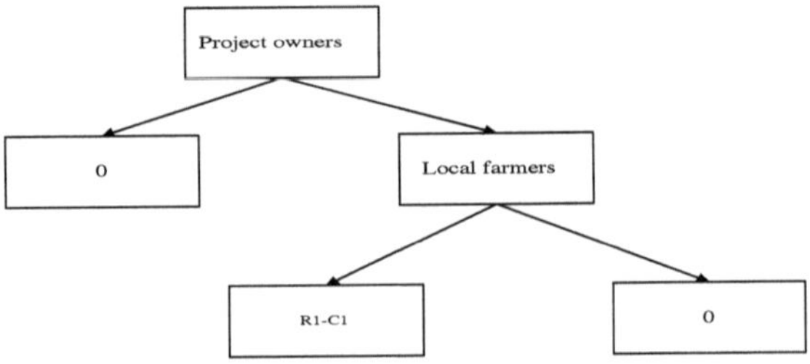

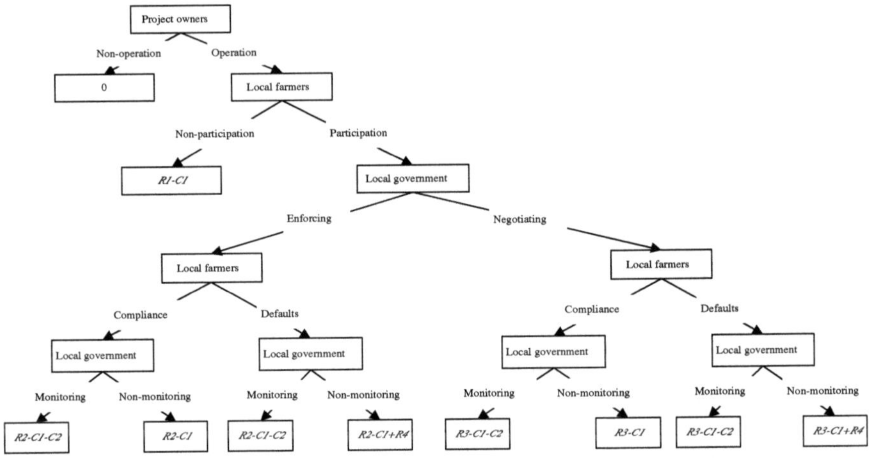

Fig. 5.13 Stakeholder game tree with government and without community intervention

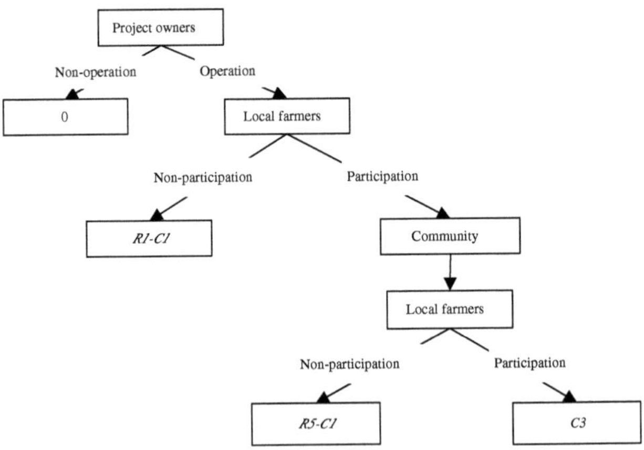

Fig. 5.14 Stakeholder game tree with government and community involvement

by the transaction costs of negotiation and management between project owners and local farmers; thus, this model mainly arises between project owners and large afforestation farmers.

With government and without community intervention. In this context, the stakeholders are project owners, local farmers, and local governments. When farmers choose to participate, local government involvement has little impact on their participation behaviour, and theoretically the involvement of local governments is unnecessary. When farmers choose not to participate, the local government intervenes and faces a choice between two behavioural strategies: one is to adopt compulsory measures to forcibly promote project development; the other is to incentivise farmers to participate in FCS projects by negotiating with them. After farmers participate in projects, they must choose whether to take up the behavioural strategy of compliance during the project operation, and the local government must choose whether to monitor such compliance.

With government and community intervention. In this case, the stakeholders are project owners, local farmers, local governments, and communities. As independent third parties, local governments provide technical and policy support and supervise and regulate the behaviour of project owners, thus not generating behavioural games with local farmers. When farmers do not participate, communities must act as intermediary organisations, and farmers are again faced with behavioural decisions about whether to participate after community intervention.

The interpretation of the indicator is as follows:

$R1$ indicates the total benefits obtained by farmers in the context of no government intervention and direct contractual relationships between project owners and farmers.

$R2$ denotes the total benefits obtained by farmers in the context of government intervention and a direct contractual relationship between local government and farmers.

$R3$ denotes the total benefits obtained by farmers in the context of government intervention and direct contract establishment between the project owner and the farmer.

$R4$ indicates the additional benefits obtained by farmers when they breach the contract.

$R5$ indicates the total benefits received by farmers in the context of community participation.

$C1$ denotes the opportunity cost of farmers' participation.

$C2$ denotes the government monitoring costs per individual farmer.

$C3$ denotes the psychological cost faced by farmers for not participating in the context of community participation.

$P1$ denotes the probability of farmer participation.

$P2$ denotes the probability of compulsory measures undertaken by the local government.

$P3$ denotes the probability of farmers' defaulting under the compulsory model.

$P4$ denotes the probability of farmers' defaulting in the context of negotiation.

$P5$ denotes the probability of the government taking supervisory actions under the compulsory model.

*P*6 denotes the probability of government adoption of supervisory behaviour in the context of consultation.

*P*7 denotes the probability of farmers' participation in the context of community intervention.

This study assumes that $R2-C1$, $R3-C1$, and $R5-C1$ are greater than 0, and $R1-C1 < 0$ due to the influence of transaction costs such as the negotiation between project owners and farmers.

5.3.2.3 Farmers' Benefits Under Different Organisational Models

First, the farmers gain under the autonomous model. The benefits of local farmers' behavioural decisions under the autonomous model are shown in Table 5.7. When farmers do not participate, the benefits are 0, but when farmers participate, their benefits are the total benefits minus the opportunity cost, that is, $R1-C1$. The study sets the probability of farmers' participation as $P1$; thus, farmers' expected benefits are $P1*(R1-C1) + (1-P1)*0-P1*(R1-C1)$. In terms of decision-making, when $R1 < C1$, farmers choose not to participate; thus, farmers' participation decision is based on the comparative cost–benefit analysis. In summary, the optimal decision of farmers in the autonomous model is not to participate, and the benefit is 0.

Second, the farmers gain under the compulsory model. In the compulsory model, the benefits of local farmers' behavioural decisions are shown in Table 5.8. Under the compulsory measures taken by the government, the benefits of farmers' compliance are $P5 \times (R2-C1-C2) + (1-P5) \times (R2-C1)$, and the benefits of default are $P5 \times (R2-C1-C2) + (1-P5) \times (R2-C1 + R4)$. Therefore, the farmers' expected benefits are $P3 \times [P5 \times (R2-C1-C2) + (1-P5) \times (R2-C1 + R4)] + (1-P3) \times [P5 \times (R2-C1 -C2) + (1-P5) \times (R2-C1)]$. In terms of behavioural decisions, under local government supervision the benefits of compliance and default are the same; however, in the context of no local government monitoring, the benefits of default are higher than those of compliance. The difference is $(1-P5) \times R4]$. Thus, in the compulsory model farmers are biased towards defaulting, and the benefit is $P5 \times (R2-C1-C2) + (1-P5) \times (R2-C1 + R4)$.

Third, farmers gain under the consultative model. The benefits of local farmers' behavioural decisions in this case are shown in Table 5.9. Under the government-driven consultative model, the benefits of compliance are $P6 \times (R3-C1-C2) + (1-P6) \times (R3-C1)$, and the benefits of default are $P6 \times (R3-C1-C2) + (1-P6) \times (R3-C1 + R4)$. Therefore, the expected benefits of farmers' participation in the

Table 5.7 Farmers' benefit matrix under the autonomous model

		Project owners	
		Operation	Non-operation
Local farmers	Participation	$R1-C1$	—
	Non-participation	0	—

Table 5.8 Farmers' benefit matrix under the compulsory model

		Local government	
		Monitoring	Non-monitoring
Local farmers	Compliance	$R2-C1-C2$	$R2-C1$
	Default	$R2-C1-C2$	$R2-C1+R4$

Table 5.9 Farmers' benefit matrix under the negotiation-based model

		Local government	
		Monitoring	Non-monitoring
Local farmers	Compliance	R3−C1−C2	R3−C1
	Default	R3−C1−C2	R3−C1 + R4

Table 5.10 Farmers' benefit matrix under the community-based model

		Community	
		Intervention	Non-intervention
Local farmers	Participation	$R5-C1$	$R1-C1$
	Non-participation	$-C3$	0

project are $P4 \times [P6 \times (R3-C1-C2) + (1-P6) \times (R3-C1 + R4)] + (1-P4) \times [P6 \times (R3-C1-C2) + (1-P6) \times (R3-C1)]$. In terms of behavioural decisions, the benefits of compliance and default are the same for farmers under local government monitoring. However, in the context of no local government monitoring, the benefits of default are higher than those of compliance [the difference is $(1-P6) \times R4$]. Thus, in the consultative model, farmers prefer to default, with a payoff of $P6 \times (R3-C1-C2) + (1-P6) \times (R3-C1 + R4)$.

Fourth, farmers gain under the community-based model. The benefits of local farmers' behavioural decisions under the community-based model are shown in Table 5.10. When the community does not intervene, its behavioural game is consistent with the autonomous, compulsory, and consultative type, and the benefits of farmers' participation in the context of community intervention are $R5-C1$, while the benefits of farmers' non-participation in AR projects are $-C3$. $C3$ indicates the benefits faced by farmers who choose not to participate. The expected benefit is $P7*(R5-C1) + (1-P7)*(-C3)$. In terms of behavioural decision, when $R5 > C1$, farmers are biased to participate, and the benefit is $R5-C1$.

5.3.2.4 Comparative Analysis of Optimal Decision of Different Models

As discussed previously, the optimal decision of farmers under the autonomous model is non-participation, under the compulsory and consultative models it is defaulting, and under the community-based model it is participation. The comparative analysis of the optimal decisions under different models is shown in Table 5.11.

First is the comparison between the compulsory and consultative types. The comparison of the benefits of optimal decisions in these cases is as follows.

$$\begin{aligned}&P5*(R2-C1-C2)+(1-P5)*(R2-C1+R4)-P6*(R3-C1-C2)\\&+(1-P6)*(R3-C1+R4)=\ [(R2-C1)-P5*C2+(1-P5)*R4]\\&-[(R3-C1)-P6*C2+(1-P6)*R4]=\ (R2-R3)+(P6-P5)(C2+R4)\end{aligned}$$

In the compulsory model, the cost of reaching a contractual relationship between local governments and farmers on land and labour supply is smaller than the total benefits that farmers obtain; thus, the local government intervenes. The total benefits derived by farmers from AR projects in the context of direct contracting between local governments and farmers is greater than the total benefits in the context of government intervention and direct contracting between project owners and farmers, that is, $R2 > R3$. Meanwhile, in the compulsory model, the township forestry is responsible for the implementation of the projects; thus, the probability of undertaking supervision behaviour is relatively low, that is, $P5 < P6$. In summary, $(R2-R3) + (P6-P5)(C2 + R4) > 0$; thus, the compulsory model is superior.

Second, there is the comparison between the compulsory and community-based types. The comparison of the optimal decision in the community-based versus the compulsory model is as follows.

$$\begin{aligned}&R5-C1-[P5*(R2-C1-C2)+(1-P5)*(R2-C1+R4)]\\&=R5-C1-[(R2-C1)-P5*C2+(1-P5)*R4]\\&=R5-R2+P5*C2-(1-P5)*R4\\&=R5-R2-R4+P5*(C2+R4)\end{aligned}$$

Due to the influence of geography, kinship, and blood, the contractual relationship between communities and farmers is closer and has lower transaction costs than that between the government and farmers. Under community involvement, the farmers' willingness to participate will be enhanced because they tend to be isolated by other community members if they choose not to participate. Thus, the total benefits gained

Table 5.11 Comparative analysis of benefits of optimal decision-making

Mode	Benefits	Sorting from largest to smallest
Autonomous	0	4
Compulsory	$P5*(R2-C1-C2)+(1-P5)*(R2-C1+R4)$	2
Consultative	$P6*(R3-C1-C2)+(1-P6)*(R3-C1+R4)$	3
Community-based	$R5-C1$	1

by farmers from AR projects in the context of community involvement are greater than those derived from government involvement, that is, $R5 > R2$. Therefore, the key to the comparison between community-based and compulsory models is $R4$. Moreover, farmers are highly dependent on the lands as they are in remote and poverty-stricken mountainous areas, so the additional benefit $R4$ obtained by farmers through defaulting is relatively high. In summary, $R5-R2-R4 + P5^*(C2 + R4) > 0$; thus, the community-based model prevails.

In sum, under the current market and policy environment, farmers' optimal decision is community-based > compulsory > consultative > autonomous, and the community-based reforestation project is an important way to ensure the effective participation of farmers and promote the sustainable operation of FCS projects.

5.3.3 *Main Problems in PAFCS*

PAFCS has seen significant development, as determined by the investigation of typical FCS projects in Sichuan, Yunnan, and Guizhou and from discussions and in-depth interviews with the heads and staff of relevant government departments, enterprises, institutions, and NGOs. It has brought new momentum to and new means of ecological poverty alleviation, green poverty reduction, and targeted poverty eradication in poverty-stricken ethnic areas in Southwest China. At the same time, however, it faces the following three outstanding problems.

5.3.3.1 Misunderstanding and Deficiency in Basic Institutional Arrangements and Poverty Benefit Mechanism

Respondents generally agree with the poverty alleviation function of FCS projects and believe that it is necessary to further expand the positive effects and actively mitigate the negative ones, making positive contributions to climate change mitigation, reducing the vulnerability of poverty-stricken areas, and enhancing the adaptability of the latter. However, compared with carbon trading in the energy and industrial sectors, the current FCS market is small, and demand is very limited. FCS projects should be promoted as a priority for carbon trading in China to continuously expand the country's market share. Wood products are to be included in the FCS measurement. However, the questions of how the AR task can be completed and how the survival and preservation rate of seedlings can be improved are often the focus of afforestation entities and forestry management departments at all levels and usually the general criteria for evaluating the performance of a project.

Many FCS projects are like traditional afforestation, leading to poverty alleviation goals being ignored. The role of FCS in promoting targeted poverty alleviation has not yet been fully played, and it is still at the stage of emphasising the participation and benefit of the community. The need to establish an FCS market, the lack of standards, and vague assessment indicators are important obstacles to expanding PAFCS. It is imperative to introduce more top-level policies and macro management and to formulate FCS standards and methodologies that are in line with international rules and China's actual situation, with an emphasis on creating benefits and development opportunities for the poor.

5.3.3.2 Insufficient Policies, Entity, Resources, and Methods for Poverty Alleviation

In the face of the huge pressure to alleviate poverty, some provinces have included carbon sink poverty alleviation in their ecological compensation effort. Some local governments have included it in their ecological poverty alleviation plan or annual work plan and are exploring new ways of carrying out personalised green poverty alleviation. While numerous administrative bodies bear responsibility for the management of PAFCS, including the Development and Reform Commission, the Ministry of Forestry, the office of poverty alleviation authorities at various local levels. In practice, however, the bulk of the work is carried out by the forestry authorities themselves. Poverty alleviation, science and education, finance, and related departments are relatively absent, and social donations and citizen participation channels are very limited. Greater emphasis is placed on policy, short-term income increases for community farmers, and material compensation, and less is ascribed to system construction, knowledge and skill training, and technical compensation. Technology poverty alleviation, financial poverty alleviation, cultural poverty alleviation, and other forms of poverty alleviation have not been given due attention. There are insufficient policies, entities, resources, and methods for poverty alleviation. The policies of finance, taxation, investment, forest ecological compensation, technical assistance, and other general matters related to industrial poverty alleviation need to be integrated. Finally, preferential policies regarding carbon tax, carbon sink mortgages and interest-bearing loans, FCS insurance, and carbon sink forest inter-logging and harvesting need to be established.

5.3.3.3 Market Failure and Weakening of Poverty Alleviation Function

As a special form of industrial poverty alleviation, PAFCS must follow the basic rules of FCS market trading. However, it faces challenges such as its wide implementation scope, long project cycle, the difficulty of nurturing and managing young forests, stricter standards, higher risks, and greater investment than general afforestation projects. The amount of carbon trading achieved by individual projects under the post-paid mechanism is very limited, and there is a shortage of funds for subsequent construction and financing channels. Even those projects which have made good progress are facing conflicts, such as insufficient endogenous motivation, and rely heavily on top-down administrative promotion. Other difficulties include underdeveloped communities, weak development capacity of farmers, inefficient organisation, very limited awareness of forest carbon market transactions, the insufficient role of the main entity, low participation of the poverty-stricken people, and rising afforestation costs. Projects also face conflicts between the long-term occupation of project land and the short-term demand for the land of landowners, large-scale afforestation and traditional livelihood of community farmers, market-oriented operation and local culture, elite-driven and poverty-stricken population being extruded, etc.

New challenges include exclusive competition for land brought by new breeding industrial projects for poverty alleviation and intensified conflicts among foresters and herders. For example, to improve the survival rate of afforestation and reduce transaction costs, professional afforestation teams have been hired to replace farmers, limiting the participation of local community farmers. As another example, the opportunity cost increased due to the increase in beef and mutton prices. Hence, the community farmers put forward new requirements to increase the compensation price, which the project owner and local government struggled to meet, resulting in the cancellation of the project and irreparable losses to the project implementation and contract fulfilment.

5.3.4 Case: The Novartis Sichuan FCS Project

In the past few years, in accordance with the new opportunities and requirements brought by the implementation of the targeted poverty alleviation strategy, various parts of the country have carried out varied explorations and experiments. Typical examples are *the Novartis model* implemented in Sichuan Province (Sichuan Provincial People's Government, 2018), *the single plant carbon sink targeted poverty alleviation platform model* in Guizhou Province (Guizhou Provincial People's Government, 2018), *the Tengchong Forest carbon sink + ecological tourism + forest economy + targeted poverty alleviation model* in Yunnan Province (Guizhou Provincial Government, 2018), *the Yanshou model of FCS + enterprise + targeted poverty alleviation* in Heilongjiang (Heilongjiang Provincial Government, 2017), and, of these, the Novartis model provides a valuable reference for solving the particular dilemmas and challenges and achieving the goal of *truly alleviating the poor and alleviating the truly poor* while satisfying the interests of multiple participants to the maximum extent.

5.3.4.1 Case Overview

As of 2018, the Novartis FCS Project had completed AR of 4,095.80 ha, established 209,000 m of forest protection fencing, and invested in 330,000 labourers engaged in seedling, afforestation, replanting, young forest management, and nurturing. In the period 2016–2018, community farmers earned more than 26 million RMB (about 3.8 million USD) from their labour and 13 million RMB (about 1.9 million USD) in seedling income through participation in the project, representing a per capita income increase of over 2,160 RMB (about 313 USD) annually. In May of that year, the Sichuan Forestry and Grassland Bureau, Sichuan Science and Technology Department, and Sichuan Poverty Alleviation and Development Bureau upgraded the Novartis model to be piloted in Yuexi, Zhajue, and Leibo Counties of Liangshan Yi Autonomous Prefecture.

News Coverage

'Government + Enterprise + University' Cooperation on PAFCS

Sponsored by the Sichuan Forestry Department, Sichuan Agricultural University, and Sichuan Dadu River Afforestation Bureau, the PAFCS Demonstration Project was launched and a seminar was held on 23 May 2018 in Chengdu.

The project creates a new model of *government + enterprise + university* cooperation. It is based on the Novartis Southwest Sichuan Forestry Carbon Sequestration Project, funded by the National Social Science Foundation's 'Policy Research on Promoting Forest Carbon Sequestration Poverty Alleviation in Southwest Ethnic Areas' and the Sichuan Provincial Science and Technology Department's special project on 'Forest Carbon Sequestration Industrial Poverty Alleviation Technology Integration and Demonstration'. Two targeted poverty-alleviation models will be implemented. One is the 'collective economy-led' type, focusing on the role of village economic organisations, collective forest land shares, etc. to improve and implement the enterprise + village collective economic organisations + farmers model. The other is the 'farmer-led' type, based on the model of enterprise + farmers, promoting the employment and technical training of poverty-stricken farmers in the vicinity. The project will be carried out in Yuexi, Zhaojue, and Leibo Counties in Liangshan Prefecture. This project provides a typical model for the gradual realisation of integrated carbon sink forest management, integration, and demonstration of intensive breeding technology in ecological but poverty-stricken counties and providing a win–win model for the realisation of the regional traditional livestock industry and the sustainable development of the project. (Reporter Jiang Yunhan).

Source Sichuan Provincial People's Government website, 24–05-2018, https://www.sc.gov.cn/10462/12771/2018/5/24/10451619.shtml.

5.3.4.2 Main Explorations

The basic idea is to expand and improve the community function of the Novartis Sichuan FCS Project, promote the effective participation of poverty-stricken farmers, and continuously improve the *farmer-led and collective economy-led* models. The two models entail the joint participation of multiple parties, organic integration of resources, complementary poverty alleviation methods, and joint promotion of poverty alleviation.

First, the FCS project should be purpose-driven. With the purpose of creating benefits and development opportunities for poverty-stricken people, the Novartis

Sichuan FCS Project consolidates the results of AR in the early stage and enhances the sustainability of the project.

Second, it is vital to implement two models of targeted poverty alleviation. Two models were introduced in the practice of the Novartis Sichuan FCS Project. One is the farmer-led model, adopted in poverty-stricken villages with clear property rights of forest land, strong market economy awareness, and rich resources of confirmed forestable land. The government department will guide poverty-stricken households to participate in technical training and give priority to those able to participate in the work of FCS seeding, breeding, planting, replanting, management, care, fence construction, and other labour activities. Thus, the 'enterprise + farmers' model with the core of farmers would be implemented and improved, promoting the employment of poverty-stricken farmers in the vicinity and effectively realising targeted poverty alleviation. The other is the collective economy-led model, where the cooperation of enterprise + village collective economic organisation + farmers will be encouraged. Collective forest land can be transferred into stakes in the company. In poverty-stricken villages with a better governance structure of collective economic organisations and richer collective forestable land resources, village-level organisations will improve the participation of poverty-stricken farmers in the development of FCS projects.

Third, the three poverty alleviation mechanisms must be improved. Firstly, as regards the poverty alleviation resources integration mechanism, a leading group and related office for the PAFCS project are to be set up to enhance coordination among the Sichuan Development and Reform Commission, Forestry and Grassland Bureau, Bureau of Poverty Alleviation and Development, Science and Technology Department, and other departments. Based on the organisational assignment, stakeholders will increase the supporting funds and policy support for the project, strengthen technical training for farmers in the community, and promote industrial poverty alleviation, relocation of farmers, and ecological afforestation. Secondly, as regards the participation mechanism of the poverty-stricken people, it is crucial to expand participation channels and degree of participation; give full play to the role of village committees, professional cooperatives, rural elites, etc., and continuously improve the organisation of poverty-stricken farmers' participation. Third, as concerns the monitoring and assessment mechanism, the case demonstrates that it is also important to identify the amount of assistance given to poverty-stricken villages, households, and people, to identify the number of technical training sessions and other assessment indicators, and to improve the monitoring and evaluation mechanism.

Fourth, four key tasks must be implemented. Firstly, FCS afforestation technologies must be integrated and demonstrated in ecologically poverty-stricken areas. Training and demonstration of new technologies, such as alpine carbon sink forest seedling cultivation, afforestation, nurturing and management, and pest control, must be provided, and advanced forestation and reforestation technologies must be continuously injected into poverty-stricken communities. Secondly, carbon sink forest management technologies must be comprehensively integrated and demonstrated, as must training and demonstration of the planting technology of mushrooms and Chinese herbs in forests to actively create conditions to improve the comprehensive management benefits of carbon sink forests and promote the long-term sustainable operation of carbon sink projects. Thirdly, intensive farming technology must be integrated and demonstrated. Advanced and applicable technologies must be introduced for cattle and sheep captive breeding, grass-farming, beekeeping, and other special livestock industries to promote the sustainable development of community livestock. In the end, monitoring and assessment technology must be applied with a focus on increasing the transparency of PAFCS, improving the compliance rate of poverty alleviation subjects, and correcting deviation from poverty alleviation actions. The comprehensive evaluation system of PAFCS performance must be improved to provide an important guarantee of greater efficiency.

5.3.4.3 Primary Experiences

The Novartis Sichuan FCS Project has accumulated valuable experiences in organisation and management, planning and design, carrying out community projects, and the innovation and construction of a targeted poverty alleviation model and its operation mechanism.

Firstly, effective organisation and support from the government are required. From the beginning, the Novartis FCS Project set up a project steering committee and an expert advisory committee which held regular meetings every year to guide the overall implementation of the project, solve and coordinate major problems, and consider annual progress. The competent departments set up three-level project coordination and management offices, clarified leadership, and appointed full-time staff to carry out the whole process of publicity, coordination, and supervision, thus effectively overcoming the barriers of policy, investment, capital, and technology in planning and design, certification and registration, organisation, construction, etc. This initiative solved many problems that the Dadu River Afforestation Bureau had encountered, such as project plot selection and boundary determination, forest land transfer, and contract signing.

Moreover, the project received funding support from the Sichuan Forestry and Grassland Bureau and the government of Liangshan Yi Autonomous Prefecture, as well as strong support from taxation authorities, foreign exchange management departments, and banks, which effectively alleviated the pressure of capital shortage and reduced the exchange rate risk. To reduce the risk that farmers would withdraw during long-term operation, a three-party contract including land use rights holders, project owners, and local county forestry authorities was signed. Annual plans and precise assessments with quantitative indicators were drawn up to ensure the participation of poverty-stricken people in labour and technical training.

Secondly, Novartis paid special attention to local knowledge, local wisdom, and local culture from the outset. A baseline survey and community survey were conducted to understand the community's economic and social development needs and to collect community farmers' opinions on plot utilisation, tree species selection, and interests. This initiative strengthened publicity as well as communication with community farmers, enabled consideration and evaluation of the potential risks of project development, and tracked community farmers' opinions and suggestions on project promotion and implementation.

Moreover, the project was implemented in a way that would increase the opportunities and extent of participation of the disadvantaged groups in the community, especially the poverty-stricken people, who are more dependent on traditional agriculture for their livelihoods. This not only provides a scientific basis for the identification of baseline scenarios, additional certification, and feasibility demonstration of projects, but also plays a key role in establishing a project operation model and its benefit linkage mechanism that is more in line with community opinion. It aims to benefit more poverty-stricken people, generate a better understanding and support, and promote long-term operation. For example, to respect the living habits of local farmers, the project clearly does not restrict the collection of firewood after the carbon sink forest is closed and takes the initiative to participate in the promotion of energy-saving wood stoves by the local government.

Thirdly, the support of science and technology must be strengthened. The project owner insists on long-term cooperation with Sichuan Agricultural University, Sichuan Academy of Forestry, Sichuan Forestry Research and Planning Institute, Sichuan Academy of Social Sciences, The Nature Conservancy of China (TNC), Beijing Landscape Nature Conservation Centre, Sichuan Greening Foundation, Liangshan Forestry Science Institute, and other scientific research institutes and NGOs to give effective scientific and technological support in the areas of the natural and social sciences. The project has achieved remarkable results in AR in alpine mountainous areas, ecologically fragile areas, and lands with poor ground conditions. It has brought additional funds, technology, investment, environment, and policies to break the poverty trap, provided external resources, mobilised internal resources, and gathered policy resources. Traditional farming and animal husbandry are the main sources of livelihood for farmers, but the economic development of the communities is relatively lagging, the production methods are relatively backward, and productivity is low. Therefore, the project invited experts from scientific research institutes to carry out technical training and guidance, presented popular science books or advanced and applicable agricultural technology manuals, and actively participated in community-specific industries, solving conflicts between forest and animal husbandry and forest and agriculture.

5.4 Summary

Our investigation shows that most AR FCS projects are in ecologically fragile and poverty-stricken mountainous areas and that PAFCS has achieved significant development and accumulated new experiences for ecological poverty alleviation. However, PAFCS is still treated as a general industrial poverty alleviation project in some cases, and the goal of poverty alleviation is not yet incorporated into planning and design, certification and registration, organisation and construction, and monitoring and evaluation. It is still in the exploratory experimental stage and led by local governments, and the policy is more of a compulsory system introduced by grass-roots governments to solve practical problems and meet the current requirements of targeted poverty alleviation. There is an insufficient supply of macro-management systems to promote the sustainable development of PAFCS, and normative systems need to be established, such as FCS standards and methodologies, and preferential policies, such as carbon tax, carbon sink mortgages and interest-bearing loans, FCS insurance, and FCS interval harvesting. Fiscal, tax, financial, and investment policies are not sufficient, and the policies are not consistent and sustainable. What is needed is to strengthen the main role of community farmers, including the poverty-stricken people. Importantly, researchers, practitioners, and policymakers should consider the overall situation of the region, emphasise targeted solutions, and continuously improve the participation of poverty-stricken farmers. This is an important direction in the future.

References

China State Council. (1952). *Outline of the Implementation of Regional National Autonomy.* http://www.cssn.cn/ddzg/xsdt/8y9r/. Accessed on 20 Aug 2022.

China State Council's Poverty Alleviation Office. (2012a). *List of 592 National key counties for poverty alleviation and development in the new era.* http://news.cntv.cn/china/20120319/115343.shtml. Accessed on 20 Aug 2022.

China State Council's Poverty Alleviation Office. (2012b). *Note on the publication of the list of sub-counties in the national contiguous special hardship areas.* http://www.gov.cn/gzdt/2012-06/14/content_2161045.htm. Accessed on 20 Aug 2022.

Guizhou Provincial People's Government. (2018). *Guizhou Province launches a pilot project of single plant carbon sink for targeted poverty alleviation.* http://www.guizhou.gov.cn/xwdt/dt_22/bm/201806/t20180615_1350909.html. Accessed on 20 Aug 2022.

Heilongjiang yanshou government. (2018). Yanshou County signs framework agreement for forestry carbon sink targeted poverty alleviation project. https://www.nlmy.com.cn/article/tz1wu0.html. Accessed on 20 Aug 2022

Jiang, B. H. (2003). *Research on Economic development of ethnic regions in Southwest China—An empirical analysis of economic development and its measures.* The Master's Thesis of Southwest University of Finance and Economics, Chengdu, China.

Lv, S. K. (2014). *Forest carbon sequestration practices and low carbon development in China.* Beijing University Press.

State Council Poverty Alleviation Office. (2021). National list of key counties for poverty alleviation and development work. Retrieved 31 December 2021, from https://www.gov.cn/gzdt/2012-03/19/content_2094524.htm

Sichuan Provincial People's Government. (2018). *"Government+Enterprise+ University" cooperation on forest carbon sink poverty alleviation demonstration project launched.* https://www.sc.gov.cn/10462/12771/2018/5/24/10451619.shtml. Accessed on 20 Aug 2022.

Yunnan Baoshan government. (2018). *Tengchong forestry: playing the symphony of green water, green mountains and golden development.* https://baijiahao.baidu.com/s?id=1623413622399427282&wfr=spider&for=pc. Accessed on 20 Aug 2022.

Zheng, C. D. (2014). *Catching up and transforming development of china's ethnic minority regions after the development.* Economic Science Press.

Zhuang, T. H. (2011). *Poverty and anti-poverty survey and assessment in the poverty-stricken counties of ethnic minorities in Southwest China.* China Agricultural University Press.

Chapter 6
Analysis of Farmers' Willingness to Participate Sustainedly in FCS Projects

With the completion of the reform of collective forest rights in China in the end of 2011, which clarifies property rights and confirms rights to households, forestation entities (enterprises) must strengthen their cooperation with farmers in FCS projects, forming the cooperative AR model of *company* + *farmers*. Therefore, community farmers are becoming the basic unit and main body of anti-poverty initiatives. However, the probability of farmers' withdrawing and the risk of default increase because projects last for a long time, sales revenues are unstable, and these projects are more stringent, demanding, and difficult than general afforestation projects (Kong et al., 2010). Accordingly, this chapter studies farmers' willingness to participate sustainedly in FCS projects and the factors influencing their willingness to do so through a sample survey of farmers who have already participated in such projects. It is expected to provide a scientific basis for formulating corresponding incentives and policies to ensure farmers' sustained and active participation and to facilitate the realisation of the goals of combating climate change and creating benefits and development opportunities for the poverty-stricken people.

6.1 Theoretical Analysis of Farmers' Willingness to Participate Sustainedly in FCS Projects

Farmers' willingness to participate in an FCS project is an important factor influencing their performance in poverty alleviation. In this study, based on a literature review, the researchers classify the influencing factors into individual characteristics, household characteristics, project characteristics, and environmental characteristics, and therefore, predict the econometric model of farmers' willingness to participate.

W. Zeng and F. Yang, *Poverty Alleviation Via Forest Carbon Sequestration*, International Research on Poverty Reduction, https://doi.org/10.1007/978-981-99-8191-5_6

6.1.1 Determination of Willingness

Some studies have addressed farmers' participation in FCS projects. Based on the theory of planned behaviour, Huang et al. (2013) constructed a theoretical model of farmers' behavioural intention to participate in FCS projects. Chen et al. (2011) analysed the factors influencing farmers' intention to participate in Sino-German cooperative afforestation projects, and the results showed that farmers' cognition of the project, average monthly household income, farthest distance of forest land from farmers' homes, project participation requirements, farmers' herd mentality, and forestry policies significantly influenced their intention to participate. Fang and Gao (2012) found from a survey in Guangdong Province that farmers' low willingness to participate and low awareness of FCS hindered project development. Tian et al. (2013) showed that age, household income level, family members working outside the home, FCS cognition, and project organisation mode significantly affected forestry workers' willingness to engage in FCS. The empirical results of Ning et al. (2014) showed that the average annual number of forestry training sessions received by household heads, forest land quality, farmers' management attitude, and regional differences were the main factors affecting farmers' willingness to participate in FCS management. The above research has important values and insights for this chapter but somewhat neglects to consider farmers' sustained participation in FCS projects.

According to rational smallholder theory (Schulz, 1987), farmers are rational economic agents who, when faced with multiple behavioural options, will choose the option that maximises their interests under certain constraints. Suppose that after participation, farmers are faced with *n* economic behaviour choices, and sustained participation in the project is only one of them. Among the remaining $n - 1$ economic behaviour opportunities, the net benefit V_i of the *i*-th one is the largest (i.e., V_i is the opportunity cost of farmers' sustained participation), and $V_i = R_i\text{-}C_i$, while R_i, *and* C_i are the benefits and costs of farmers' choice of the *i*-th economic behaviour, respectively, and R_1 and C_1 are the expected benefits and costs of farmers' sustained participation in the FCS project, respectively. Assuming that there is temporal mutual exclusivity between continuing to participate in FCS projects and choosing the *i*-th economic behaviour, the decision model of farmers' sustained participation can be expressed as $D = P(F = R_1 - C_1 - R_i + C_i)$, F is the expected net benefit of farmers' sustained participation, *P* is the probability of farmers' sustained participation, and *D* is the decision function of farmers' sustained participation. From the decision model, it can be seen that farmers are willing to participate in the project only when $F > 0$ and are unwilling to participate when $F < 0$.

The value of *F* is determined by R_1, C_1, R_i, and C_i, and the values of R_1, C_1, R_i, and C_i are influenced by the combination of farmers' own and the external environment. Using *j* to denote farmers' internal factors, such as gender, age, years of education, and household characteristics, and *l* to denote external environmental factors, such as project characteristics, institutional environment, and natural environment, and $g(j, l)$, $h(j, l)$ to denote the forces that positively and negatively affect R_1, C_1, R_i, and C_i, respectively, then $F = R_I - C_I - R_i + C_i = f[g(j, l), h(j, l)] = g(j, l) - h(j, l)$, so that the value of F is greater than zero only when $g(j, l) > h(j, l)$ and farmers are willing to participate sustainedly in the project.

In summary, farmers' willingness to participate sustainedly in FCS projects is not only influenced by internal factors, such as farmers' individual characteristics and family characteristics, but also by external factors, such as project characteristics and environmental characteristics, and farmers' willingness is a rational choice under the joint action of internal and external factors.

6.1.2 *Identification of Factors Influencing Farmer's Willingness to Engage in Sustained Participation*

Based on the above theoretical analysis, this study divides the factors influencing farmers' willingness to engage in sustained participation into four groups: individual characteristics, family characteristics, project characteristics, and environmental characteristics. The following hypotheses are proposed regarding farmers' willingness to participate sustainedly in FCS projects and their influencing factors.

As regards individual characteristics, both behavioural and behavioural economics theories, as well as numerous empirical studies on individual behaviour, show that individual behavioural decisions are closely related to demographic characteristics. In this section, three variables, namely gender, age, and years of education, were selected as individual characteristic variables of farm households. Generally, the implementation of FCS projects, including afforestation and forest management, requires high physical strength. In this respect, men have more advantages than women. Thus, their willingness to sustain participation may also be stronger than that of women. In addition, the older farmers are, the more likely they are to become risk-avoiders and make conservative choices based on the comparison between their individual life cycle and the project period, and their willingness to participate sustainedly will also gradually decrease. The higher the education level, the higher the ability to analyse economic profit and loss rationally, and the higher the cognitive ability of the dialectical relationship between economy and ecology. The former may lead to negative decisions when the benefits are lower than expected, while the latter may lead to the inclination to make positive decisions regardless of the benefits.

Regarding household characteristics, four variables, namely contributed land area, household income level, labour force quantity, and part-time occupation, are selected to reflect the resource endowment status of household production factors such as land, capital, and labour force for the FCS project. Generally speaking, the size of the contributed land area reflects, to a certain extent, the dependence of farming households on forestry income and the size of expected returns. The larger the contributed land area, the higher the dependence of farming households on forestry income; the larger the expected return value, the stronger the willingness of farming households to continue to participate.

Farming households must invest certain costs if they want to obtain higher potential returns from FCS projects. The higher the level of household income, the more money they can spend on forestry investment and the more willing farmers are to participate sustainedly in the project. The higher the number of labourers, the more abundant the labour capital that can be used and the stronger the willingness to sustain participation. This section defines the degree of part-time employment as the proportion of non-farm income to total household income. The higher the degree of part-time employment the farmers have, the less dependent households are on agricultural income, including FCS income, and the less likely they are to be willing to sustain participation.

Project characteristic variables are closely designed around the income, cost, and contractual constraints of farmers' sustained participation in the project, including the project organisation model, upfront benefit satisfaction, later benefit expectation, and the opportunity cost of sustained participation. Theoretically, different project organisation models may lead to differences in farmers' participation methods, benefit linkage mechanisms, benefit and risk guarantees, and perceptions of their rights and obligations, which, in turn, affect farmers' willingness to participate sustainedly. The higher the farmers' satisfaction with the initial participatory benefits and the higher the expectation of long-term benefits, the stronger their willingness to continue to participate; if the farmers think that the opportunity cost of continuing to participate is high, their willingness to continue to participate may be weak.

Regarding environmental characteristics, since the suitability of the natural environment would have been investigated during pre-project site selection, environmental characteristics in this section mainly refer to the influence of the institutional environment on farmers' willingness to sustain participation. Individual farmers' behavioural intentions are always nested in and interact with a certain institutional environment. In this section, three variables, namely government support, ease of access to forestry information, and road traffic conditions, are selected to reflect the institutional environment. In general, the stronger the government support for FCS projects, the greater the incentive for farmers to participate sustainedly. The ease of access to forestry information is measured by the timeliness and accuracy of information released by the government and afforestation entities (enterprises). The timely and accurate release of information on project-related policies and measures, weather forecasts, and disaster warnings by the government and afforestation entities (enterprises) can improve farmers' policy perception and disaster prevention ability, which is conducive to their sustained participation. The better the condition of the roads, the better the management and marketing of forest products, and the stronger farmers' willingness to participate in the project.

6.1.3 Measurement Model Construction and Indicator Quantification

The explained variable of farmers' willingness to participate sustainedly in FCS projects refers to farmers' willingness to continue to fulfil their obligations of not changing land use and carrying out afforestation and management after the completion of the initial afforestation projects. Farmers' willingness to participate sustainedly can be classified into five levels on a Likert scale: very reluctant, not very willing, neutral, willing, and very willing. Since the explained variable belongs to multi-categorical ordered variables and the explanatory variables are mainly discrete data, the ordered Probit model is suitable as the measurement analysis method. The general form of the ordered Probit model is $Y_i = \beta X_i + \varepsilon_i$, Y_i is the latent variable, X_i is the set of explanatory variables, β is the coefficient to be estimated, and ε_i is the random disturbance item.

The basic model is formulated as follows: farmers' willingness to participate sustainedly in FCS projects = F (individual characteristics, household characteristics, project characteristics, environmental characteristics) + random disturbance item. Based on the above analysis, the information on model variables is shown in Table 6.1.

6.1.4 Investigation of Farmers' Willingness to Participate Sustainedly in FCS Projects

In order to verify the above econometric model, a research sample of farmers in the area around the Novartis Sichuan FCS Project was selected for this study. This section introduces the survey methodology, the results of the survey on farmers' willingness to sustain participation, and the basic characteristics of these farmers.

6.1.5 Data Description

In order to exclude the influence of the project organisation model and its interest linkage mechanism, regional socioeconomic and cultural differences, and other factors, this chapter selects the Novartis Sichuan FCS Project implementation communities as the research target. The sample area was located in the project counties (districts) where afforestation is mainly carried out on lands subcontracted to families, specifically in Meigu, Ganluo, and Yuexi Counties (Fig. 6.1). All the project lands in the three counties are forestable, they have been confirmed to be household-owned, and they have completed the preliminary afforestation and entered the stage of cultivation and management. Farmers participated in the project by signing a contract with the project owner for 30 years. During the contract period, the project

Table 6.1 Variable descriptions and expected direction of influencing

Variable name	Variable meaning and assignment	Expected direction of influence
Explained variables		
Farmers' willingness for sustained participation (Y)	Very reluctant = 1; *not* very willing = 2; neutral attitude = 3; willing = 4; very willing = 5	
Explanatory variables		
Variables of Individual characteristics		
Gender (X_1)	Male M = 1; Female F = 0	+
Age (X_2)	Actual observation	−
Years of education (X_3)	Actual observation	?
Variables of family characteristics		
Contributed land area (X_4)	Actual observation(Ha)	+
Household income (X_5)	Per capita net household income (RMB/month) in 2013: below 2,000 = 1; 2,000–4,000 = 2; 4, 000–6,000 = 3; 6,000–8,000 = 4; above 8,000 = 5	+
Labour force quantity (X_6)	Actual observation	+
Degree of part-time employment (X_7)	Non-farm income as a share of total household income: 0–20 = 1; 20–40 = 2; 40–60 = 3; 60–80 = 4; 80–100 = 5 (%)	−
Project organisation mode (X_8)	Home production according to individual contract = 1; others = 0	?
Front-end return satisfaction (X_9)	very dissatisfied = 1; dissatisfied = 2; average = 3; satisfied = 4; very satisfied = 5	+
Post-period benefit expectation (X_{10})	Very low = 1; low = 2; average = 3; high = 4; very high = 5	+
Opportunity cost of sustained participation (X_{11})	Very low = 1; low = 2; average = 3; high = 4; very high = 5	−
Variables of environmental characteristics		
Degree of government support (X_{12})	Very low = 1; low = 2; average = 3; high = 4; very high = 5	+
Information accessibility (X_{13})	Very difficult = 1; difficult = 2; average = 3; easy = 4; very easy = 5	+
Road traffic condition (X_{14})	Very poor = 1; poor = 2; fair = 3; good = 4; very good = 5	+

Note ① +, −, and ? denote the direction of influence of the explanatory variables on the explained variables as positive, negative, and uncertain, respectively; ② considering the specificity of the study area, the labour force population is defined as the population aged 14–65 with working ability

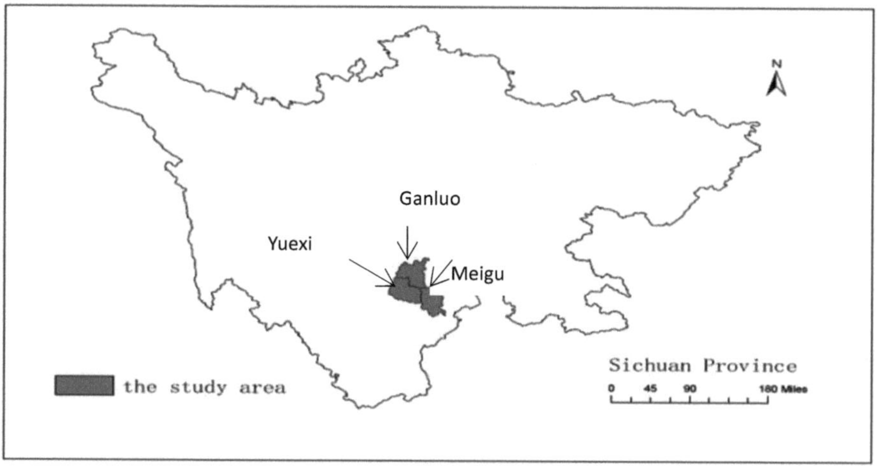

Fig. 6.1 Location of the study area in Sichuan Province

owner, Sichuan Dadu River Afforestation Bureau, is responsible for organising the project, while the farmers are responsible for afforestation, forest cultivation, and management. The farmers' participation income mainly includes the income derived from labour during the project period and the income from forest products and forest by-products, while the carbon sequestration income generated from the afforestation during the contract period is all owned by the afforestation entities. The participation cost of farmers mainly consists of labour and the opportunity of participating in exploiting the lands, and the participation condition is that the forestable land whose household ownership is confirmed must meet the site selection requirement of carbon sequestration afforestation.

6.1.6 Degree of Farmers' Willingness to Participate

Farmers'sustained participation in FCS projects involves the input of various factors, such as land and labour, and there are often significant differences in the degree of farmers' willingness to participate because of the differences in their resource endowments, such as physical capital, natural capital, and social capital. The statistical results of the questionnaires collected from 397 farmers show that 46.35% have a neutral attitude in the decision to participate sustainedly in the FCS project, 6.80% are very willing to participate, and 28.97% are willing to participate, while 24.69% of farmers are not willing to participate sustainedly. In general, the willingness of the sample farmers to participate sustainedly in the FCS project is normally distributed (Fig. 6.2), with the largest number of farmers having a neutral attitude, accounting for 46.35% of the total sample. The mean value of willingness to participate sustainedly

Fig. 6.2 Willingness of sample farmers to participate sustainedly

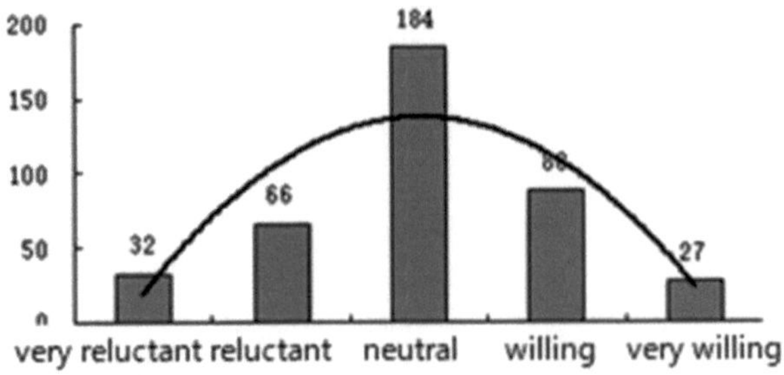

Table 6.2 Results of descriptive statistics of variables

Variables	Min	Max	Mean	S. D
Farmers' willingness to engage in sustained participation (Y)	1	5	3.03	0.99
Gender (X_1)	0	1	0.66	0.47
Age (X_2)	17	65	42.11	11.62
Years of education (X_3)	0	15	4.15	3.53
Contributed land area (X_4)	0.05	0.53	0.12	0.05
Household income (X_5)	1	5	2.77	0.76
Labour force quantity (X_6)	1	7	3.83	1.21
Degree of part-time employment (X_7)	1	5	2.80	0.76
Project organisation mode (X_8)	0	1	0.49	0.50
Front-end return satisfaction (X_9)	1	5	2.97	0.95
Post-period benefit expectation (X_{10})	1	5	2.96	0.80
Opportunity cost of sustained participation (X_{11})	1	5	2.62	1.10
Degree of government support (X_{12})	1	5	2.74	0.96
Information accessibility (X_{13})	1	5	2.71	0.86
Road traffic condition (X_{14})	1	5	2.90	0.70
Sample size	397			

is 3.03 (Table 6.2), which is between *neutral* and *very willing*, indicating that farmers' willingness to sustain participation was not high and needed to be improved.

6.1.7 Basic Characteristics of Farming Households Participating in the FCS Project

The age of participating farmers is mainly concentrated in the ranges of 40–49 (31.49%) and 30–39 (20.48%). On average, farmers have had only 4.15 years of education. Each family has an average of 3.83 labourers. The majority of families had a per capita net income of 4000–6000 RMB (about 600–870 USD) in 2013

(57.68%), which is lower than the average level of Sichuan Province (7895 RMB or about 1144 USD). The source of income is mainly traditional agriculture and outworking, and the degree of part-time employment is low (2.80 or about 36%). The basic situation of the sample and the survey found that the farmers in the survey area have a low level of education, a high rate of young adults left behind, a low level of household income, a narrow source of income, and a strong desire to escape from poverty.

6.1.8 Empirical Analysis of Farmers' Willingness to Participate Sustainedly in FCS Projects

Based on the above data, this study examines the influence of individual characteristics, family characteristics, project characteristics, and environmental characteristics on farmers' willingness to participate sustainedly in FCS projects. The study also proposes countermeasures to improve farmers' willingness to participate in the project.

6.1.9 A Multicollinearity Test

To ensure the accuracy and stability of the model, a multicollinearity test is conducted on each variable. The method is to use one of the independent variables as the dependent variable and the rest of the variables as the independent variables for regression analysis. The criterion for determining the presence of multicollinearity is the tolerance or variance inflation factor (VIF). A smaller value of tolerance indicates that the dependent variable is explained by other variables to a greater extent, and therefore, the more likely it is that there is serious covariance. A reasonable range of tolerance is (0.1 to $+\infty$); VIF is the inverse of tolerance, and if its value is $\geq$10, there may be serious covariance between independent variables.

Table 6.3 shows the results of the multicollinearity test with gender as the dependent variable and other variables as independent variables. Of these, the minimum value of tolerance (Tolerance) is $0.606 > 0.1$, and the maximum value of VIF is 1.650 < 10, which shows that there is no serious multicollinearity. Similarly, all other variables can be tested for multicollinearity. Due to limitations on space, the test process is not presented in its entirety. The results of all the runs show that the tolerance of the multicollinearity test and the VIF are within a reasonable range. Therefore, there is no serious multicollinearity problem among the respective variables in the regression equation.

Table 6.3 Model multicollinearity test

		Covariance statistics	
		Tolerance	VIF
Gender (X_1)	Ln (Age) (X_2)	0.715	1.399
	Years of education (X_3)	0.914	1.095
	Ln (Contributed land area) (X_4)	0.864	1.158
	Household income (X_5)	0.899	1.112
	Labour force quantity (X_6)	0.804	1.243
	Degree of part-time employment (X_7)	0.944	1.059
	Project organisation mode (X_8)	0.726	1.377
	Front-end return satisfaction (X_9)	0.606	1.650
	Post-period benefit expectation (X_{10})	0.607	1.647
	Opportunity cost of sustained participation (X_{11})	0.932	1.073
	Degree of government support (X_{12})	0.790	1.266
	Information accessibility (X_{13})	0.704	1.420
	Road traffic condition (X_{14})	0.763	1.310

6.1.10 *Probit Model Estimation*

With Stata 12.0, the data were processed by ordered Probit regression. To ensure the stability of the model operation, age and contributed land area were taken into the model after being taken logarithm. The results are shown in Table 6.4: the overall fit of the model is good, and the impact of independent variables is also basically in line with expectations.

6.1.11 *Analysis of Model Results*

As can be seen from Table 6.4, the variables of age, contributed land area, household income level, degree of part-time work, project organisation mode, satisfaction with the initial income, expectation of the later income, government support, difficulty in obtaining forestry information, and road traffic condition passed the significance test. It can also be seen that these variables are the key factors that significantly affect farmers' willingness to participate sustainedly in FCS projects. The specific analyses are as follows.

First, for the influence of individual characteristics, the coefficient of *age* is negative and significant at the 1% level, indicating that the older the age, the lower the willingness of farmers to participate, which is consistent with the previous hypothesis about farmers' physical ability and risk preference. The coefficient of *gender*

Table 6.4 Regression results of ordered Probit model

Independent variable	Coefficient	Standard error	Z	P
Individual characteristics				
Gender (X_1)	0.190	0.129	1.470	0.141
Ln (Age) (X_2)	−1.034***	0.242	−4.270	0.004
Year of education (X_3)	0.014	0.018	0.800	0.424
Household characteristics				
Ln (contributed land area) (X_4)	0.528**	0.268	1.970	0.048
Household income (X_5)	0.322***	0.084	3.810	0.003
Labour force quantity (X_6)	0.076	0.054	1.400	0.161
Degree of part-time employment (X_7)	−0.227***	0.079	−2.870	0.004
Project characteristics				
Organisationalmodel (X_8)	−0.831***	0.142	−5.870	0.000
Satisfaction with the initial income (X_9)	0.387***	0.079	4.890	0.000
Expectation of the later income (X_{10})	0.628***	0.097	6.490	0.005
Cost of sustained participation (X_{11})	−0.028	0.055	−0.510	0.612
Environmental characteristics				
Government support (X_{12})	0.228***	0.068	3.350	0.001
Information accessibility (X_{13})	0.377***	0.083	4.550	0.007
Road traffic conditions (X_{14})	0.370***	0.095	3.890	0.000
Log likelihood	−347380			
Pseudo R^2	0.312			
P	0.000			

Note ***, **indicate significance at the 1% and 5% levels, respectively

failed in the significance test, indicating that men do not have a stronger willingness to participate than women. The coefficient of *years of education* failed in the significance test, probably because the surveyed farmers generally had few years of education, and the significance was not statistically significant.

Second, as regards the influence of household characteristics, the coefficient of the *contributed land area* is positive and significant at the 5% level, indicating that the larger the land area involved, the stronger the household's willingness to continue to participate, which is consistent with the previous hypothesis. The coefficient of *household income* is positive and significant at the 1% level, as expected, indicating that the higher the farm household income, the stronger their willingness to participate sustainedly. This finding also reflects that the lower the household income is, the higher the likelihood that farmers will withdraw from participation. The reason may be that the household income is a reflection of the household's economic ability and economic status, and a lower household income indicates a lower economic ability and economic status, which also means a lower public political status and weaker public voice. Such farming households are more likely to be marginalised in the negotiations for project benefit distribution and have a higher likelihood of having their interests eroded, and their sustained participation is more likely to be discouraged.

The coefficient of *degree of part-time employment* is negative and significant at the 1% level, which is in line with the expectation, indicating that farmers who are less dependent on agricultural income, including income from FCS projects, are more likely to withdraw from participation, which provides a basis for targeting FCS projects, namely carrying out projects in poverty-stricken areas that are farming-(especially forestry-) based. The reason that the coefficient of *labourers* failed to pass the significance test may be that the labour force requirements for sustaining FCS projects in their middle and late stages are not harsh; thus they can be met by the average household.

Third, for the influence of project characteristics, the coefficient of *project organisation mode* is negative and significant at the 1% level. Based on the characteristics of the project, organisation mode is divided into two types, according to the difference in contracting mode, namely individual household contract mode and joint contract mode. The former refers to the direct signing of project contracts between afforestation entities (enterprises) and individual farmers. In this case, the afforestation entities (enterprises) pay funds, single-family farmers participate in the project on their own contributed land, and farmers receive income from afforestation, forest care labour, and forest by-products according to the project contracts. Under the latter mode, the forestation entity (enterprise) and multiple farmers jointly sign the project contract, the forestation entity (enterprise) pays the funds, and multiple farmers jointly participate in the project construction on their contributed land, the implementation of revenue sharing and risk sharing. In this case, multiple farmers are tied to form a community of interest, and one defaulting will damage the interests of all, giving rise to the formation of a mechanism of mutual supervision and control. The empirical tests show that farmers' willingness to continue to participate is stronger under the joint contracting model.

The coefficient of *satisfaction* with the initial income and that of *expectation of the later income* are both positive and significant at the 1% level, as expected, indicating that economic income is an important factor in farmers' consideration of whether to participate sustainedly in the project or not. The higher the satisfaction and expectation of economic returns, the stronger the willingness of farmers to participate sustainedly.

The coefficient of *opportunity cost of sustained participation* failed in the significance test because the study area is poor in resources, has inconvenient transportation, and is economically backward, and the farmers have limited ways to generate income, so their perception of the opportunity cost of sustained participation in FCS projects is low.

Fourth, as regards the influence of environmental characteristics on farmers' sustained participation in FCS projects, the coefficient of *government support* is positive and significant at the 1% level, which is in line with the expectation. This indicates that the government should provide moderate policy, financial, material, and technical support to motivate farmers' willingness to participate in the project. The coefficient of *forestry information accessibility* is positive and significant at the 1% level, as expected, indicating that the more timely and accurate the release of forestry information related to the project, the stronger the willingness of farmers to participate sustainedly. The coefficient of *road traffic condition* is positive and significant at the 1% level, which is in line with expectation, indicating that the improvement of road traffic conditions can enhance the willingness of farmers to participate sustainedly.

6.2 Summary

Based on the cross-sectional data from the field survey, this chapter empirically analysed farmers' willingness to participate sustainedly in FCS projects and their influencing factors. The results show that the willingness of farmers to participate sustainedly in FCS projects is not strong and needs to be improved; age, contributed land area, household income level, degree of part-time employment, project organisation mode, satisfaction with early benefits, expectation of later benefits, government support, ease of access to forestry information, and road traffic conditions have significant effects on farmers' willingness to participate sustainedly.

The above findings have obvious policy implications for enhancing farmers' enthusiasm to participate sustainedly in and promote the sustainable operation of FCS projects and poverty reduction. For afforestation entities (enterprises), an important strategy is to improve the revenue incentive mechanism. First, they should implement multiple targeting of regions, land, and population, focusing on ecologically fragile and forestry-dependent poverty-stricken areas, pay close attention to the willingness of poverty-stricken households to participate, and avoid the marginalisation of poverty-stricken households that could hinder the sustainable development of the project. Second, they should select and optimise the project organisation model according to local conditions, the key is to realise benefit sharing and risk sharing. Enterprises are encouraged to choose the organisation mode of signing project contracts with multiple farmers who are eligible for this contractor signing tripartite contracts with farmers and county forestry authorities or township governments to reduce the risk of farmers' withdrawal. Third, they should issue relevant subsidies and relevant expenses for carbon sequestration forest management, etc., in full and on time to guarantee farmers' short-term income. Fourth, they should select local tree species with high economic value as much as possible, combine the forest sequestration project with eco-tourism, use the forest resources for eco-tourism development after the project period, increase farmers' long-term income expectation, and establish a long-term incentive mechanism.

As far as the government is concerned, an important strategy is to improve the policy protection system. First, government support should be increased, strengthening the integration of poverty alleviation resources, methods, and policies and providing preferential policy, financial subsidies, material assistance, and technical guidance. Second, the government should establish and improve forestry information dissemination and disaster warning mechanisms and improve the forest insurance system. Third, it should financially support the construction of roads, irrigation, and other infrastructure in forest areas to overcome the natural conditions that limit project development and constrain the participation of farmers in their communities.

References

Chen, K., Chen, W. T., Wang, Y. M., et al. (2011). An empirical analysis of factors influencing farmers' willingness to participate in Sino-German cooperative afforestation projects—An example from Chaoyang City, Liaoning Province. *Agricultural Economics, 5*(5), 32–35.

Fang, X. L., & Gao, L. (2012). Research on the influencing factors and countermeasures of forest carbon sequestration projects in Guangdong. *Jiangsu Agricultural Science, 40*, 6–8.

Huang, Y. L., & Nie, J. (2013). A theoretical analysis framework of forest farmers' intention to participate in forest carbon sequestration behavior—A perspective based on TPB model. *Economist,* (11), 24–25+34.

Kong, F. B. (2010). Research progress of forestry in response to global climate change and research direction of policy mechanism in China. *Issues in Agricultural Economics, 7*, 105–109.

Ning, K., Shen, Y. Q., & Zhu, Z. (2014). Analysis of farmers' perceptions of forest carbon sequestration and willingness to operate carbon sink forests-based on a survey of farmers in Zhejiang, Jiangxi, and Fujian provinces. *Journal of Beijing Forestry University (social Science Edition), 13*(2), 63–69.

Tian, Q., Ke, S. F., Du, X., et al. (2013). Analysis of factors influencing willingness to supply forestry carbon sinks in China--an empirical investigation based on the Tizi Mountain Forestry in Hubei Province *Ecological Economics (Academic Edition), 2*(2), 168–173.

Chapter 7
Analysis of Constraints on Farmers' Participation in FCS Projects

Farmers are the main suppliers, the most direct participants, and the ultimate beneficiaries of FCS projects, and their participation not only directly affects the smooth implementation of FCS projects and carbon transactions, but also determines the performance of PAFCS (Gong & Zeng, 2018). The status quo of farmers' participation and the barriers they face have been extensively studied in the domestic and international literature (Benitez et al., 2004; Nhung, 2009; Antle & Stoorvogel, 2009; Chen, 2010; Benessaiah, 2012; Ding & Ma, 2013; Ming et al., 2015; Yang et al., 2016; Zhu et al., 2013), but the vast majority of studies have simply started from the internal factors of farmers or households, ignoring the premise that the current poverty alleviation development of FCS projects is mainly driven by the government. In the current context of China's grassroots governance system and carbon market development, local governments or village and community collectives are still the main drivers of FCS projects. It is thus necessary to include external factors such as government actions and project characteristics in the analysis of farm household participation. To this end, this chapter draws on the theory of embedding social structure and uses micro-sampling data to empirically study the participation of farmers and its influencing factors under the established CDM FCS standards. It thus provides a useful reference for clearing or reducing barriers to and formulating FCS support policies to enhance farmers' participation.

7.1 Theoretical Basis of Farmers' Participation in FCS Projects

This section first defines the scope of farmers' participation in FCS projects based on the four stages of planning and design, organisation and execution, sustainable operation, and monitoring and evaluation; then it selects specific indicators that affect farmers' participation based on embedding social structure theory.

W. Zeng and F. Yang, *Poverty Alleviation Via Forest Carbon Sequestration*, International Research on Poverty Reduction, https://doi.org/10.1007/978-981-99-8191-5_7

7.1.1 Definition of Farmers' Participation in FCS Projects

The participation of farmers in FCS projects is a comprehensive concept involving not only the input of factors such as land, labour, advice, and consultation, but also the four stages of FCS project planning and design, organisation and execution, sustainable operation, and monitoring and evaluation. In the planning and design stage, farmers' participation mainly takes the form of *cooperating* and *suggesting* during the project background survey and project planning and design phases, which is a typical input of intangible factors. The organisation and execution stage involves participation in afforestation, forestry, technical training, and other project activities, including FCS land developing such as land transfer or lease, shareholding, and other forms, and providing support and cooperation for project execution. In the monitoring and evaluation stage, due to the strict professionalism of measurement and monitoring standards, farmers' participation is very restricted, principally the input of intangible elements such as the right to know and the right to suggest. In summary, this study concludes that the participation of farmers in FCS projects refers to the actual behaviour of farmers in the three stages of the FCS project planning and design, execution and operation, and monitoring and evaluation, and nine specific behaviours (see Table 7.1).

Table 7.1 Participatory behaviour of farmers in FCS projects

Stages	Behaviours
B1: Planning and design	B11: Whether to participate in or know about the background survey (questionnaire, discussion, etc.)
	B12: Whether to put forward their or village representatives' opinions
B2: Execution and operation	B21: Whether to publicise (announce) the project plan or sign a contract
	B22: Whether to participate in training
	B23: Whether to acquire shares in FCS projects or gain benefits from transferring the forestable land
	B24: Whether to obtain income from labour, management, grazing, and ecological compensation for public welfare forests
	B25: Whether to comply with the contract
B3: Monitoring and evaluation	B31: Whether to participate in the project stages of acceptance, monitoring, and evaluation
	B32: Whether suggestions of means to avoid conflicts and disputes are adopted

7.1.2 Indicator Selection for the Analysis of Constraints on Farmers' Participation in FCS Projects

According to embedding social structure theory (Granovetter, 2018), economic actions are not completely atomised and isolated, but embedding in the social structure, which is constrained by social relations. Farmers' participation in FCS projects is not only influenced by autonomous factors, such as individual farmers' characteristics, household resource endowment, and social capital, but constrained by embedding factors, including project characteristics such as organisational mode, property rights ownership of FCS projects, and village and community environment characteristics, such as economic development and villagers' interrelationship in the project area.

First of all, under the household contract responsibility system, the household is still the main body of agricultural production and management decisions. Against the background of reform of the collective forest rights system, in particular, the status of farmers or households as the main body of FCS supply has been reflected and enhanced. Individual characteristics such as the education level and gender of household heads, as well as household characteristics such as household resource endowment and social capital, will influence farmers' willingness and ability to participate and then their participation decisions and form of participation.

Second, China's long-standing household contract responsibility system has led to the fragmentation of rural land; however, FCS requires large-scale centralised production, which makes it impossible for a single farmer to supply it. Moreover, a farmer is inevitably influenced by the behaviour of the local government, the collective, other farmers, and the economic, social, and natural environment of the project. Finally, China's grassroots governance system determines the prominent role of local governments or village and community collectives in the execution of FCS projects.

Since China's carbon trading market is not yet well developed, the government will remain the main driving force of FCS projects for some time. In this context, farmers' participation has two main aspects: the cooperation behaviour under involuntary choice and the participation decision under autonomous choice. Therefore, this chapter draws on embedding social structure theory to explore the influence of three autonomous factors, namely individual farmers' characteristics, household business characteristics, and household social capital, and two embedded factors, namely project characteristics and community characteristics, on farmers' participation in FCS projects.

The individual characteristics, as the autonomous factor, refer to the education level, age, and gender of the household head or household decision maker, which influence awareness, willingness to participate, and decision-making. In this chapter, education, gender, and age are used as individual characteristics. Household characteristics are factors such as labour force structure, forest land area, and income owned by the household, which have an influence on the participation mode and degree of participation.

Social capital characteristics of households are their relationship networks and social trust, which influence the acknowledgement of the information related to the decision, proposal, and monitoring and evaluation in FCS projects. In this chapter, the social network, trust in government, and mutual trust among villagers are used as secondary indicators.

As an embedded factor, community characteristics should be included in the study. Since the production cost, registration, and management of FCS projects are borne by local governments or third-party enterprises, the degree of farmers' participation is inevitably influenced by the political, economic, and natural environmental characteristics of the project area. In this chapter, the degree of dependence on livestock, the degree of government support for FCS, and the degree of community harmony are used to reflect the community characteristics.

The project characteristics, as another embedded factor in this study, mainly refer to the source of labour, ownership of forest land, and organisation mode of the FCS project, which affect the participation mode, the benefit mechanism, the risk guarantee of benefits, and the perception of farmers' rights and obligations. In this chapter, the researchers use labour source, forest land ownership, and organisation model as the secondary indicators of project characteristics.

The degree of participation mainly depends on both the willingness to participate and the ability to participate. Thus, the influence of autonomous and embedded factors on the degree of participation of FCS farmers includes two paths. The first one is to indirectly influence the degree of participation of farmers by influencing their willingness to participate. Existing studies have shown that autonomous factors, such as farmers' characteristics and family business characteristics, have significant effects on farmers' willingness to participate (Ming et al., 2015; Zhu et al., 2013). Moreover, embedded factors, such as government behaviour and project model, affect farmers' expectations of the benefits and risks of FCS projects, which, in turn, affect their willingness to participate (Huang & Nie, 2013).

The second path is to directly *control* farmers' participation and affect their ability to participate (Huang & Nie, 2013). Therefore, this chapter uses willingness to participate as an intermediate variable to analyse the influence of autonomous and embedded factors on the degree of participation, on the one hand, and the level of farmers' autonomous choice in the current government-driven context, on the other hand. The influence path is shown in Fig. 7.1.

7.1.3 Investigation and Analysis of Farmers' Participation in FCS Projects

In order to test the specific effects of the factors influencing farmers' participation in FCS projects, the researchers selected farmers' participation in the *Northwest Sichuan FCS Project and the Novartis Sichuan FCS Project* and conducted an extensive survey. This section describes the survey sample, methodology, and results.

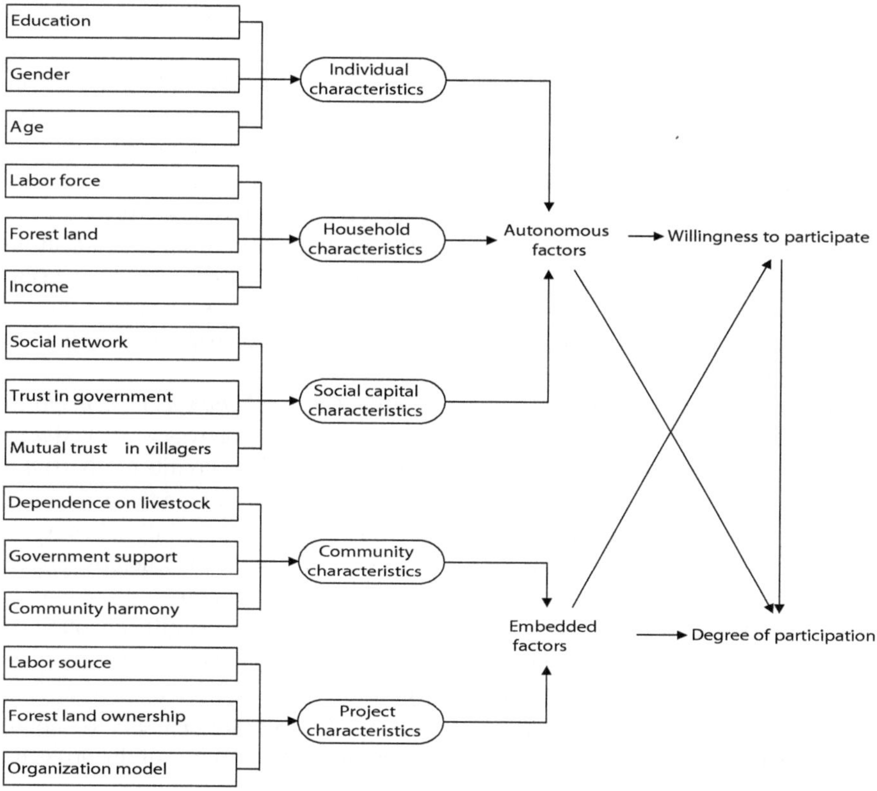

Fig. 7.1 Framework for analysing farmers' participation in FCS projects

7.1.4 Overview of the Survey Area

The data in this chapter were collected from field questionnaires and case interviews conducted by the research team in the counties and villages where the two projects mentioned above are located and many farmers participate in FCS. A total of 728 valid samples were generated.

Both projects are AR under the Kyoto rules (CDM/AR) and are consistent in terms of land eligibility, carbon baseline, boundary, carbon sequestration measurement, FCS monitoring standards and methods, etc. The project areas are mainly located in the poverty-stricken areas of Sichuan, and the main afforestation body is the Dadu River Afforestation Bureau of Sichuan Province. However, the differences in the organisational and management modes and the economic, social, and cultural environments of the project areas have led to there being significant differences in the participation mode, willingness, and degree of participation of farmers.

During the visit, the researchers found that the *Northwest Sichuan FCS Project—Pingwu County Subproject* (henceforth, *Pingwu FCS Project*) is a unilateral project executed by the Dadu River Afforestation Bureau with its own funds before any revenue is generated. The Pingwu Forestry Bureau assists in afforestation and management, provides part of the afforestation seedling funds, and jointly carries out afforestation technology training with the Dadu River Afforestation Bureau and other units which initiates pilot demonstration projects of edible mushroom cultivation and biomass solid fuel processing. At the end of the project period, farmers receive all the timber revenue and 30% of the carbon sequestration revenue, and the Forestry Bureau receives 20% of the carbon sequestration revenue. With strong support from the government, village rules and regulations are established to ensure the smooth implementation of the project, while the Dadu River Afforestation Bureau has installed wooden fences in the FCS area to ensure the survival rate and standard of afforestation.

The Novartis Sichuan FCS Project—Ganluo County Subproject (*henceforth, Ganluo FCS Project*) is a bilateral project in which the revenue was generated prior to execution. The preliminary funds mainly come from carbon credits partly paid in advance by Novartis, which greatly reduces the financial pressure in the early stage. To ensure the survival rate and specification of afforestation, it mainly relies on the Leibo Afforestation Company to carry out afforestation. The means whereby farmers can participate in the project are limited, and farmers can only obtain little benefit. At the same time, the FCS land mainly consists of forestable barren hills or rotational rest land, with prominent contradictions between forestry and animal husbandry, so the farmers have high resistance to the project. Moreover, the Forestry Bureau is an auxiliary unit that does not have an interest in benefit distribution and is not highly motivated, resulting in a large deviation from farmers' expectations of the projected income (see Table 7.2).

Table 7.2 Project characteristics of Ganluo and Pingwu FCS projects

Project	Ganluo FCS Project	Pingwu FCS Project
Source of funding	Novartis 57%; government 43%	State-funded afforestation seedlings, self-financing by Dadu River Afforestation Bureau
Source of forest land	Farmers' barren hills and rotational land, collective forest land	Villagers' group collective forest land, farmers' own forest land
Revenue distribution	100% of timber revenue goes to farmers; 100% of carbon sink revenue goes to Novartis	100% of timber revenue goes to farmers, 30% of carbon sink revenue to farmers, 25% to Dadu River Afforestation Bureau, 20% to Pingwu County Forestry Bureau, 25% to management expenses
Operation and management	Managed by Dadu River Afforestation Bureau, and some local farmers are employed	Dadu River Afforestation Bureau and Pingwu Forestry Bureau; some afforestation construction teams of local farmers are employed
Forestation entity	Leibo afforestation company; local farmers	Local, pre-construction technical training provided by protected area authorities; on-site guidance by technical personnel

7.1.5 Results and Analysis

In terms of the overall participation percentage, farmers' participation in the Ganluo FCS Project (26.34%) and Pingwu FCS Project (30.03%) was generally low, and the degree of participation in the three stages was significantly differentiated, with the lowest participation in the monitoring and evaluation stage. From the perspective of welfare improvement, the participation that can bring non-economic welfare, such as the right to advise and the right to obtain information, is relatively low compared with the participation that can bring direct economic welfare, such as labour input and forest land input. Comparing Ganluo FCS Project with Pingwu FCS Project, the latter has a higher overall participation level than the former, and there are obvious differences in behaviours at each stage, especially in the behaviours of *whether to participate in training* and *whether to fulfil the contract.*

On the one hand, the Pingwu FCS Project is led by the government and carries out many initiatives to benefit farmers, and the frequency and scope of agricultural production skills training are better. Ganluo County, where the *Novartis FCS Project is implemented,* belongs to the Yi ethnic group area, and the perennial pastoral culture weakens farmers' demand for agricultural production skills. On the other hand, farmers in Ganluo County have a high reliance on livestock farming, and the forest–pastoral conflict resulting from the FCS project is serious, which causes low willingness and low farmer participation. However, it is worth noting that indirect farmer participation in the Ganluo FCS Project is significantly better than that in the Pingwu FCS Project, because a large number of farmers in Pingwu County, especially young adults, have left the farm and gone out to work.

More male respondents were interviewed than females, accounting for 69.61% of the total sample. The age of the interviewed farmers was high, with 61.49% of the respondents being over 45 years old. In the Ganluo FCS Project sample, the proportion of Yi ethnic group is high (71.79%), as is the proportion of female household heads (31.09%).[1] Agriculture is mainly grazing, and the dependence on forest land is greater; therefore, the conflict between forest and livestock is obvious. The education level is relatively low, and most farmers have only 1–2 years of education. The income level is low. The carbon sequestration forest is mainly on the farmers' barren hills and wastelands or rotational land, and professional afforestation companies are used to create and manage it. The level of trust in the government is relatively low, but the level of trust among villagers is high. The willingness to participate indirectly in the local survey, planning, monitoring, and evaluation stages of the project is high, but the willingness to participate directly in labour and forest land input is relatively low.

The majority of the samples in the Pingwu FCS Project are Han Chinese, with a relatively high level of education that 96.09% of the households have members with elementary school level or above. Forest land resources and labour force are relatively abundant. Household production is mainly agriculture, but many farmers work outside the farm, so the dependence on pastoralism is weak, and the income

[1] In rural China, the head of the household is usually male.

level is high. Carbon sequestration forests are mainly collective forest land. The labour force input of farmers in the project area is relatively high (88.83%), and they gain partly benefits from the carbon sequestration. They have a high degree of trust in the government and a high willingness to participate directly and indirectly (see Table 7.3).

7.1.6 Econometric Analysis of the Constraints on Farmers' Participation in FCS Projects

To confirm the correlations between the aforementioned influencing factors and farmer participation in FCS projects, an econometric approach is adopted. In other words, the structural equation model was used to examine the relationships after data quality inspection and testing.

7.1.7 Reliability and Validity Tests

In order to test the stability and reliability of the questionnaire, SPSS 20.0 was used in this chapter to test the reliability and validity of the sample. The overall Cronbach α was 0.801, indicating that the questionnaire had good internal consistency. The results of factor analysis showed that the KMO value was 0.802, Bartlett's test approximate chi-square was 2798.308, and the significance level was 0.000, indicating the internal consistency is high. The contribution of cumulative variance after rotation is high, indicating that the data have good structural validity. The explanatory variable, Cronbach's α, was 0.572 (greater than 0.5), with a total of two valid factors extracted, and the distribution was consistent with direct and indirect participation, indicating that the division of participation into direct and indirect participation coincided with reality (see Table 7.4).

7.1.8 Model Estimation

In this chapter, AMOS 17.0 software was used as a tool for SEM analysis, and the confirmatory factor analysis of the overall sample was conducted using data from 728 households in the Pingwu FCS Project and Ganluo FCS Project, with the Comparative Fit Index (CFI) and the General Fitness Index (GFI) greater than 0.90 and Root Mean Square Error of Approximation (RMSEA) less than 0.08, indicating a good fit of the model to the data. The normalised results of the path coefficients are shown in Table 7.5.

Table 7.3 Evaluation criteria and descriptive analysis of explanatory variables of participation

Latent variables	Explanatory variables and quantification (yes = 1; no = 0 if no marks)	Pingwu		Ganluo	
		Mean	S.D	Mean	S.D
B Degree of participation	B11:Whether or not to participate in or understand the survey (questionnaire, discussion, etc.)	0.922	0.072	0.949	0.049
	B12:Whether to give their or village representatives' opinions on the planning design	0.341	0.225	0.327	0.220
	B21:Whether the project plan or contract is publicised	0.810	0.154	0.955	0.043
	B22:Whether to participate in training	0.849	0.128	0.455	0.248
	B23:Whether or not to obtain shareholding or transfer proceeds from forestable land	0.675	0.249	0.481	0.250
	B24:Whether to obtain income from labour, management, grazing, and ecological compensation	0.754	0.185	0.310	0.082
	B25:Whether to comply with the contract	0.788	0.167	0.173	0.143
	B31:Whether to participate in project monitoring and evaluation	0.151	0.128	0.256	0.191
	B32:Whether opinions to avoid conflicts and disputes were adopted	0.073	0.067	0.096	0.087
	Comprehensive Participation in FCS projects	0.300	0.022	0.263	0.037
SF Project characteristics	SF1:Labour source—a professional company is used for afforestation and management = 0; local farmers are employed = 1	0.888	0.099	0.276	0.200
	SF2:Forest land ownership—collectively owned = 0; individually owned = 1	0.318	0.217	0.776	0.174
	SF3:Organisational mode—contract signed by individual farmer = 0; contract signed by a group of farmers = 1	0.875	0.103	0.421	0.135

(continued)

Table 7.3 (continued)

Latent variables	Explanatory variables and quantification (yes = 1; no = 0 if no marks)	Pingwu		Ganluo	
		Mean	S.D	Mean	S.D
SE Community characteristics	SE1:Livestock dependence—very dependent = 1;not very = 2;unimportant = 3; non-dependent = 4;totally non-dependent = 5	3.626	0.826	2.718	0.869
	SE2:Government support—very supportive = 5;not very = 4;moderately = 3;not = 2;very unsupportive = 1	3.341	0.716	3.365	1.450
	SE3:Degree of village community harmony—very harmonious = 5; harmonious = 4; fair = 3; non-harmonious = 2; very non-harmonious = 1	3.000	1.520	3.333	1.107
PF Farmer's characteristics	PF1:Education—high school or above = 5; middle to high school = 4; elementary to middle school = 3; below elementary school = 2; no schooling = 1	3.143	0.500	2.365	0.693
	PF2:Age of head of household—(0.8,1] = 5; (0.6,0.8] = 4; (0.4,0.6] = 3; (0.2,0.4] = 2; (0,0.2] = 1	3.123	1.192	3.947	1.113
	PF3:Gender—female = 0, male = 1	0.897	0.093	0.689	0.214
PE Household business characteristics	PE1:Income level—(0.8,1] = 5; (0.6,0.8] = 4; (0.4,0.6] = 3; (0.2,0.4] = 2; (0,0.2] = 1	3.034	0.513	2.032	1.454
	PE2:Forest land area—(0.8,1] = 5; (0.6,0.8] = 4; (0.4,0.6] = 3; (0.2,0.4] = 2; (0,0.2] = 1	3.654	0.852	2.660	1.288
	PE3:Labour force ratio—(0.8,1] = 5; (0.6,0.8] = 4; (0.4,0.6] = 3; (0.2,0.4] = 2; (0,0.2] = 1	2.765	0.772	3.468	1.044
PS Family social capital	PS1:Social network—whether any relative is a village head or above: yes = 1, no = 0	0.092	0.084	0.074	0.068

(continued)

Table 7.3 (continued)

Latent variables	Explanatory variables and quantification (yes = 1; no = 0 if no marks)	Pingwu		Ganluo	
		Mean	S.D	Mean	S.D
	PS2: Level of trust in government—fully trust = 5; partially = 4; neutral = 3; distrust = 2; complete distrust = 1	3.847	1.255	2.481	1.249
	PS3:Level of trust in villagers—full trust = 5; partial = 4; neutral = 3; distrust = 2; complete distrust = 1	1.788	1.167	3.941	0.817
BI Willingness to participate	BI1:Indirect willingness to participate—very willing = 5; willing = 4; not necessarily = 3; not willing = 2; very unwilling = 1	3.877	0.253	3.045	0.645
	BI2:Forest land input—very willing = 5; willing = 4; not necessarily = 3; not willing = 2; very unwilling = 1	3.872	0.195	2.840	0.571
	BI2:Forest land input—very willing = 5; willing = 4; not necessarily = 3; not willing = 2; very unwilling = 1	3.849	0.318	2.872	0.432

Note The segment nodes corresponding to age, income level, forest land area, and labour force ratio are the results of normalising the Initial value to the values of [0–1]

7.1.9 Analysis of Statistical Results

The structural model analysis shows that all five latent variables, including farmers' individual characteristics, household business characteristics, and household social capital, directly affect or indirectly influence the degree of farmers' participation through the willingness to participate, which is consistent with the findings of Zhu et al. (2013), Huang and Nie (2013), Ming et al. (2015). Community characteristics and project characteristics have the largest effect on the degree of participation with influence coefficients of 1.154 and 1.031, respectively. Hence, whether farmers can or are willing to participate in the project is not only an autonomous economic choice, but also a choice influenced by the local government and the project owner and one which is more influenced by the latter. The coefficient of *the willingness to participate* on *the degree of participation* is relatively small (0.357), and the path significance is not high ($p = 0.137 > 0.05$), further indicating the significance of the study on the degree of participation.

However, the effect of the willingness to participate on the degree of participation in different projects is significant, mainly because farmers' independent choices in

Table 7.4 Test results of each variable and index

Latent variables	Secondary indicators	SFL	Factor N	KMO	Bartlett
SF Project characteristics	SF1:Labour resource	0.607	1	0.572	242.483***
	SF2:Forest land ownership	0.741			
	SF3:Organisational mode	0.719			
SE Community characteristics	SE1:Livestock dependence	0.693	1	0.567	49.307***
	SE2:Government support	0.743			
	SE3: Degree of village community harmony	0.772			
PF Farmer's characteristics	PF1: Education	0.768	1	0.591	113.489***
	PF2: Age of head of household	0.559			
	PF3: Gender	0.534			
PE Household business characteristics	PE1:Income level	0.572	1	0.491	63.480***
	PE2: Forest land area	0.603			
	PE3:Labour force ratio	0.579			
PS Family social capital	PS1: Social network	0.733	1	0.542	57.361***
	PS2: Level of trust in government	0.699			
	PS3: Level of trust among villagers	0.807			
BI Willingness to participate	BI1: Indirect willingness to participate	0.839	1	0.726	749.050***
	BI2: Forest land input	0.908			
	BI3: Labour force input	0.898			
B Degree of participation	B11: Survey participation	0.684	1	0.572	285.764***
	B12: Opinions on the planning design	0.684			
	B21: Project openness	0.717			
	B22: Participation in training	0.721			
	B23: Obtaining shareholding land benefits	0.754			
	B24: Obtaining incomes	0.677			
	B25: Complying with the contract	0.703			
	B31: Participation in monitoring	0.725			
	B32: Avoiding conflicts	0.693			

Note ***Indicates significance at 1% level

Table 7.5 SEM analysis results

Paths	M1 general model	M2 Ganluo	M3 Pingwu	Paths	M1 general model	M2 Ganluo	M3 Pingwu
B <—SF	0.855**	0.915**	0.890**	SE1 <—SE	0.892	0.903	0.317
B <—SE	0.783*	0.866*	0.701**	SE2 <—SE	0.884	0.780	0.988
B <—PF	0.318*	0.357**	0.291*	SE3 <—SE	0.687	0.538	0.718
B <—PE	0.297*	0.333**	0.274**	PF1 <—PF	0.875	0.801	0.863
B <—PS	0.757**	0.743***	0.777**	PF2 <—PF	0.886	0.839	0.894
B <—BI	0.357	0.277**	0.197**	PF3 <—PF	0.671	0.699	0.605
BI <—SF	0.494*	0.451**	0.502*	PE1 <—PE	0.915	0.844	0.972
BI <—SE	0.538*	0.687**	0.494*	PE2 <—PE	0.848	0.819	0.875
BI <—PF	0.535**	0.570**	0.505**	PE3 <—PE	0.895	0.921	0.814
BI <—PE	0.511**	0.483**	0.512**	PS1 <—PS	0.933	0.852	0.960
BI <—PS	0.507**	0.537*	0.495*	PS2 <—PS	0.947	0.863	0.991
SF1 <—SF	0.719	0.972	0.668	PS3 <—PS	0.959	0.936	0.978
SF2 <—SF	0.734	0.683	0.899	BI1 <—BI	0.894	0.891	0.910
SF3 <—SF	0.607	0.577	0.615	BI2 <—BI	0.816	0.855	0.738
				BI3 <—BI	0.870	0.894	0.841

Note *Indicates significant at 10% level, ** indicates significant at 5% level, ***indicates significant at 1% level

the two projects are different. The Ganluo FCS Project mainly relies on farmers' barren hills and wastelands or rotational resting land, and farmers have the right to operate forest land, so they can choose their own forest land input. In contrast, Pingwu FCS project is dominated by collective forests, and farmers do not have an autonomous choice in forest land input, but do in regard to labour input. Thus, the effect of the willingness to participate on the participation behaviour is not the same in different projects, resulting in an overall insignificant effect of the willingness to participate on the participation behaviour, but a significant effect in some projects individually.

In terms of project characteristics, the source of labour, forest land tenure, and contracting mode of FCS projects directly determine the possible extent of farmers' participation. In order to ensure that the afforestation in the Ganluo FCS Project meets the CDM standard, professional afforestation companies are utilised for afforestation, which limits farmers' labour input. Thus, the labour participation rate in the Ganluo FCS Project (31.03%) is lower than that in the Pingwu FCS project (75.42%), and the impact of labour source on participation is higher (0.972). The carbon sequestration forests in Pingwu FCS project are mainly collectively-owned forest land, which makes it less feasible for farmers with management rights only to participate in the project through forest land shareholding and transfer. Thus, the method of forest land input limits farmers' participation and has a higher marginal impact on the degree of participation (0.899).

In addition, from the perspective of contracting mode, farmers often form informal organisations based on kinship and neighbourhood relations that jointly own forest land or have high decision-making power. The leaders of these organisations usually have a certain influence in the village group, so they have more say in the selection of tree species, forest land management, distribution of carbon sequestration earnings, etc., and a stronger sense of ownership of the FCS project. Thus, they tend to have a higher degree of cooperation in the projects. On the whole, labour and forest land inputs are the most important ways for farmers to participate in FCS projects, and restrictions on either type of participation will greatly weaken farmers' participation. Meanwhile, with the formation of the agricultural land ownership system after 2015 and the increase in labour transfer, the fragmentation of farmers and forest land is more and more obvious. Hence, the feasibility of individual farmers' participation is reduced.

In terms of community characteristics, Ganluo is a county with a large number of Yi ethnic residents, who rely more on livestock farming and livestock products, such as cattle and sheep. These animals are not only a source of income for Yi people, but have multiple other functions, such as status symbols and wedding gifts. Thus, the forest–pastoral conflict is significant in the Ganluo FCS Project area and has a higher impact on the degree of participation than in the Pingwu FCS Project area. Thus, the marginal contribution of dependence on livestock farming to the degree of participation is higher (0.903). However, in terms of the degree of village and community harmony, since the Pingwu FCS Project has a high proportion of migrant workers and is a mixed multi-ethnic area, the daily communication between villagers is weak. In contrast, the Ganluo FCS Project area is dominated by the Yi ethnic group, with the existence of informal organisations such as Bimo[2] and clans. Thus, the degree of harmony between villagers is higher, and the homogeneity in project participation decisions is greater, which is more favourable to farmers' participation. Consequently, the positive contribution of villager harmony to the degree of participation is higher than in the Pingwu FCS Project area ($0.988 > 0.780$).

In both the FCS project areas, the degree of government support is the main factor influencing the degree of participation, which is related to the current grassroots governance structure in China. Meanwhile, the village committee has lived with the villagers for a long time and established a certain trust relationship, and the government is more credible than the enterprises. Thus, the higher the government support, the higher the degree of farmers' participation. For the present and future development of FCS projects, grassroots organisations always assume an important promotion role, especially in the current situation where farmers are increasingly dispersed into different occupations and regions around the country. It is important to rely on grassroots organisations to optimise the environment for FCS development and reduce the negative impacts, so as to increase farmers' support for FCS projects. A comparison of the two project areas indicates that informal organisations have a

[2] Bimo refers to the priests in the traditional religion of the Yi in China.

significant role in mitigating the community environment on farmers' participation. However, it is worth noting that informal organisations have a significant impact on enhancing farmers' right to voice their opinions and alleviating the constraints of the community environment on farmers' participation. It is undeniable that the function of informal organisations is limited, especially in recent years. The mass exodus of rural labour has led to the increasingly obvious division of farmers, and the improvement of the grassroots governance system based on two village committees has also, to some extent, weakened the function of informal organisations. Thus, it is not enough to rely on informal organisations to enhance farmers' participation.

In addition, autonomous factors such as individual farmers' characteristics, family business characteristics, and family social characteristics are important aspects that affect the degree of farmers' participation, especially family social characteristics. The population distribution in rural China is still tied by relationships such as the same surname and kinship, and villagers mostly have direct or collateral kinship relationships with each other. Thus, the behaviour is more homogeneous. The wider the relationship network and the closer the relationship with the government, the greater the participation of farm households. Meanwhile, gender, literacy, forest land size, and income level also have significant effects on participation, indicating that women, people with less land, poverty-stricken and indigenous people, etc., tend to be marginalised and excluded from project participation. This is not only because of the existence of unequal political games, but also because women, landless people, and poverty-stricken people often do not have afforestation and forestry skills, and their forest land is scattered and small. Hence, they have few ways to participate in FCS projects.

7.2 Summary

This chapter quantified farmers' FCS participation in three stages: project planning and design, execution and operation, and monitoring and evaluation. It found that under the current government-driven FCS projects, farmers' participation is weak overall, and three autonomous factors, namely farmers' individual characteristics, farmers' household characteristics, and household management characteristics, and two embedded factors, namely project characteristics and community characteristics, have a significant influence on such participation.

Project characteristics and community characteristics are the main aspects that restrict farmers' participation. Project characteristics mainly restrict the possibility of farmers' labour and land inputs through mandatory project contracts, and community characteristics are manifested in two aspects: the influence of group decision-making on farmers' choices and the constraint of regional development on the project itself. Household social and material capital and individual household characteristics are considered a reflection of farmers' household resource endowment, and they directly determine farmers' ability to participate. From the factor perspective, farmers' participation is mainly the input of three types of factors: labour, forest land, and intangible factors such as suggestions, decisions, and choices. It is difficult for farmers to meet the requirements of projects with their backward forestry and silvicultural technology. Moreover, strict standards for forest land restrict the input of large amounts of forest land resources, and informal organisations have weaker functions than in the past, while formal organisations are still underdeveloped. All the above are the main reasons for the low level of farmers' participation.

The key to improving such participation is therefore to enhance the input capacity of farmers in these three factors. Furthermore, farmers' modern forestry management techniques must be improved, and farmers must be helped to improve their forest land standards to input more forest land resources. Lastly, it is important to promote the development of formal organisations and give rise to their role in helping farmers to voice their opinions.

Therefore, attention should be paid to the following aspects to clear the barriers to participation. First, farmers' capacity to provide more labour inputs must be improved. With the government taking a leading role and support from the market, a grassroots education system should be established to provide training on practical techniques such as afforestation, forestry, and carbon sequestration monitoring and accounting, so as to enhance the feasibility of farmers who are willing to participate, especially women, people with less land, and poverty-stricken households.

Second, farmers should be helped to provide more forest lands. The execution and monitoring standards of FCS projects should be optimised and the restrictions on land eligibility and tree species relaxed as much as possible under the premise of meeting the carbon sequestration measurement and monitoring standards. Farmers should be given more independence in choosing forest land investments. The roles of government and farmers in the execution of FCS projects should be changed towards a model of 'government-led + market main + voluntary farmers'.

Third, the government must focus on farmers' capacity in intangible factor input, accelerate the development of organisations such as forestry share cooperatives, forestry professional cooperatives, forestry technical associations, and other farmers' organisations, and play an important role in promoting farmers' participation in FCS execution, management, and supervision. In some areas, forestry cooperatives could be established whose main business focus is on FCS.

References

Antle, J. M., & Stoorvogel, J. J. (2009). *Payments for ecosystem services, poverty and sustainability: The case of agricultural soil carbon sequestration* (pp. 133–161). Springer.

Benitez, P., McCallum, I., Obersteiner, M., et al. (2004). *Global supply for carbon sequestration: Identifying least-cost afforestation site sunder country risk considerations*. Laxenbur: II A-SA, IR-04-022.

Benessaiah, K. (2012). Carbon and livelihoods in Post-Kyoto: Assessing voluntary carbon markets. *Ecological Economics* (77), 1–6.

Chen, C. Y. (2010). Forestry carbon sequestration and farmers' livelihoods: The example of the first global forestry carbon sequestration project. *World Forestry Research, 23*(5), 15–19.

Ding, Y., & Ma, P. P. (2013). Research on forest carbon sequestration and economic development in ethnic minority areas in West Sichuan Province: A case study of Yuexi County, Liangshan Yi autonomous prefecture, Sichuan Province. *Rural Economy, 5*, 38–41.

Gong, R. F., & Zeng, W. Z. (2018). A study on the constraints of farmers' participation in FCS projects in the context of government promotion. *Resource Science, 40*(5), 201–211.

Granovetter, M. (2018). Economic action and social structure: The problem of embeddedness. *The sociology of economic life* (pp. 22–45). Routledge.

Huang, Y. L., & Nie, J. (2013). A theoretical analysis framework of forest farmers' intention to participate in forest carbon sequestration behavior—A perspective based on TPB model. *The Economist* (11), 24–25+34.

Ming, H., Qi, Y. B., Li, Y. M., et al. (2015). Do foresters have the willingness to participate in forestry carbon sequestration projects?–Taking the CDM forestry carbon sink pilot project as an example. *Agricultural Technology Economics, 7*, 102–113.

Nhung, N. T. H. (2009). Optimal forest management for carbon sequestration: A case study of Eucalyptus Urophylla and Acacia Mangium in Yen Bai Province, Vietnam. EEPSEA Final Report.

Yang, F., Zeng, W. Z., & Zhang, W. K., et al. (2016). Forest farmers' willingness to sustained participation in forest carbon sequestration projects and its influencing factors. *Forestry Science, 52*(7), 138–147.

Zhu, Z., Shen, Y. Q., Wu, W. Q., et al. (2013). Optimal decision-making of farmers' forest management and carbon sequestration supply capacity under carbon sequestration objectives - a survey based on Zhejiang and Jiangxi provinces. *Journal of Ecology, 8*, 2577–2585.

Chapter 8
Analysis of Farmers' Willingness to Accept Ecological Compensation for Participating in FCS Projects

To save land rent and reduce labour cost, FCS projects are generally implemented in remote and poverty-stricken rural areas, and farmers are the most important participating party. Due to the imperfect market mechanism of FCS, the opportunity cost of farmers' participation in a project exceeds their participation benefit, and farmers mostly participate passively and are led by enterprises and the government. The execution period of FCS projects is long, generally between 20 and 40 years, and the sustained participation of farmers is a guarantee of the sustainable and healthy development of projects (Yang et al., 2016; Zhu et al., 2013). Therefore, in addition to the market mechanism, it seems necessary to establish another set of compensation mechanisms, led by the government, to stimulate the active participation of farmers. In this chapter, the researchers investigate the willingness of farmers to accept ecological compensation for transforming their land from grazing land to forested land through a field survey of farmers participating in the Novartis Sichuan FCS Project. The ecological compensation amount is also measured, and factors affecting the compensation amount received by farmers are analysed through a Tobit regression model. The chapter aims to provide a reference for the formulation of an ecological compensation policy for FCS and the integration of FCS resources for poverty alleviation.

8.1 Theory, Methodology, and Data

The researchers use a Hicks analysis to analyse the balance between the government's willingness to compensate and farmers' willingness to accept the compensation in FCS projects. Moreover, they determine the measurement of this compensation by comparing different methods of calculating ecological compensation standards. Lastly, they explain the characteristics of surveyed samples in the study.

W. Zeng and F. Yang, *Poverty Alleviation Via Forest Carbon Sequestration*, International Research on Poverty Reduction, https://doi.org/10.1007/978-981-99-8191-5_8

8.1.1 Theory

In this chapter, the researchers analyses the willingness of the government to compensate, and of farmers to accept ecological compensation, with the help of a Hicks analysis (Hicks, 1942). In the specific application of the method, the researchers refer to its introduction by Zhang (2006) and Lai (2007) and its application by Cao and Guo (2002) and Yu and Cai (2015). Hicks divides consumer surplus (CS), which measures welfare changes, into compensating variation (CV) and equivalent variation (EV). As shown in Fig. 8.1, it is assumed that the production possibility curves of farmers are l_1 and l_2, the utility function curves are u_1 and u_2, the Marshall demand curve is j, the Hicks demand curves are j_1 and j_2, the horizontal axis Y_1 denotes the area of grazing land, and the vertical axis Y_2 denotes other production factor inputs in grazing livestock. Assuming that the initial utility level of the farmer is point E_1 on u_1, the change of land use from grazing to afforestation now leads to the reduction of grazing land area, which, in turn, leads to the reduction of grazing volume and grazing livestock quantity or quality. Then, the utility level of the farmer decreases to point E_3 on u_2, and the reduced grazing land area A_1A_2 is transformed into forested land. Hence, by creating a carbon sink forest, the amount of vegetation on the land has an increased capacity to absorb carbon dioxide, the ecological function of the land is enhanced, the ecological environment of the region is improved, and the utility level of other members of society is increased.

CV represents the subsidy that must be given to farmers to restore their utility level to u_1 (point E_2). *CV* is thus a measure of the subsidy required to avoid a decrease in the utility level of the farmers when they convert their grazing land to forested land, that is, the willingness to accept (*WTA*). From the Hicks demand curve j_1, $S_{P1E1'E2'P2}$ is the size of WTA and also the size of *CV*. From the Marshall demand curve j, $S_{P1E1'E3'P2}$ is the magnitude of *CS*, that is, the increased cost to the farmer to restore the level of utility to that before the change of land use.

Similarly, the amount that other social members are willing to pay for the ecological improvement due to afforestation can be derived. Assuming that the initial utility level of other members of society is E_3 on u_2, a farmer's grazing land changes to forested land, and the ecological environment is improved through afforestation, the utility level rises to the level of u_1 (point E_1); thus *EV* can measure the highest price that other members of society are willing to pay for the improved ecological environment caused by forest carbon afforestation, that is, willingness to pay (*WTP*), whose magnitude can be expressed as $S_{P1E4'E3'P2}$.

From Fig. 8.1, it can be seen that $S_{P1E1'E2'P2} > S_{P1E1'E3'P2} > S_{P1E4'E3'P2}$, that is, $WTA = CV > CS > EV = WTP$, or the amount of compensation for farmers > opportunity cost of farmers' participation in FCS projects > the amount of compensation that other social members are willing to pay. The compensation that social members are willing to pay is not enough to cover the amount of ecological compensation for farmers, which thus needs to be compensated by government subsidies.

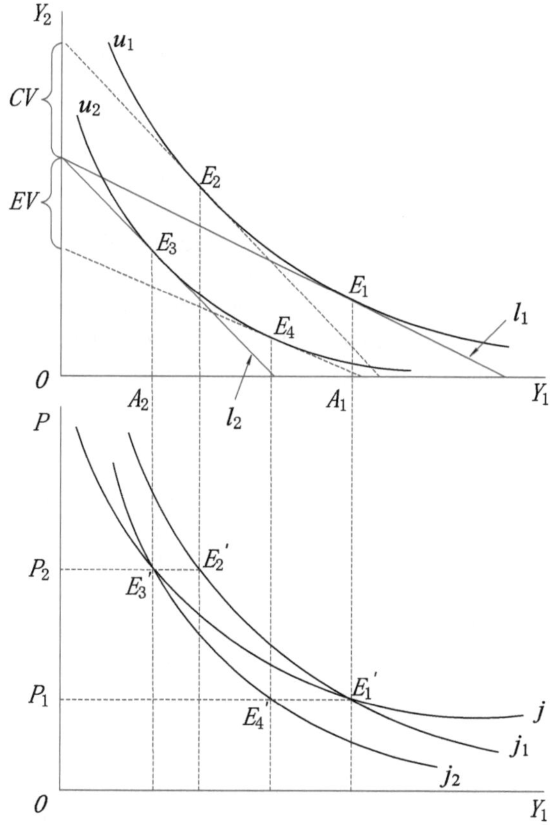

Fig. 8.1 Hicks analysis of ecological compensation for FCS afforestation

8.1.2 Methods

The main existing methods for determining ecological compensation standards are the opportunity cost method, willingness survey method, market value method, alternative market method, shadow price method, non-equilibrium state economic thermodynamic method, and resource physics method (Su 2007; Yu & Cai, 2015). Of these, the latter three are rather recent, and their theoretical basis and application research lack maturity. As the market or alternative market for ecological values other than carbon sequestrations has not yet been formed, the market value method and alternative market method are not applicable either. The opportunity cost method and willingness survey method are the mainstream methods for determining ecological compensation standards, and most existing studies adopt one of these (Li et al., 2013, 2009). The willingness survey method is based on the subjective will of the respondents (compensators or recipients), while the opportunity cost method may lack the subjectivity of the compensators or recipients in its purely objective accounting. Therefore, this chapter combines the willingness survey method and the opportunity cost method to determine the range of ecological compensation for FCS, which can effectively compensate for the limitations of these methods and validate the estimation results.

8.1.2.1 Opportunity Cost Method

Opportunity cost, also known as the cost of a choice, refers to the maximum income of the various other uses foregone when a certain resource is used for a certain purpose. Applied to the ecological compensation of FCS, it is the amount of reduction in livestock income caused by the change of the original land use of grazing in order to participate in afforestation. Its calculation formula is as follows:

$$V_i = M_i - C_i \tag{8.1}$$

where i denotes the i-th household participating in afforestation, $i \in (1, n)$, V_i is the opportunity cost of the household participating in afforestation, M_i is the annual income from livestock grazing on the original grazing land, and C_i is the carbon sequestration-related income that the household has received from the enterprise, including labour costs (L_i) and enterprise grazing subsidies (K_i). In the current research, M_i, L_i, and K_i are the incurred values in 2013. As a result:

$$V_i = M_i - (L_i + K_i) \tag{8.2}$$

In formula (8.2), V_i is the opportunity cost of the i-th farmer's participation in afforestation in 2013, and the left and right sides of the equation are divided by the participation area of the household's forested land (R_i), which is the average opportunity cost per mu of the i-th farmer's participation in afforestation in 2013.

$$\frac{V_i}{R_i} = \frac{M_i - (L_i + K_i)}{R_i} \tag{8.3}$$

The average opportunity cost per mu of all farmers' participation in afforestation in 2013 is summed up and averaged to obtain the average opportunity cost per mu of all farmers' participation in afforestation in 2013:

$$\frac{\sum_{i=1}^{n} \frac{V_i}{R_i}}{n} = \frac{\sum_{i=1}^{n} \frac{M_i - (L_i + K_i)}{R_i}}{n} \tag{8.4}$$

8.1.2.2 Willingness Survey Method

The willingness survey method, also known as the conditional valuation method (CVM), uses the principle of utility maximisation to directly ask potential compensators about their willingness to pay for non-market goods or potential recipients about their willingness to accept compensation in a hypothetical market (Boxall et al., 1996; Cai & Zhang, 2011). Since the state does not provide financial compensation for the ecological benefits of afforestation, and the current market for FCS only includes carbon sequestration products without other ecological values, this study uses the willingness survey method to directly ask the interviewed farmers about the amount of compensation that they are willing to accept, the average of which (*P*, equivalent to the WTA in the previous section) is calculated by the following formula, where i is the *i-th* surveyed household, p_i is the amount of compensation, and n is the total number of surveyed households:

$$P = \frac{\sum_{i=1}^{n} p_i}{n} \tag{8.5}$$

8.1.3 Data Description

In this section, the researchers will describe the study area on which the survey is conducted, the questionnaire used in this study, and the sample profile, respectively.

8.1.3.1 Overview of the Research Region

The data in this chapter derive from the field research undertaken around the Novartis Sichuan FCS Project, where the surveyed farmers have transformed their contracted grazing land into forested land. This project was officially launched in 2010, creating 4,196.80 ha of multifunctional forests in five counties (Yuexi, Ganluo, Meigu, Zhaojue, and Leibo) and three nature reserves (Shenguozhuang, Maanshan, and Mamizhe) in Liangshan Yi Autonomous Prefecture, in Sichuan Province. It is expected to generate 1.206 million tonnes of long-term certified carbon dioxide equivalent emissions reductions over the 30-year project crediting period, with an average of 40,000 tonnes of carbon dioxide equivalent per year. The project is located in the basin of the Yangtze River and Dadu River, the upper reaches of the Yangtze River, and a secondary tributary of the Jinsha, respectively, where soil erosion is severe. The project area is also one of the 32 priority areas for biodiversity conservation in China, namely the southern section of the Hengduan Mountains, which, as an important habitat for rare and endangered species such as the giant panda, was selected in the 2010 China Biodiversity Conservation Strategy Action Plan.

This project can not only help to absorb carbon dioxide and slow climate change by creating forests, but also enhance biodiversity conservation and its adaptability to climate change by improving the connectivity of the forest ecosystem landscape around the protected area. At the same time, it can improve the ability of soil and water conservation in the upper reaches of the Yangtze River, realising the multiple benefits and values of the forest. However, participating farmers, especially those whose land has changed from grazing to carbon sequestration forested land, have experienced different degrees of reduction in livestock revenue due to the reduction in grazing land, resulting in different degrees of reduction in the quantity and quality of livestock. According to internal information from the project, a total of 1716.70 ha changed from grazing land to forested land, accounting for 40.9% of the total project land.

8.1.3.2 Questionnaire Design, Amendment, and Survey

The questionnaire is divided into three parts. The first part mainly investigates the basic characteristics of farmers, including individual characteristics, family characteristics, project characteristics, and community characteristics. The second part focuses on farmers' willingness to receive ecological compensation and the amount of compensation required for the use of grazing land. The researchers use open-ended questions to ask farmers directly about the amount they want to be paid, which can effectively compensate for the psychological anchoring effect that payment cards or dichotomous questionnaires may induce in farmers, thus increasing the credibility of the results (Tang et al., 2014). The third part involves the opportunity cost of changing the use of grazing land, which mainly concerns the reduction in the number of grazing animals and in grazing income.

The survey was conducted by a combination of a questionnaire and in-depth case interviews, and the questionnaire was administered on a randomly sampled basis. To ensure the validity of the questionnaire, several pilot surveys were conducted before the formal survey was carried out, and the questionnaire was amended according to the result of the pilot survey. Moreover, training for the surveyors was carried out in August 2015 before the formal survey was conducted. In order to overcome the difficulty that some Yi farmers do not speak Chinese, local youths and students who had studied Chinese were invited to act as interpreters. A total of 200 questionnaires were distributed, and 183 valid questionnaires were collected, yielding a sampling efficiency of 91.50%.

8.1.3.3 Basic Characteristics of the Sample

The basic characteristics of the surveyed samples are shown in Table 8.1. In regard to gender, most of the respondents were male. Their ages ranged from 30 to 50; their education levels were generally low; their household labour force was sufficient, with three to four members; their per capita net household income in 2013 was low, with

Table 8.1 Basic characteristics of households

Variables		Frequency	Relative frequency (%)	Accumulated frequency (%)
Gender	Male	135	73.77	73.77
	Female	48	26.23	100.00
Age	≦30	33	18.03	18.03
	30–40	76	41.53	59.56
	40–50	45	24.59	84.15
	50–60	19	10.38	94.54
	>60	10	5.46	100.00
Years of education (year)	≦6	146	79.78	79.78
	7–9	32	17.49	97.27
	≧10	5	2.73	100.00
Household labour force (persons)	1–2	26	14.21	14.21
	3–4	148	80.87	95.08
	5–6	9	4.92	100.00
Average household income ($\times 10^3$ RMB)	≦2	54	29.51	29.51
	2–4	74	40.44	69.95
	4–6	33	18.03	87.98
	>6	22	12.02	100.00
Source of income	Agriculture	130	71.04	71.04
	Others	53	28.96	100.00
Forested land ($\times 10$ mu)	≦1	123	67.21	67.21
	1–2	21	11.48	78.69
	2–3	39	21.31	100.00
Reduction of livestock	≦5	11	6.01	6.01
	5–10	160	87.43	93.44
	>10	12	6.56	100.00
Distance from home to market ($\times 10$ km)	≦2	4	2.19	2.19
	2–5	152	83.06	85.25
	>5	27	14.75	100.00
Village collective economy	Very weak	37	20.22	20.22
	Week	122	66.67	86.89
	Average	24	13.11	100.00

most households earning 2,000–4,000 RMB (about 290–580 USD), which was much lower than the average level of Sichuan Province (7,895 RMB or about 1,144 USD) in the same period; and their source of household income was mainly agriculture. The forested land is generally less than 10 mu; the number of livestock reduced due to participation in the project is generally between 5 and 10 (only); the distance of households from the nearest market is long, generally between 20 and 50 km; and the village collective economy is generally weak.

8.2 Empirical Analysis

Using survey data from the farms around the Novartis Sichuan FCS Project, this study measures the compensation amount that farmers are willing to accept and uses Tobit regression analysis to analyse the influence of their individual characteristics, household characteristics, project characteristics, and environmental characteristics on the compensation amount. Then, the researchers propose corresponding countermeasures for the government to promote FCS projects.

8.2.1 *Calculation of the Ecological Compensation that Farmers are Willing to Accept*

Formulae (8.4) and (8.5) are used to measure the opportunity cost and the amount of compensation that the farmers are willing to accept. The average opportunity cost of farmers' participation in an FCS project is 1,758 RMB (about 255 USD)/(mu-years), and the average compensation amount that they are willing to accept is 2,025 RMB (about 293 USD)/(mu-years), which is consistent with the theoretical expectation of $CV > CS$ before. To a certain extent, these findings confirm the credibility of the research data. It is important to note that the average compensation amount is high; for example, it is much higher than the national standard of public forest compensation (15 RMB or about 2.2 USD)/(mu-years) determined in 2013. The reasons for this are, first, that the land was mainly used for grazing, resulting in the change having a high opportunity cost; second, the surveyed area is a poverty-stricken and remote mountainous area with a high dependence on agriculture and animal husbandry, and farmers have few opportunities to obtain other benefits; and third, this study does not exclude that, in the context of the national policy of vigorously implementing subsidies for agriculture, farmers are dependent on national agricultural subsidies.

8.2.2 *The Factors Influencing the Compensation Amount that Farmers are Willing to Accept*

In this part, the researchers use four types of explanatory variables, namely individual characteristics, family characteristics, project characteristics, and environmental characteristics, and apply an econometric Tobit model to confirm their influence on the amount of farmers' expected compensation.

8.2.2.1 Empirical Model Selection

Econometric analysis of the factors affecting the expected compensation amount is one of the keys to verify the effectiveness of this compensation as well as the basis for setting the ecological compensation standard of afforestation. In the actual survey, all farmers are willing to be compensated and the amount of compensation is $\in (0, +\infty)$. If OLS is used for estimation, the parameters will produce serious bias and inconsistency; therefore, the Tobit model is used to verify the factors affecting the amount of farmers'expected compensation (Zhou & Li, 2012). The Tobit model is a standard censored model which is applicable to regression analysis and in which the observations of the dependent variable are approximately continuously distributed in positive values but contain some missing observations due to some limitations. Its general form is as follows:

$$\begin{cases} Y^* = \beta^T X_i + \varepsilon_i \\ Y = \max(0, Y^*) \end{cases} \quad \varepsilon_i \sim N(0, \sigma),\ i = 1, 2, \ldots, n \tag{8.6}$$

where Y^* is the latent variable, Y is the dependent variable, X_i is the explanatory variable, β is the coefficient to be estimated, and ε_i is the random error.

Based on the literature and the purpose of the research, this book has identified four types of explanatory variables: individual characteristics, family characteristics, project characteristics, and environmental characteristics, which are defined in Table 8.2. The Tobit model is designed in the following form:

$$y_i^* = \beta_0 + \beta_i x_i + u$$

$$y_i = \max(0, y_i^*) \tag{8.7}$$

where y_i^* is the latent variable, y_i is the dependent variable (i.e., *WTA*) answered by the surveyed farmers), x_i is the explanatory variable, β_0 is the intercept term, β_i is the parameter to be estimated, u is the residual, and i is the number of explanatory variables, of which 10 are selected in this chapter.

8.2.2.2 Model Multicollinearity Test

In order to ensure the accuracy and stability of the model, it is necessary to test the multicollinearity of the respective variables. As has been done in Chap. 6, the method is to use one of the independent variables as the dependent variable and the remaining variables as the independent variables for regression analysis (Ma, 2008). Table 8.3 shows the results of the multicollinearity test with gender as the dependent variable and other variables as independent variables. Of these, the tolerance value is $0.780 > 0.1$ and the maximum VIF value is $1.283 < 10$, which shows that there is no serious multicollinearity.

Table 8.2 Variable definitions and values

Variable			Variable definition	Min	Max	Avg.	S.D.	Expected correlation
Dependent	Compensation (WTA value, y)	Actual observed value ($1000/ mu-yr)	0	1	0.74	0.44		
Explanatory	Individual characteristics	Gender (x_1)	Male = 1; Female = 0	17	70	39.42	10.52	Negative
		Age (x_2)	Actual observations (years)	0	12	4.15	3.18	?
		Education (x_3)	Number of years of schooling (years)	1	6	3.99	0.42	Negative
	Family characteristics	Num. of Labourers (x_4)	Actual observations (persons)	1.5	8	3.83	2.11	Negative
		Income (x_5)	Per capita net family income in 2013 ($\times 10^3$ RMB)	0	1	0.71	0.45	Negative
		Source of income (x_6)	Agriculture = 1; Other = 0	0.2	3	1.49	0.88	Positive
	Project characteristics	Afforestation area (x_7)	Actual observation value ($\times$ 10 mu)	4	50	10.39	6.36	Positive
		Livestock reduction (x_8)	Number of livestock reduced due to participation in FCS, actual observed value (numbers)	2	12	5.44	1.70	Positive

(continued)

Table 8.2 (continued)

Variable			Variable definition	Min	Max	Avg.	S.D.	Expected correlation
	Environmental characteristics	Distance to market (x_9)	Actual observation value (× 10 km)	1	3	1.93	0.57	Positive
		Village collective economy status (x_{10})	Very poor = 1; relatively poor = 2; average = 3 relatively affluent = 4; very affluent = 5	0.3	5	2.02	1.09	Negative

Note +, −, and ? denote the direction of influence of the explanatory variables on the explained variables as positive, negative, and uncertain, respectively

Table 8.3 Multicollinearity test between variables

Model		Covariance	
		Tolerance	VIF
x_1	x_2	0.932	1.073
	x_3	0.924	1.083
	x_4	0.956	1.046
	x_5	0.780	1.283
	x_6	0.895	1.117
	x_7	0.851	1.175
	x_8	0.922	1.084
	x_9	0.957	1.045
	x_{10}	0.889	1.125

Similarly, the other variables can be tested for multicollinearity. Due to limitations on space, the test process is not presented in its entirety in this chapter. The results of all the runs show that the tolerance of the multicollinearity test and the VIF are within a reasonable range. Therefore, there is no serious problem of multicollinearity among the respective variables in the regression equation.

8.2.2.3 Analysis of Empirical Results

Tobit regression model analysis was conducted using Eviews 7.2 to analyse the factors influencing the expected compensation amount (see Table 8.4). From the regression results, factors such as years of education, main source of household income, area of forestation land, livestock reduction, distance of household from the nearest market, and village collective economic status significantly affected the expected compensation amount. Specific analyses are as follows.

From the regression results for individual characteristics of the surveyed farmers, the coefficient of *education* is negative and significant at the 5% level. This means that, all other conditions being equal, the higher the education level of the farmers, the lower the amount of ecological compensation that farmers are willing to accept, for two principal reasons. First, the higher the education level, the richer the human capital of farmers and the stronger their ability to look for jobs and generate income; thus their dependence on ecological compensation decreases. Second, the higher the education level, the stronger the cognitive ability of individuals and the stronger their awareness of the ecological benefits of FCS. In terms of the socialisation function of education, in general, the more educated an individual is, the higher the cognition of their social role and social responsibility will be. The citizen's sense of social responsibility may be enhanced as a result, and their behaviour may shift from self-interest to altruism, that is, benefiting the public, thus reducing the amount of compensation for ecological FCS.

From the regression results for the household characteristics, the coefficient of the *main source of income* is positive and significant at the 1% level, implying that the more households are dependent on farming as the main source of income, the higher the level of willingness to accept compensation. The fact that agriculture is the main source of income indicates that farming households have few ways to generate non-farm income, are strongly dependent on agroforestry income, and therefore, have a strong willingness to obtain income from agroforestry.

From the regression results for project characteristics, the coefficient of *afforested land area* and *livestock reduction* are both positive and significant at the 1% level.

Table 8.4 Tobit model Regression Results ($n = 183$)

Independent Variables	Coefficient	Std. Error	z-Statistic	Prob.
x_1	0.063	0.141	0.447	0.655
x_2	0.003	0.006	0.459	0.646
x_3	–0.041**	0.020	–2.050	0.040
x_4	0.107	0.147	0.731	0.465
x_5	–0.033	0.032	–1.017	0.309
x_6	0.420***	0.140	2.999	0.003
x_7	0.248***	0.074	3.351	0.001
x_8	0.044***	0.010	4.364	0.000
x_9	0.177***	0.037	4.825	0.000
x_{10}	–0.597***	0.111	–5.370	0.000
C	0.805	0.765	1.052	0.293
R-squared	0.801			
Significance	0.0000			

Note *** denotes significance test value $p < 0.01$, ** denotes $p < 0.05$

The former implies that the larger the afforested land area, the higher the amount of expected compensation for farmers, as farmers hope that the incremental output of FCS will exceed the incremental input, which is consistent with the hypothesis of rational homo economics and the rational small farmer theory proposed by Schultz (1966) in *Transforming Traditional Agriculture.* The latter implies that the more a farmer reduces the number of livestock, the more compensation they expect to receive. The greater the reduction in livestock numbers, the greater the reduction in income from livestock farming, the greater the opportunity cost of participating in the FCS project, and therefore, the greater the compensation desired by the farmer.

The regression analysis of environmental characteristics reveals that both the coefficient for the distance from the nearest market and the one for the village collective economic status were statistically significant at the 1% level. The coefficient for market distance was positive, indicating that households situated further away from markets anticipate higher compensation. This suggests that the proximity to marketplaces, to some extent, influences the market participation opportunities for farm households. As the likelihood of these households to benefit from their own economic transactions diminishes with distance, they require greater financial assistance to compensate for this disadvantage.Conversely, the negative coefficient for the village collective economic status implies that in villages with stronger collective economies, the surrounding economic environment is more conducive to the well-being of farmers. In such contexts, farmers are more inclined to accept lower levels of compensation because their economic prospects are bolstered by the benefits they derive from the local economic environment. As their ability to generate income from the community's economic activities increases, their reliance on financial subsidies correspondingly decreases.

8.3 Summary

Based on a theoretical Hicks analysis of ecological compensation for FCS, this chapter surveyed the farmers participating in the Novartis Sichuan FCS Project, calculated the opportunity cost caused by the change of land use from grazing to afforestation, and measured the compensation amount that farmers are willingness to accept. The opportunity cost method and willingness survey method were used to calculate the range of ecological compensation for FCS, and the Tobit model was used to analyse the factors influencing the amount of expected ecological compensation. The study showed that the surveyed farmers were willing to accept ecological compensation ranging from 1,758 RMB (about 255 USD)/(mu-years) to 2,025 RMB (about 293 USD)/(mu-years). Factors such as years of education, main source of household income, area of afforested land, livestock reduction, distance from the nearest market, and village collective economic status had significant effects on the expected compensation amount.

This chapter argues both theoretically and empirically that it is necessary to give ecological compensation for afforestation at the present stage. This need is related to both the current imperfection of the carbon sequestration market and the task of poverty alleviation presently. As President Xi Jinping pointed out in his speech at the Central Poverty Alleviation and Development Work Conference in November 2015, it is necessary to increase ecological protection and restoration in poverty-stricken areas, increase transfer payments to key ecological areas, and expand the scope of policy impact. The financial subsidies for FCS projects constitute a specific practice for targeted poverty alleviation and poverty eradication. Although FCS projects must focus on poverty, given that the government's financial resources are limited, the following policy recommendations are proposed to reduce the dependence of farmers on financial subsidies.

First, since education significantly affects the amount of compensation received by farmers, it is necessary to strengthen investment in education for farmers, especially those in poverty-stricken areas and the next generation, and enhance their awareness of eco-environmental protection through school education, propaganda, and lectures and training. The latter means of education have positive implications for promoting the local ecological environment and reducing farmers' over-reliance on financial subsidies. Second, opening up diverse channels for farmers to increase their income and helping them to move beyond their dependence on agriculture, forestry, farming, and animal husbandry is conducive to reducing the pressure to pay for ecological subsidies from the public purse. Third, in addition to economic compensation, it is necessary to increase technical compensation to farmers, for example, by providing them with afforestation and management techniques and intensive breeding techniques, improving the output efficiency of FCS projects and animal husbandry, and increasing their income. Fourth, participation opportunities must be created in poverty-stricken areas. Fifth, the rules of the carbon sequestration trading market must be improved: curbing buyer pricing behaviour and helping farmers benefit directly from FCS market transactions are the fundamental ways to reduce farmers' dependence on financial subsidies.

References

Boxall, P. C., Adamowicz, W. L., Swait, J., Williams, M., & Louviere, J. (1996). A comparison of stated preference methods for environmental valuation. *Ecological Economics, 18*(3), 243–253.

Cai, Y. Y., & Zhang, A. L. (2011). The measurement of ecological compensation amount of farmland based on farmers' willingness to be compensated—a survey in Wuhan City as empirical evidence. *Journal of Natural Resources, 2*(2), 177–189.

Cao, J. H., & Guo, X. P. (2002). The application of willingness survey method in evaluating the environmental value of forest resources. *Journal of Jiangxi Agricultural University (Natural Science Edition), 5*, 645–648.

Hicks, J. R. (1942). Consumers' surplus and index-numbers. *The Review of Economic Studies, 9*(2), 126–137.

Li, G. P., Li, X., & Xiao, D. J. (2013). Exploring the theoretical criteria and measurement methods of ecological compensation. *The Economist, 2*, 42–49.

Li, X. G., Miao, H., & Zheng, H. (2009). The application of opportunity cost method in determining ecological compensation criteria—Taking the central mountainous region of Hainan as an example. *Journal of Ecology, 9*(1), 4875–4883.

Lai, Y. Y. (2007). *Microeconomics* (p. 155). Higher Education Press.

Ma, X. W. (2008) Diagnosis of multiple cointegration in linear regression equations and its empirical analysis. *Journal of Huazhong Agricultural University (Social Science Edition)* (2), 78–81+85.

Schultz, T. W. (1966). Transforming traditional agriculture: Reply. *Journal of Farm Economics, 48*(4), 1015–1018.

Su, G. S. (2007). Research on the value of natural resources and its assessment method. *Academic Forum, 4*, 77–80.

Tang, W. H., Xu, X. H., Wang, M., et al. (2014). The preconditions of the anchoring effect and its influencing mechanism. *Psychological Science,* (5), 1060–1063.

Yang F, Zeng W Z, Zhang W K, et al. (2016). Foresters' willingness to sustainedly participation in forest carbon sequestration projects and its influencing factors. *Forestry Science, 52*(7), 138–147.

Yu, L. L., & Cai, Y. Y. (2015). Ecological compensation of farmland based on farmers' willingness to be compensated–Jingshan County, Hubei Province as an example. *Journal of Applied Ecology, 26*(1), 215–223.

Zhang, S. M. (2006). *Intermediate microeconomics tutorial* (pp. 106–116). China Economic Press.

Zhou, H. L., & Li, X. S. (2012). Tobit model estimation methods and applications. *Economics Dynamics, 5*, 105–119.

Zhu Z, Shen Y Qn, Wu W G, et al. (2013). Optimal decision-making for household forest management and carbon sink supply capacity under carbon sink targets - based on surveys in Zhejiang and Jiangxi Provinces. *Acta Ecologica Sinica, 8*, 2577–2585.

Chapter 9
Analysis of Farmers' Cultural Adaptation in FCS Projects

FCS projects have strict regulations for maintaining carbon stock growth and avoiding carbon leakage, especially AR projects under the CDM mechanism. Such projects usually require farmers to comply with contractual agreements over a fairly long project period (generally 20–40 years), give up or partially give up their rights to use forest land and ownership of forest trees, and continuously carry out FCS cultivation and management. This has led to restrictions on traditional livelihood activities such as fuel wood harvesting and grazing (Chen, 2010; Jindal et al., 2008; Yang et al., 2018). There are potential contradictions between certain institutional arrangements and contractual terms originating from international FCS emission reduction standards and project trading rules, on the one hand, and traditional agricultural and pastoral cultural norms and ethnic cultural practices rooted in poverty-stricken areas of China, on the other hand. If not handled properly, such contradictions will not only impede the sustainable development of FCS projects, but even hinder the harmony of ethnic groups and endanger social stability. Therefore, the market mechanism-oriented FCS projects in remote poverty-stricken areas, especially in poverty-stricken ethnic areas, must deal with the question of ecological compensation from the perspective of economics, the question of farmers' incentives from the perspective of management, and the question of adaptation and integration between two different cultures: farmers' deep-rooted emotional identification with the business culture carried by FCS projects and their adaptation to it.

In this chapter, using the theoretical framework of Berry's (1980) two-dimensional cultural adaptation, the strategies the participating farmers adopt to adapt traditional and business cultures will be investigated empirically. To promote the cultural adaptation of the indigenous people in the project area, the influencing factors on farmers' strategies of cultural adaptation will be explored, in particular as regards established contracts. The purpose of this study is to provide suggestions to promote the implementation and sustainable development of FCS projects from the perspective of

W. Zeng and F. Yang, *Poverty Alleviation Via Forest Carbon Sequestration*, International Research on Poverty Reduction, https://doi.org/10.1007/978-981-99-8191-5_9

the cultural adaptation of the indigenous people and to refer to the development of internationally funded eco-industrial poverty alleviation and agriculture-related projects.

9.1 A Scale for Identifying Cultural Adaptation Strategies

Several studies have begun to discuss the association between FCS and culture. According to Guan et al. (2012), the construction of the FCS system has an important role in guaranteeing the sustainable development of Yunnan ethnic culture. Wen (2013) assessed the impact of Tibetan culture on economic development by establishing an indicator system and analytical model of such impact on agriculture, concluding that agriculture dominated by Tibetan culture can naturally integrate with the carbon sink economy. A survey by Ding and Ma (2013) of FCS projects in Yuexi County, Liangshan Prefecture, Sichuan Province showed that farmers in the project area have high beliefs in Yi Bimo culture, and the influence of Bimo culture is fully used to facilitate the propagation, implementation, and later management of AR FCS projects. A study by Walker et al. (2014) of indigenous people in the Amazon River Basin found that they have historically valued forests and are opposed to outsiders' control of local forestry resources. Existing research offers important values and insights for FCS cultural adaptation study, but studies have largely neglected the cultural adaptation of farmers in forest carbon trading, and empirical studies based on micro-survey data are lacking.

Since its introduction in 1980 (Berry, 1980), the framework of two-dimensional cultural adaptation theory has become the theoretical basis and an important paradigm in the field of cultural adaptation research. This analytical framework classifies adaptation styles into four categories based on the attitudes of individuals facing two basic questions in adaptation: whether they tend to maintain their original cultural traditions and identity or whether they tend to engage with and participate in the new culture. The four categories are integration, assimilation, separation, and marginalisation.

Integration refers to the individual's emphasis on maintaining the source culture as well as on the absorption and acceptance of the new culture (i.e., high source culture identity, high new culture identity), assimilation refers to the individual's unwillingness to maintain the source culture and willingness to identify with and pursue the new culture (i.e., low source culture identity, high new culture identity), separation refers to the individual's emphasis on their original culture and their desire to avoid contact with other cultures (i.e., high source culture identity, low new culture identity), and marginalisation refers to the individual's lack of interest in both maintaining the source culture and absorbing the new culture (i.e., low source cultural identity, low new cultural identity). Of the four adaptation strategies, integration is usually considered the most adaptive, while the least adaptive is marginalisation, with assimilation and separation strategies ranking between the two (Matsudaira, 2006; Phinney & Navarro, 1997) (Table 9.1).

Table 9.1 Two-dimensional cultural adaptation theory analysis framework

		Source culture	
		High identification	Low identification
New culture	High identification	Integration	Assimilation
	Low identification	Separation	Marginalisation

From the analytical framework of two-dimensional adaptation theory, it can be seen that an individual's adaptation is closely related to their identification with the source culture and the new culture. Moreover, the difference between an individual's identification with the source culture and the new culture influences their choice of adaptation strategies in cross-cultural communication. According to Graves (1967), Murphy (1987), and Schwartz et al. (2006), a complete cultural identity includes three aspects: value identity, attitude identity, and behavioural identity. Therefore, this chapter draws on relevant research to develop a traditional cultural identity scale for farmers and a business culture identity scale for FCS projects (Tables 9.2 and 9.3), which requires farmers interviewed to score the questions according to a Likert scale and rate each questionnaire item as best fits their own opinions or attitudes. The scores ranged from 1 to 5, indicating strongly disagree, relatively disagree, neutral, relatively agree, and strongly agree, respectively. Then, the scores of each item were summed to find the average value, and the scores of the traditional culture and commercial culture of the participating farmers were obtained.

Table 9.2 Scale of identification with traditional culture

Latent variable	Items
Value identification	The Yi are a nature-worshipping people, and shifting cultivation is an important production practice
	The Yi are a people who worship freedom, and herding cattle and sheep is an important traditional custom
	The Yi are fire worshippers, and fire pits are sacred places and torch festivals national customs
	Bimo culture is the core of Yi culture and cultural wealth
Attitude identification	Rotational farming is better than intensive farming
	Free-range is better than captive livestock
	Burning the fire hall is the only way to ward off evil and pray for peace
	Bimo is the embodiment of wisdom, justice, and selflessness
Behaviour identification	I will continue to cultivate in shifting hibernation
	I will keep free-range livestock
	I will continue to burn fire pits with firewood
	I prefer to follow the advice of Bimo or village cadres to deal with contractual disputes over FCS projects

Table 9.3 Scale of identification with business culture through FCS projects

Latent variable	Items
Value identification	One of the cores of business culture is the spirit of contract
	Compliance with a contract is the basis for achieving contractual goals
	Violation of contract should be punished
	Commercial contracts are binding on both parties to the contract
Attitude identification	Planting or grazing under the carbon sink forest has a negative impact on the achievement of the contract goals
	Harvesting firewood in the forest has a negative impact on the achievement of the contract goals
	The use of fire in and around the forest is a potential risk to the achievement of the contract objectives
	There are no 'bullying' clauses in the project contract
Behaviour identification	I will not plant or graze in the forest before the afforestation project ends
	I will not harvest firewood before the afforestation project ends
	I will not engage in dangerous burning practices in and around the forest I will accept disputes in accordance with the project contract

9.1.1 Influencing Factors of Cultural Adaptation Strategies

Cultural adaptation is a fundamental concept that reflects cultural identity and cultural function. Established research shows that cultural adaptation is influenced by internal and external factors of individuals. Internal factors include cognitive evaluation styles, coping styles, personality, culturally relevant knowledge and skills, and demographic factors, while external factors can be life changes, social support, exposure, cultural distance, and discrimination and prejudice (Chen et al., 2003; Wang, 2013). It has been specified that adaptation occurs under the circumstances of spatial displacement of individuals or groups, as well as whenever individuals or groups encounter different cultural contacts in their daily lives (Koneru et al., 2007; Tang et al., 2011).

In this study, the farmers' adaptation of traditional and commercial cultures did not involve their spatial displacement, but was generated by an incoming foreign culture. Therefore, external factors, other than the degree of exposure, which require demographic spatial displacement to occur, do not apply to this study. Moreover, because internal factors such as farmers' cognitive evaluation style, coping style, and personality are implicit factors, they require a long period of continuous observation before accurate data can be obtained. Therefore, this chapter examines the impact of external factors such as exposure and internal factors such as culture-related knowledge and skills and demographic factors on farmers'adaptation of the two cultures. Exposure is characterised by two observed variables, namely participation time and land area; culture-related knowledge and skills are characterised by the distance between home and the nearest marketplace; and demographic factors are characterised by gender, age, education level, and main source of household income.

The surveyed sample was predominantly Yi (96.86%), mostly male (57.59%), with a low education level (4.47 years), little experience of working outside the home (only 28.27%), and large household size, namely of four to six persons (69.63%). Family income sources were mainly traditional farming and animal husbandry (85.86%). The average per capita disposable income of households in 2015 was 5,705 RMB (about 827 USD), which was less than the average level of 9,422 RMB (about 1366 USD) in 2015 in Liangshan Prefecture, and 19.37% of the farm households were poverty-stricken households with a higher incidence of poverty than the average level in Liangshan Prefecture (13.40%). The average distance between households and the nearest market was 29.82 km, and 91.62% of the land involved in the project was between 0.1 and 1 ha.

In the survey, it was found that due to historical, natural, and social factors, the villages are generally located in remote areas with many difficulties, such as inconvenient transportation, poor living environment, weak infrastructure, backward public services, a scattered population, and problems caused by difficult road access, schooling, medical care, income generation, and housing for farming households. Farmers generally have a strong reliance on traditional production and living methods, such as rough land management, livestock grazing, and firewood burning. Due to the constraints of education level and language, their opportunities to go out to work are very limited, and the rate of young and strong farmers staying in the villages is high (Table 9.4).

Table 9.4 Basic information of the valid sample

	Min	Max	Avg.	A.D.
Gender (1 = male; 0 = female)	0	1	0.58	0.45
Age (years)	18	61	39.09	12.30
Education level (years)	0	12	4.47	3.63
Whether working outside the home (1 = yes; 0 = no)	0	1	0.28	0.45
Number of family members (persons)	2	8	5.00	1.40
Household disposable income per capita (RMB/year)	2500	9800	5704.71	377.29
Main source of household income (1 = agriculture; 0 = non-agriculture)	0	1	0.86	0.35
Distance between home and the nearest marketplace (km)	5	60	29.82	16.74
Land area involved in the project (ha)	0.10	1.20	0.46	0.27

9.1.2 Investigation and Analysis of FCS Participants' Cultural Adaptation Strategies

To investigate participants' cultural adaption strategies in FCS projects, the researchers carried out several formal survey interviews in the villages in which the Novatis Sichuan FCS Project is implemented, sampling 12 villages and more than 200 participants. Based on the data generated by these surveys, the researchers compare the differences in FCS participants' attitudes towards business and traditional culture, identify their models of acculturation strategies statistically, and highlight their differences from several perspectives.

9.1.3 Data Description

The data in this chapter are derived from field questionnaires and case interviews in the project areas of the Novartis Sichuan FCS Project. In order to ensure the quality of the survey data and to better reflect the farmers' adaptation to commercial and traditional culture, the project team designed the questionnaire and interview outline and conducted pilot survey visits to forestry authorities at all levels in Sichuan Province and Liangshan Prefecture, the Dadu River Afforestation Bureau of Sichuan Province, and the Chengdu office of the Beijing Shanshui Nature Conservation Centre. The formal research adopted a combination of stratified and random sampling methods. First, three counties, namely Meigu, Ganluo, and Yuexi, were selected among the five project counties; then, two to four project villages were selected in the main afforestation areas of each county, and finally, 20–30 farmers were randomly selected in the selected project villages for the survey. A total of 210 farmers were surveyed, 191 valid samples were obtained, and the sampling efficiency was found to be 90.95%.

9.1.4 Farmers' Identification with the Two Cultures

The statistical results show that the average score of traditional cultural identity is 4.05, and the average business cultural identity score is 3.48. The results of a paired sample T-test show that the traditional cultural identity was significantly higher than the business cultural identity ($t = 11.10, p = 0.00 < 0.01$). In this regard, it was found in the survey that most of the afforestation areas in the project are located in the mid-hill slopes or alpine mountains, where the benefit is distributed in such a way that farmers or village collectives provide the land for free, and all the forest revenue goes to the land contractor or owner at the end of the 30-year project period. While local farmers experience modest economic gains, including income from labor, subsidies for grazing losses, and forest compensation, the majority maintain a favorable outlook on the project. This positive sentiment can be attributed to the cultural heritage of the Yi people, which emphasizes the protection and reverence for forests.

Due to the inadaptability of the farmers, before the forestation starts, it is necessary to build fences and take other measures to close off the mountain. However, squeezing the feeding and exercise space for cattle and sheep herds under the traditional custom of free grazing increases the cost of breeding and makes cattle and sheep more prone to disease. Second, part of the afforestation land occupies the rotation rest land, challenging the traditional customs of rotation planting. Third, policies on forest planting, firewood harvesting and fire prevention, and other harsh regulations restrict Chinese herbal medicine planting and traditional firewood collection, as well as the torch festival and other traditional fire activities. The contract provisions do not reasonably provide for the fire culture and heritage. Fourth, rights and interests cannot be fully protected, and the legal system procedures to deal with contract disputes are complicated. Local Bimo, village cadres, and other village elites need to become involved to solve contract disputes.

9.1.5 Identification of Farmers' Cultural Adaptation Strategies

According to the identity scores, traditional cultural identities and business cultural identities were grouped into high and low categories, with scores lower than 3 being classified as low identity, and those of 3 and above being classified as high identity. The traditional cultural identity and commercial cultural identity were then matched to form four cultural adaptation strategies: integration, assimilation, separation, and marginalisation. As seen in Table 9.5, the distribution of farmers in each category is as follows: integration (39.27%), separation (26.70%), assimilation (18.32%), and marginalisation (15.71%). The results of the nonparametric chi-square test showed significant differences in the distribution of the number of people corresponding to the four cultural adaptation strategies ($\chi^2 = 25.78$, $p = 0.00 < 0.01$).

9.1.6 Analysis of Farmers' Adaptation Strategies

The distribution of the number of people with different adaptation strategies is shown in Table 9.6, and the cross-tabulation difference significance test is conducted for the distribution of the number of people with two adaptation strategies correspondingly. The analysis of the potential main factors affecting their cultural adaptation strategies is summarised below.

In terms of gender, nearly half of the female farmers mainly applied an integration strategy to traditional and commercial culture (49.38%), while male farmers adopted integration and separation strategies in equal numbers, with a combined share of more

Table 9.5 Judgement criteria and number distribution of cultural adaptation strategies (person, %)

Cultural adaptation strategy	Judgement criteria	Frequency	Relative frequency
Integration	High traditional cultural identity, high business cultural identity	75	39.27
Assimilation	Low traditional cultural identity, high business cultural identity	35	18.32
Separation	High traditional cultural identity, low business cultural identity	51	26.70
Marginalisation	Low traditional cultural identity, low commercial cultural identity	30	15.71

than 60%, and one fifth of male farmers chose a marginalisation strategy. There is a significant difference between male and female farmers' choice of cultural adaptation strategies ($\chi^2 = 9.80, p = 0.02 < 0.05$). Overall, female farmers were more adaptable and had a higher degree of adaptation, which was similar to the findings of Ding and Gao (2015). Females are more likely to use risk-taking adaptation strategies and accept cultures coming from outside, thus showing higher adaptation (Wang, 2002). In the Yi family branch culture, males dominate economically and politically, and females are in an economically and politically subordinate position due to their ethical tradition (Hao, 2015). As a carrier of business culture, FCS projects emphasise the fairness and equality of participants, and men and women are equally paid. This practice tends to make males feel frustrated as they lose economic dominance, causing them to be uncomfortable with the business culture brought about by the project.

In terms of age, the integration strategy shows an inverted U shape as age increases, indicating that middle-aged people between 25 and 55 years old have a higher rate of adopting the integration strategy than young people under 25 years old and older people of 55 years old and above. In terms of assimilation strategy, young people under 25 years old have the highest rate (28.89%); as regards the separation strategy, the highest percentage was among older people aged 55 and above (65.63%); in the marginalisation strategy, the highest percentage was among young people aged under 25 (24.44%). There is a significant difference among age groups in terms of choosing adaptation strategies ($\chi^2 = 42.81, p = 0.00 < 0.01$). The distribution pattern outlined above confirms Beiser's (1988) conclusion that both adolescents and older people tend to choose an extreme adaptation strategy. Older people are deeply influenced by their original culture and are prone to form solidified beliefs; thus, they are hesitant to accept alien cultures that are different from their own and adopt separation strategies. Young people, meanwhile, have fewer implicit cultural beliefs and are more capable of identifying with and accepting alien cultures, and are even more easily assimilated into alien cultures than older people.

Table 9.6 Distribution of cultural adaptation strategies of farmers in FCS projects (persons, %)

		Cultural adaptation strategy							
		Integration		Assimilation		Separation		Marginalisation	
		F	F%	F	F%	F	F%	F	F%
Gender	Male	35	31.82	18	16.36	35	31.82	22	20.00
	Female	40	49.38	17	20.99	16	19.75	8	9.88
Age	18–25 years	11	24.44	13	28.89	10	22.22	11	24.44
	25–40 years	37	56.92	10	15.38	10	15.38	8	12.31
	40–55 years	21	42.86	10	20.41	10	20.41	8	16.33
	55 yrs. or above	6	18.75	2	6.25	21	65.63	3	9.38
Education	No schooling	7	16.67	10	23.81	16	38.10	9	21.43
	Elementary school	35	49.30	11	15.49	15	21.13	10	14.08
	Junior high school	25	51.02	6	12.24	13	26.53	5	10.20
	High school (secondary school) and above	8	27.59	8	27.59	7	24.14	6	20.69
Source of income	Agriculture and livestock	67	40.85	28	17.07	43	26.22	26	15.85
	Non-farm	8	29.63	7	25.93	8	29.63	4	14.81
Distance between home and nearest market	Less than 10 km	11	36.67	2	6.67	12	40.00	5	16.67
	10–30 km	33	49.25	14	20.90	11	16.42	9	13.43
	30–50 km	21	33.87	11	17.74	21	33.87	9	14.52
	50 km and above	10	31.25	8	25.00	7	21.88	7	21.88
Participation time	Within 2 years	14	35.90	8	20.51	10	25.64	7	17.95
	2–4 years	17	26.56	12	18.75	21	32.81	14	21.88
	4 years and above	44	50.00	15	17.05	20	22.73	9	10.23
Land area	Less than 0.25 ha	8	15.38	10	19.23	21	40.38	13	25.00
	0.25–0.5 ha	19	34.55	6	10.91	21	38.18	9	16.36

(continued)

Table 9.6 (continued)

		Cultural adaptation strategy							
		Integration		Assimilation		Separation		Marginalisation	
		F	F%	F	F%	F	F%	F	F%
	0.5–0.75 ha	22	53.66	4	9.76	7	17.07	8	19.51
	0.75 ha and above	26	60.47	15	34.88	2	4.65	0	0.00

In terms of education level, as the education level gradually increases, so does the number of people who adopt integration strategies. There is a significant difference in levels of education ($\chi^2 = 18.70, p = 0.03 < 0.05$). This may be because education is closely linked to cognitive ability, knowledge skills, and socioeconomic wealth, and the level of education is positively correlated with cultural adaptation (Chen et al., 2003). The higher the level of education of farmers is, the better they are able to recognise the ecological and social benefits of FCS projects and that participation in training and practical activities in afforestation and forestry technology enables them to better grasp the new technology and effectively migrate to other economic activities, such as employment and entrepreneurship.

In terms of source of household income, the largest number of farmers for whom agriculture was the main source of income chose the integration strategy (40.85%), while for those farmers whose main income is not from farming, the difference in terms of choosing integration, assimilation, and separation strategies is not significant. The difference in income sources is not significant ($\chi^2 = 1.85, p = 0.60 > 0.10$). This may be related to the relatively backward economic and social development of the project area, where traditional farming is still the main source of household income, and non-farm income is still generally limited.

There is no significant difference in the distances between the home and the marketplace ($\chi^2 = 13.18, p = 0.16 > 0.10$). A possible explanation is that the FCS project is located in a poverty-stricken and remote rural area with poor transportation and underdeveloped markets, and the marketplace is only used for the exchange of daily production and necessities. As the marketplace has a limited role in the formation and dissemination of business culture, farmers whose homes are closer to it do not acquire more business culture in the primary market transactions.

In terms of participation time, farmers who had participated in the project for four years or more were most likely to choose the integration strategy (50.00%), while those who had participated in the project for two to four years were more likely to choose the separation strategy (32.81%), and those who had participated in the project for less than two years also tended to choose the integration strategy (35.90%). However, there is no significant difference in the length of participation for the four adaptation strategies ($\chi^2 = 10.30, p = 0.11 > 0.10$). The main reason for this finding is that it takes individuals a long time to find their own adaptation equilibrium and adopt suitable adaptation strategies. The adaptation to business culture takes even longer for those farmers living in relatively remote and poverty-stricken mountainous areas.

In terms of the land area involved in the FCS project, as this increases the number of people who choose the integration strategy also gradually increases, and the number of people who choose the separation strategy gradually decreases. There is a significant difference in land areas for the four cultural adaptation strategies ($\chi^2 = 50.32$, $p = 0.00 < 0.01$). A possible explanation is that the larger the land area, the larger the size of the household and the higher the economic and political status of the family in the region; thus the family is more willing to comply with the terms of the project contract as a way to consolidate its authority and reputation within the community.

9.2 Summary

Based on survey data from the Novartis Sichuan FCS Project, the researchers empirically analysed the adaptation strategies, namely integration, assimilation, separation, and marginalisation, and their influencing factors in the Yi settlement area. The results show that (1) the traditional culture identity of the surveyed farmers is significantly higher than the business culture identity, and the formal system alone is not sufficient to regulate farmers' compliance, (2) integration and separation are the two main strategies adopted by the farmers, and (3) gender, age, education level, and land area are important factors influencing the choice of cultural adaptation strategies. Specifically, the degree of adaptation was higher among female, middle-aged, and more educated farmers and farmers with larger land areas.

Based on these findings, the following insights can be given. First, the traditional customs and traditions of the project area should be taken into account in the selection of FCS project plots, environmental assessment, community assessment, and planning. More attention should be given to men, the elderly, the less educated, and farmers with smaller land areas to enhance cultural adaptability. Second, women, middle-aged people, and people with higher education levels and larger land areas should be promoted as role models to encourage other community members to adapt to the new culture. Third, it is necessary to give full play to the role of informal systems, local elites, and villagers' self-governance by integrating the institutional arrangements of FCS projects into informal institutions such as village rules and regulations gradually, thus promoting the integration of business culture with regional traditional culture. More importantly, it is essential to ensure villagers' autonomy to secure the support of formal and informal systems, offset the interest-driven nature of forestation enterprises, and safeguard the legitimate rights and interests of farmers through village organisations.

References

Beiser, M. (1988). Influences of time, ethnicity, and attachment on depression in Southeast Asian refugees. *American Journal of Psychiatry, 145*(1), 46–51.

Berry, J. W. (1980). Acculturation as varieties of adaptation. *Acculturation: Theory, models and some new findings* (pp. 9–25). Westview Press.

Chen, H., Che, H. S., & Zhu, M. (2003). A review of research on factors influencing cross-cultural adaptation. *Advances in Psychological Science, 11*(6), 704–710.

Chen, C. Y. (2010). Forestry carbon sequestration and farmers' livelihoods: A case study of the first global FCS project. *World Forestry Research, 23*(5), 15–19.

Ding, F. Q., & Gao, J. J. (2015). Cultural adaptation of population migration of ecological migrants in western ethnic minority settlements—Taking the central arid belt region of Ningxia as an example. *Agricultural Economic Issues, 6*, 75–82.

Ding, Y., & Ma, P. P. (2013). Research on forest carbon sequestration and economic development of ethnic minority areas in western Sichuan province: Taking Yuexi County of Liangshan Yi autonomous prefecture in Sichuan province as an example. *Rural Economy, 5*, 38–41.

Graves, T. D. (1967). Psychological acculturation in a tri-ethnic community. *Southwestern Journal of Anthropology, 23*(4), 337–350.

Guan, B., Shi, Z., & Ning, P. (2012). The role of forest carbon sequestration system in guaranteeing the sustainable development of ethnic culture in Yunnan. *Journal of Kunming University of Technology (Social Science Edition), 12*(4), 60–64.

Hao, Y. (2015). The status of women in Yi family branch culture. *Journal of Southwest University for Nationalities (Humanities and Social Sciences Edition), 11*, 55–59.

Jindal, R., Swallow, B., & Kerr, J. (2008). Forestry-based carbon sequestration projects in Africa: Potential benefits and challenges. *Natural Resources Forum, 32*(2), 116–130.

Koneru, V. K., Mamani, A. G. W. D., Flynn, P. M., et al. (2007). Acculturation and mental health: Current findings and recommendations for future research. *Applied & Preventive Psychology, 12*(2), 76–96.

Matsudaira, T. (2006). Measures of psychological acculturation: A review. *Transcultural Psychiatry, 43*(3), 462–487.

Murphy, J. W. (1987). *Culture, identity, and politics* (pp. 3–17). Cambridge University Press.

Phinney, J. S., & Navarro, D. M. (1997). Variation in bicultural identification among African American and Mexican American adolescents. *Journal of Research on Adolescence, 7*(1), 3–32.

Schwartz, S. J., Montgomery, M. J., & Briones, E. (2006). The role of identity in acculturation among immigrant people: Theoretical propositions, empirical questions, and applied recommendations. *Human Development, 49*(1), 1–30.

Tang, X. Q., Qian, J. X., & Chen, L. X. (2011). Cultural adaptation and reconstruction of ethnic minority festivals under the influence of tourism: An analysis based on the evolution of the hani long street feast. *Geographical Studies, 30*(5), 835–844.

Walker, W., Baccini, A., Schwartzman, S., et al. (2014). Forest carbon in Amazonia: The unrecognized contribution of indigenous territories and protected natural areas. *Carbon Management, 5*(5–6), 1–26.

Wang, T. (2013). Cultural adaptation of the Li ethnic group: Characteristics, influencing factors and theoretical models (pp. 15–30). PhD dissertation of East China Normal University.

Wang, Y. P. (2002). *Ethnic identity, cultural adaptation and psychological alienation of Tibetan college students* (pp. 5–12). Master's Thesis of Northwest Normal University.

Wen, Y. L. (2013). Tibetan cultural agriculture and carbon sequestration economic development. *Forestry Economics, 3*, 112–115.

Yang, F., Paudel, K. P., Cheng, R. Z., Qiu, L., Zhuang, T., & Zeng, W. (2018). Acculturation of rural households participating in a clean development mechanism forest carbon sequestration programme: A survey of Yi ethnic areas in Liangshan, China. *Journal of Forest Economics, 32*, 135–145.

Chapter 10
Analysis of the Impact of Livelihood Capital on Farmer Earnings in FCS Projects

The Novartis Sichuan FCS Project and the Tengchong FCS Project are typical projects implemented in poverty-stricken areas populated by Yi, Dai, Hani, and Bai ethnic minorities. The two projects are executed by the Sichuan Dadu River Afforestation Bureau and the Sujiang Forestry Administration of Yunnan Tengchong Forestry Bureau, respectively. Both projects have adopted *the government + enterprise + farmer operation model*, in which the benefit derived from forest trees goes to the landowners at the end of the project in accordance with CDM AR standards. In the study, it was found that the farmers in both project communities received some earnings, but there were significant differences in the earnings received by farmers with different capital resources (Qiu et al., 2018). It is inferred that the difference in capital owned by farmers will influence the choice of how farmers participate in FCS projects and thus their earnings. Therefore, this chapter attempts to draw on livelihood capital theory, farmer behaviour theory, and behavioural economics theory to empirically test the relationship between farmers' livelihood capital and earnings in FCS projects implemented in different areas, with a view to providing useful references for the optimisation of FCS and poverty alleviation.

10.1 Theoretical Analysis and Research Hypothesis

Livelihood capital is both an important guarantee of livelihood and a necessary condition for achieving different livelihood strategies (Moser, 1998). People's choices in livelihood strategy are not completely free (Ellis, 2000), capital status determines the choice of livelihood strategy, and the ability to achieve different livelihood strategies depending on the livelihood capital one possesses. The amount and type of livelihood capital play a key role in the number of available options, and the diverse composition of capital and assets, both quantitative and qualitative, can present very different livelihood situations (Xing & Ge, 2013). In their investigation of multidimensional

W. Zeng and F. Yang, *Poverty Alleviation Via Forest Carbon Sequestration*, International Research on Poverty Reduction, https://doi.org/10.1007/978-981-99-8191-5_10

poverty among farm households in the Wuling Mountains, Li and Wang (2014) found that the scarcity of farm households' livelihood capital led to a transformation of their livelihood capital stuck in a low-level equilibrium trap. In studies on poverty alleviation in tourism, it has been argued that lack of capital severely constrains the participation mode of the poverty-stricken people in the tourism industry and hinders their benefitting from tourism development (Li & Li, 2015; Wang, 2007). Studies on labour participation suggest that human capital enhancement due to investment in education, health, and migration has a positive impact on labour participation capacity (Cai & Wang, 2004; Shi & Yang, 2012). The above studies suggest that, in general, for a single economic activity, the greater the livelihood capital of potential participants, the greater their ability to choose to participate in that economic activity.

When both supply and demand have free choice, the labour market is more uniform and competitive, and workers with richer human capital can obtain higher earnings by virtue of their capital endowment (Wan, 2005), i.e., they will prefer projects with higher earnings to capital. Therefore, when multiple economic activities are available at the same time, the competitive rate of return of economic activity becomes the basis for workers to choose whether and how to participate based on their livelihood capital. For instance, farmers rich in livelihood capital can earn more from non-farming activities (Liang et al., 2011; Zhang et al., 2012). These studies show that, in general, for multiple competitive economic activities, the richer the livelihood capital of potential participants, the greater their ability to choose to participate in economic activities with higher earnings from investment.

From the above analysis, it is clear that livelihood capital has a significant impact on people's livelihood strategies, in a complex manner. Generally, the livelihood capital people have is positively correlated with their choice of livelihood strategy, and the richer this livelihood capital, the larger the range of livelihood strategy opportunities. Therefore, on the one hand, the more livelihood capital farmers have, the stronger their ability to participate in livelihood activities, including FCS projects, and the more likely they are to participate heavily in the projects, which may bring them greater earnings (through their participation). On the other hand, farmers with more livelihood capital may, on the contrary, have greater choice in livelihood strategies and be more likely to obtain higher earnings in non-farming activities. Thus, farmers with more livelihood capital may be more likely to have a negative attitude towards FCS projects. Based on the above analysis, the following hypothesis is derived:

Hypothesis 1: Farmers' livelihood capital has an impact on their earnings in FCS projects.

Different forms of livelihood capital have different impacts on people's livelihood strategies. An increase in natural capital usually makes farmers more active in agriculture-based livelihood activities, and thus farmers earn more from them (Su et al., 2009; Zhao et al., 2015). Analysing the livelihood activities of farmers and herders on the Gannan Plateau, Zhang et al. (2012) found that farmers with a lack of natural capital tended to choose other, non-farming livelihood activities, while the diversification of farmers' livelihoods was influenced by a combination of factors.

Similarly, based on the quantitative analysis of farmer livelihood strategies, the findings of scholars such as Su and Shang (2012), Meng et al. (2013), Hao et al. (2014), Zhao et al. (2016), and Wu (2016) show that most farmers engaged in agricultural production activities have rich natural capital and physical capital. It is noteworthy that the FCS projects investigated in this chapter are located in remote mountainous areas, and the projects rely on resources such as rural labour and forest land; thus they are agricultural livelihood activities. Accordingly, the following hypotheses are proposed:

Hypothesis 2: Natural capital positively affects farmers' earnings in FCS projects.

Hypothesis 3: Physical capital positively affects farmers' earnings in FCS projects.

Through empirical analysis, Meng et al. (2013) and Hao et al. (2014) found that, in contrast, farmers who use non-farming livelihood activities as their main livelihood strategy have richer human and financial capital than others do. That is, farm households with richer human and financial capital are more likely to rely on non-farming livelihood strategies as their main source of earnings (Wu, 2016; Zhao et al., 2015, 2016). In general, the richer their financial capital is, the more farm households will prefer a non-farming sector (Su et al., 2009). Some scholars have also found through studies that as the human capital of farm households increases, their main livelihood strategy is highly likely to make a shift from agriculture to non-agriculture (Guo et al., 2017). Accordingly, the following hypotheses are proposed:

Hypothesis 4: Human capital negatively affects farmer earnings in FCS projects.

Hypothesis 5: Financial capital negatively affects farmer earnings in FCS projects.

Meanwhile, Su and Shang (2012) empirically analysed the livelihood strategies of farm households in Ganzhou District, Zhangye City, and found that farm households with richer social capital were more likely to have off-farm employment. Similarly, Wu (2016) and Zhao et al. (2016) argue that farm households richer in social capital generally lack interest in agricultural livelihood activities and are prone to focus their attention on non-farming industries. Therefore, the following hypothesis is proposed:

Hypothesis 6: Social capital negatively affects farmer earnings in FCS projects.

10.1.1 Data, Sample, Variables, and a Theoretical Model

To test the theoretical hypotheses proposed in the previous section, the researchers take two typical FCS projects from Southwest China and sample the participants conforming to the empirical method. In this section, the researchers discuss the sampling method, data acquisition access, sample characteristics, and the theoretical model underlying the empirical study.

10.1.2 *Data and Sample Characteristics*

The data were collected through several surveys in southwest China where two typical FCS projects were implemented. Data and sample characteristics are presented below.

10.1.2.1 Data

The data used in this chapter is derived from surveys conducted by the research team in Sichuan and Yunnan provinces. The sample areas were Ganluo County and Yuexi County in Liangshan Yi Autonomous Prefecture, Sichuan Province, and Tengchong City, Yunnan Province. Since some of the items in the questionnaire were based on historical data, the survey was conducted based on data in 2016 in order to make the research data more realistic and scientific. It should be emphasised that all research data were at the household level, not at the individual level. In addition, in order to avoid interference caused by endogeneity among variables, the data on farmers' livelihood capital were collected from the beginning of 2016, and the data on participation earnings were aggregated from January to August 2016.

10.1.2.2 Sample Characteristics

This chapter examines the impact of the livelihood capital held by farm households on their participation in FCS projects, using rural households rather than individual farmers as the research target. There were two requirements for the interviewees: that they were well informed about their household and that they agreed to represent their household. The number of households interviewed for the Tengchong FCS project in Yunnan Province was significantly higher than that in the Liangshan Yi Autonomous Prefecture project area in Sichuan Province, mainly because the project area in Yunnan has a higher density of households and more participants. The second reason was that the project area in Yunnan Province borders on tourist attractions and has a relatively higher level of socio-economic development, farmers in the community were better able to express their opinions clearly, and some could clearly evaluate the development of FCS projects, making the survey more efficient. In terms of age distribution, middle-aged farmers aged 30–60 years were the main, accounting for 89.95%, indicating that middle-aged farmers in the project area are still the main force participating in various activities. In terms of education, most of the interviewees had an education level of junior high school or below, accounting for 96.46%, indicating that the overall education level of participating farmers in the project area is low (see Table 10.1 for details).

The researchers found that the majority of farmers obtained economic earnings in two ways: one was to obtain compensation such as forest compensation; the other was to obtain labour earnings in seedling, breeding, land preparation, afforestation,

Table 10.1 Basic characteristics of interviewees

Type	Items	Number (households)	Proportion (%)
Area	Liangshan, Sichuan	105	28.61
	Tengchong, Yunnan	262	71.39
Gender	Male	254	69.02
	Female	113	30.98
Age	30 years and below	57	15.53
	31–45 years	158	43.05
	46–60 years	116	31.61
	61 years and above	36	9.81
Education	Primary school or below	197	53.68
	Junior high school	158	43.05
	High school or above	12	3.27

Table 10.2 Participation methods of farmers in FCS projects

Participation method	Sample size (households)	
	Sichuan province	Yunnan province
Work	74	150
Land transfer	60	130
Work and land transfer	29	18

replanting, management and care, fence construction, etc. Among them, the number of people participating in the project in the form of labour and in the form of land transfer was roughly equal, and a few farmers participated in the project through both land transfer and labour (see Table 10.2). The average daily wages of farmers were about the same, usually about 120 RMB (about 17 USD)/day.

10.1.3 Variable Selection and Description

First, the dependent variable is the earnings of farmers participating in the FCS projects, which is a fixed ratio variable. The participation earnings are the sum of the economic earnings of farmers participating in FCS projects.

Second, this chapter focuses on the influence of livelihood capital on the earnings obtained by a farmer in FCS projects. Based on the research hypotheses, the independent variables are set as follows.

Natural capital. Natural capital refers to natural resources such as farmland, water, and forest land that farmers can use in carrying out their livelihood activities. After taking into account the specific conditions of the FCS projects, two indicators are used to describe a household's natural capital status. Generally speaking, the more arable land a household has, the richer its natural capital is; the larger the area of forest land included in the FCS projects, the more natural capital it has. Other conditions being held constant, the more forestable land is included in the FCS projects, the more earnings the household will get from participating in the FCS projects, and the expected impact is therefore positive. The larger the farming area of a farmer's household, on the one hand, the busier farmers may be with agricultural livelihood activities outside FCS projects, thus affecting their participation benefits. On the other hand, farmers with more farmland may pay more attention to new projects in the community and actively seek opportunities to participate in FCS projects for more earnings. It can be seen that the effect of household farming areas on farmer participation earnings is uncertain.

Physical capital. Physical capital refers to the infrastructure and physical composition required for livelihood activities. Two indicators are used to describe the physical capital status of farmers, namely the amount of fixed assets owned by the household and the transportation accessibility of the home. In general, the more fixed assets a household owns and the better the road access to the home, the richer its physical capital. All other conditions being equal, the better the road access to a farmer's home, the easier their obtaining information and opportunities to participate in FCS projects, and the more earnings they can acquire from participation. Thus, the impact of transportation accessibility on their participation earnings is expected to be positive.

For households with a significant amount of fixed assets, on the one hand, they may leverage their material capital advantages to participate more actively in FCS projects, thereby gaining greater benefits. On the other hand, due to their possession of abundant material capital, they might prefer non-agricultural livelihood activities and show less interest in FCS projects, resulting in reduced earnings from their involvement in such projects.

Human capital. Human capital refers to the human resources, knowledge, and technology that farmers possess when they engage in livelihood activities. The quantity and quality of the labour force can influence a household's earnings in livelihood activities, mediated by their behaviours in livelihood activities. Therefore, five indicators were chosen to describe human capital status: the number of adult labourers in the household, the education level of the labourers in the household, major diseases or chronic illnesses of family members, whether members in the household have non-farming production skills, and the number of times family members participated in labour skills training organised by the village (community) in a year. Generally speaking, the larger the number of adult labourers, the higher their education level, the greater the frequency of participating in labour skills training, and the higher the number of non-farming production skills, the richer the human capital of the household. At the same time, if there are family members with major illnesses or chronic diseases in the household, then the household's human capital will be weakened to some extent. In general, all other things being equal, the more adult labourers in the household, the more human capital it will have, and the higher its participation earnings are likely to be, so the expected impact is positive. The more educated the labourers in the household, the more they will tend to participate in non-farming livelihood activities, thus reducing their participation in FCS projects, so the expected impact is negative. If there is a family member with a major disease or chronic illness, the labourers will have to devote time and energy to take care of the sick person, which will reduce their participation in the FCS project and eventually lead to a decrease in their participation earnings, so the expected impact is negative.

Financial capital. Financial capital refers to the adequacy and availability of funds for the livelihood activities of farmers, and to some extent reflects the ability of farmers to sustain these activities. In this chapter, two indicators describe the financial capital status of farm households: household debt and ease of borrowing. In general, if farmers have debts in the household, they are relatively short of financial capital; if it is easier to borrow money, they are richer in financial capital. All other things being equal, if farmers have debt, on the one hand, they are likely to participate more in livelihood activities that require labour on a permanent basis, and less in FCS projects that require labour on a seasonal or short-term basis, so their participation earnings will be lower. On the other hand, farmers who are indebted are likely to increase their participation in all livelihood activities (including FCS projects) as much as possible because they have a greater financial burden, so their participation earnings in FCS projects are likely to be higher. Therefore, the impact of household debt on the participation earnings of FCS projects is complicated, and the impact is uncertain. If farmers can borrow more easily, they are more likely to invest in livelihood projects with a higher return on investment, and less likely to participate in FCS projects with lower returns, and thus their return on participation in FCS projects is likely to be lower.

Social capital. Social capital refers to social resources and networks that can be used in livelihood activities, such as relationships with neighbours and village officials and ease of access to information. In this chapter, the social capital status of farmers is reflected by three indicators: the relationship between farmers and their relatives and neighbours, the relationship between farmers and village cadres, and the distance between the homes and the nearest town centre. Generally speaking, the better the relationship between farmers and their neighbours and the better the relationship with village cadres, the richer their social capital; the farther the farmer's home is from the nearest town, the more difficult it is for them to interact with others, i.e., the less social capital they will have. In general, the better the relationship between farmers and their neighbours and that between farmers and village cadres, the more livelihood opportunities will be available to them, i.e., the more diverse their livelihood activities will be. In this case, FCS projects will lack comparative advantages, resulting in less motivation for farmers to participate in them, and ultimately, fewer earnings from participation. Therefore, the influence of relationships with neighbours and village officials on farmer participation in FCS projects is expected to be negative. In addition, the distance between the home and the nearest town centre can influence the participation behaviour of farmers, mediated by the convenience of accessing information, which ultimately affects their participation earnings. For farmers whose home is farther away from the market, it is more difficult for them to access information related to livelihood activities, which leads directly to a narrower range of these activities. In this case, FCS projects will be more attractive; farmers will choose to participate more actively in the projects, and their participation earnings will be higher. Therefore, the influence of the distance between the home and the nearest town on farmers' participation in FCS projects is expected to be negative.

Control variables. Since the surveyed FCS projects are located in two regions, geographical dummy variables were introduced as control variables in the empirical analysis to study the effects of differences in project institutional rules and socio-economic development levels of different project communities on the dependent variables. Definitions, values, and descriptive statistics for variable indicators are shown in Table 10.3.

10.1.4 Measurement

In determining the weights of the indicators, the more widely used method is hierarchical analysis, while some scholars also use a weighting formula. In the calculation of livelihood capital values, most scholars use the extreme difference standardisation method to make dimensionless the indicator values and then calculate the values.

To calculate weights. The hierarchical analysis method was adopted to determine the weight of each livelihood capital indicator. The main steps were as follows: establish the hierarchical structure model; assign values to the judgement matrix; calculate the weight vector and conduct a consistency test; weight the arithmetic average of each expert's ranking vector through group decision and experts assignment method; finally obtain the weights of the livelihood capital indicators. The calculation results are shown in Table 10.4.

To calculate livelihood capital. Because of the unit differences among the indicators, the researchers first standardised the indicators dimensionlessly through the extreme difference standardisation. The formula for calculating the extreme difference standardisation is as follows:

$$Z_{\mathrm{ij}} = (X_{ij} - \min X_{ij})/(\max X_{ij} - \min X_{ij}) \tag{10.1}$$

In formula (10.1), Z_{ij} represents the value of the j-th indicator of the i-th category of livelihood capital after standardisation, and X_{ij} represents the original value of the j-th indicator of the i-th category of livelihood capital, i.e., the unstandardised value.

After standardisation of the livelihood capital indicators, the livelihood capital values were calculated based on the standardised value of each indicator and its weight. The calculation formula is as follows:

$$L = \sum_{i=1}^{5} \sum_{j=1}^{n} W_{ij} Z_{ij} \tag{10.2}$$

Table 10.3 Definitions, values, and descriptive statistics of variables

Variable		Definition and value	Mean	S.D.
Benefit (y)		Actual earnings from FCS project (RMB)	2650.905	4227.243
Natural capital	Area of forest land (x_1)	Area of forest land included in FCS project (ha)	0.183	0.1251
	Area of farmland (x_2)	Actual arable land area (ha)	0.456	0.465
Physical capital	Household fixed assets (x_3)	Household fixed assets	2.4387	0.8337
	Transportation accessibility (x_4)	Poor = 1; Fair = 2; Good = 3	2.4768	0.6219
Human capital	Number of adult labourers (x_5)	Number of actual labours greater than or equal to 16 years old (persons)	3.0272	1.3628
	Education (x_6)	Average years of education greater than or equal to 16 years old (years)	3.8011	1.1505
	Major or chronic disease in family (x_7)	None = 0; Yes = 1	0.4687	0.4997
	Non-farming skills possessed (x_8)	None = 0; Yes = 1	0.9591	0.1983
	Skills training attended in a year (x_9)	Participation (times)	0.6049	1.5056
Financial capital	Debt (x_{10})	No = 0; Yes = 1	0.5204	0.5620
	Difficulty of borrowing money (x_{11})	Very difficult = 1; Difficult = 2; Fair = 3; Easy = 4; Very easy = 5	3.2262	1.1945
Social capital	Relationships with relatives and neighbours (x_{12})	Very bad = 1; Not very good = 2; Fair = 3; Better = 4; Very good = 5	4.3924	0.7233
	Relationships with village officials (x_{13})	Very bad = 1; Not very good = 2; Fair = 3; Better = 4; Very good = 5	2.3406	0.8691
	Distance to nearest town (x_{14})	Actual distance between home and nearest downtown (km)	7.2240	12.1654
Geographical differences (Sichuan Province as reference group)		Designating Sichuan Province as a reference group; Yunnan Province is assigned a value of 1	0.7139	0.4526

Table 10.4 Weights of livelihood capital indicator

Variable indicator		Weight
Natural capital	Area of forest land (x_1)	0.3037
	Area of farmland (x_2)	0.0506
Physical capital	Household fixed assets (x_3)	0.0144
	Transportation accessibility (x_4)	0.0575
Human capital	Number of adult labourers (x_5)	0.1148
	Education (x_6)	0.0301
	Major or chronic disease in family (x_7)	0.0400
	Non-farming skills possessed (x_8)	0.0643
	Skills training attended in a year (x_9)	0.0643
Financial capital	Debt (x_{10})	0.0271
	Difficulty of borrowing money (x_{11})	0.0813
Social capital	Relationships with relatives and neighbours (x_{12})	0.0380
	Relationships with village officials (x_{13})	0.0380
	Distance to nearest town (x_{14})	0.0301

In formula (10.2), L is the livelihood capital value, W_{ij} is the weight of the j-th indicator of the i-th category of livelihood capital, and Z_{ij} is the standardised value of the j-th indicator of the i-th category of livelihood capital.

10.1.5 Theoretical Model

The dependent variable y in this chapter is a farmer's earnings in FCS projects, which is a fixed ratio variable. Therefore, the researchers used general least squares to construct a multiple linear regression model to analyse the impact of farmers' livelihood capital on their earnings. First, the model of the impact of five types of livelihood capital on participation earnings is as follows:

$$y = c + \alpha_n x_n + \alpha_m x_m + \alpha_h x_h + \alpha_f x_f + \alpha_s x_s + \alpha_{con} x_{con} + \mu \quad (10.3)$$

In formula (10.3), x_n, x_m, x_h, x_f, x_s, x_{con} are natural capital, physical capital, human capital, financial capital, social capital, and control variables, respectively; α_n, $\alpha_m . \alpha_h$, α_f, α_s, α_{con} are regression coefficients of the six variables, reflecting the effects of each variable on participation earnings; c is the constant term and μ is the random error term.

Second, the multiple linear regression model of the effects of the specific livelihood capital variables x_1, x_2, x_3, *and*—x_n on the dependent variable, a farmer's earnings, is as follows:

$$y = c + \beta_1 x_1 + \beta_2 x_2 + \beta_3 x_3 + \ldots + \beta_n x_n + \beta_{con} x_{con} + \varepsilon \quad (10.4)$$

In formula (10.4), x_n stands for the n-th livelihood capital variable; x_{con} is the control variable; β_n stands for the regression coefficient of the n-th livelihood capital variable; and β_n is the regression coefficient of the control variable; the random error term and the constant term are ε and c, respectively. In conducting the empirical analysis, the values of all variables were standardised for extreme differences to avoid the influence of extreme values.

10.2 Discussion of Empirical Results

After data collation and measurement testing, the researchers performed a fit test of the empirical model. In this section, the researchers discuss the empirical results.

10.2.1 Impact of Livelihood Capital on Participation Earnings

Although FCS project participants are required to bear opportunity costs (Ickowitz et al., 2017; Ndjondo et al., 2014; Rakatama et al., 2017), it is undeniable that forestry projects such as REDD+ and AR projects can have a positive impact on community livelihoods (Sunderlin et al., 2018). Awono et al. (2014) showed that REDD+ projects are able to reduce CO_2 emissions and protect and improve the livelihoods of communities near the projects. However, due to differences in socioeconomic status, farmers have varying degrees of participation, which ultimately results in differentiated participation earnings (Shrestha & Shrestha, 2017). Some studies have shown that farmers with different livelihood strategies differ in their ownership of livelihood capital and that their choice of livelihood strategy varies according to their livelihood capital (Hua et al., 2017). As shown in Table 10.5, all five categories of livelihood capital significantly affected FCS participation earnings from FCS projects, and Hypothesis 1 was tested.

Natural and physical capital had a positive effect on the earnings from FCS participation, indicating the richer the natural and physical capital of farmers, the higher their earnings, and thus Hypothesis 2 and Hypothesis 3 were verified. Similarly, Fang et al. (2014) and Hua et al. (2017) found through empirical analysis that natural capital has a positive effect on agricultural livelihood activities, implying that farm households richer in natural capital tend to derive more earnings from agriculture-based livelihood strategies. Possible explanations are that the greater the natural and physical capital of farm households, the higher their dependence on agricultural livelihood activities (Fang et al., 2014; Jiao et al., 2017), which means they are more likely to stay in the countryside for prolonged periods to participate in agricultural

Table 10.5 Effects of five types of livelihood capital on FCS participation earnings

Variables	Mean	S.D	T value	Coefficient
Natural capital (x_n)	0.0064	0.0066	2.6594	0.0564***
Physical capital (x_m)	0.0476	0.0183	1.7912	0.0211**
Human capital (x_h)	0.1463	0.0345	1.7463	−0.0960***
Financial capital (x_f)	0.0137	0.0065	2.3700	−0.0239**
Social capital (x_s)	0.1626	0.0654	−1.634	−0.0264**
Control variables (x_{con})	0.0516	0.0171	−3.4662	−0.0486**

Note *, **, and *** indicate significance at 10%, 5%, and 1%, respectively

activities. Plus, the sustained operation of FCS projects requires regular maintenance and replanting; therefore, a relatively stable labour supply is essential. Households with relatively abundant natural and physical capital tend to have a stronger dependency on land and are more likely to live in rural areas long-term. This situation conveniently meets the labor demands of FCS projects. As a result, these households are more likely to deeply engage in FCS projects and reap greater benefits from their participation.

Human capital, financial capital, and social capital had negative impact on the benefits that households gain from participating in FCS projects. This indicates that the more abundant the human, financial, and social capital of a household, the lower the benefits they gain from project participation. Hypotheses 4, 5, and 6 are confirmed. Similarly, studies by Su et al. (2009), Meng et al. (2013), Zhao et al. (2016), and Hua et al. (2017) showed that human capital and financial capital have a positive effect on farmers' non-farming livelihood strategies, i.e., farmers who are richer in these types of capital will earn more from non-farming industries. Furthermore, through empirical studies, Fang et al. (2014) found that financial capital and social capital are catalysts for farm households to engage in non-farming livelihood activities, but at the same time, they concluded that human capital is positively associated with agricultural livelihood activities. Some scholars also argue that financial capital does not have a significant effect on the farmers' choice of livelihood strategies (Wu et al., 2017). Possible explanations for these results are as follows: farmers with higher human capital, financial capital, or social capital are more likely to choose to participate in non-farming livelihood activities with higher marginal earnings, and these farmers are more willing to move outside the village and tend to be more mobile, so their participation in FCS projects is lower, as are the earnings from participation.

10.2.2 *Impact of Livelihood Capital Indicators on Participation Earnings*

To further investigate the impact of the specific components of livelihood capital on the participation benefits of FCS projects, the researchers analysed the relationship between specific indicators and FCS participation benefits. The results are discussed below.

10.2.2.1 Model Fitting Results and Comparison

Using the multiple linear regression model, eight regression equations were constructed by two methods: stepwise entry and stepwise screening. As showed in Table 10.6 Eqs. 1–7 were constructed by the stepwise entry method, and Eq. 8 was constructed by the stepwise screening method. Among them, Eqs. 1–5 are models that examine the independent effects of independent variables on dependent variables without considering other factors, i.e., in Eqs. 1–5, only natural capital, physical capital, human capital, financial capital, and social capital was introduced as an independent variable, respectively. In Eq. 6, natural capital, physical capital, human capital, financial capital, and social capital were introduced together. Equation 7 is a model where control variables were introduced on the basis of Eq. 6, reflecting the effect of independent variables on dependent variables after controlling for the influence of geographical factors. Equation 8 is a model constructed after introducing control variables, considering the multicollinearity among variables, and gradually eliminating nonsignificant variables.

From the regression results for each variable in Table 10.6, it can be seen that the probabilities corresponding to the F statistics of the eight regressions model are all less than the significance level of 5%, indicating that the linear relationship between the independent variables and the dependent variable is significant, and all of them can build a linear model. Meanwhile, the adjusted R-squared of Eqs. 1–7, without considering the influence of other factors, are 0.090, 0.059, 0.198, 0.063, 0.065, 0.216, and 0.346, respectively. For the goodness of fit of the equations, human capital has the strongest explanatory power, followed by natural capital, social capital, financial capital, and physical capital. The combination of the five types of capital has stronger and better explanatory power after controlling for geographical factors. The adjusted R-squared of Eq. 8 is 0.451, which is better than that of other equations. It can be seen that the regression constructed by the stepwise screening method has the best explanatory power among all equations.

10.2.2.2 Analysis of Regression Results

Equation 8 includes five types of livelihood capital as independent variables, controls for geographical factors, takes into account the multicollinearity among variables,

Table 10.6 Multiple linear regression

Livelihood capital	Variable	Equation 1	Equation 2	Equation 3	Equation 4	Equation 5	Equation 6	Equation 7	Equation 8
Natural capital	x_1	0.047*** (2.197)					0.042** (2.135)	0.039** (2.124)	0.038** (2.310)
	x_2	0.029 (1.048)					0.031 (1.003)	0.030 (1.214)	
Physical capital	x_3		−0.042 (−1.488)				−0.028 (−0.932)	−0.012 (−0.411)	
	x_4		0.011** (2.442)				0.014** (2.047)	0.019** (2.965)	0.017** (2.354)
Human capital	x_5			0.072** (2.446)			0.062*** (2.064)	0.053*** (2.764)	0.061*** (2.453)
	x_6			−0.108** (−2.704)			−0.114** (−2.746)	−0.102** (−2.498)	−0.092** (−2.342)
	x_7			−0.028** (−2.421)			−0.024** (−2.025)	−0.031** (−2.639)	−0.034** (−2.964)
	x_8			−0.050* (−1.732)			0.044 (1.510)	−0.061** (−2.099)	−0.066** (−2.295)
	x_9			−0.344*** (−4.503)			−0.288** (3.664)	−0.272** (−3.519)	−0.078** (−3.695)
Financial capital	x_{10}				−0.045 (−1.074)		−0.056 (−1.310)	−0.021 (−0.496)	
	x_{11}				−0.057** (−2.889)		−0.035* (−1.768)	−0.032* (−1.613)	−0.037* (−1.924)
Social capital	x_{12}					−0.018* (−2.711)	−0.017** (−2.289)	−0.016** (−2.796)	−0.017** (−2.267)

(continued)

Table 10.6 (continued)

Livelihood capital	Variable	Equation 1	Equation 2	Equation 3	Equation 4	Equation 5	Equation 6	Equation 7	Equation 8
	x_{13}					−0.028** (−2.397)	−0.021** (−2.530)	−0.025** (−2.007)	−0.024** (−2.579)
	x_{14}					0.503* (4.051)	0.045* (3.050)	0.054* (1.829)	0.044** (1.907)
	$x_{control}$							−0.061** (−3.659)	−0.058*** (−3.885)
	c	0.078	0.001	0.041	0.045	0.049	−0.039	−0.028	0.011
	R^2	0.066	0.103	0.207	0.071	0.072	0.352	0.483	0.569
	Adjusted R^2	0.090	0.059	0.198	0.063	0.065	0.216	0.346	0.451
	F值	5.167	8.280	4.977	6.499	3.268	4.201	4.915	9.129
	P值	0.002	0.000	0.007	0.000	0.036	0.000	0.000	0.000

Note Values in parentheses are t-statistics, and *, **, and *** indicate significance at 10%, 5%, and 1%, respectively

and eliminates nonsignificant variables. The following analysis revolves around the regression results of Eq. 8.

First, after controlling for geographical factors, the coefficient of *the area of forest land* is positive and significant at the 5% level, indicating that the area of forest land had a significant positive effect on participation earnings. Specifically, for every 1% increase in the area of forest land, participation earnings increased by 0.038%.The larger the area of forest land included in the FCS project within a farmer's household, the greater the forest land rent income they receive. The acquisition of forest land rent income will positively encourage the household to participate in the project through labor or other means, ultimately leading to greater participation benefits.

Second, after controlling for geographical factors, the coefficient of *transportation accessibility* is positive and significant at the 5% level, indicating that transportation accessibility had a significant positive effect on participation earnings. In particular, the more convenient road access near the home, the more earnings the farmer obtained. The fact that most AR projects are implemented in remote, poverty-stricken areas means that many participating farmers are inevitably restricted by geographical factors in their contact with the outside world (Ding & Ma, 2013; Yang et al., 2018; Zeng et al., 2016). In this case, the more convenient the road access to farmers' homes, the better they can access information about the FCS projects, i.e., they tend to know more about the FCS projects, and therefore, have more opportunities to participate, and the larger the earnings they can acquire from participation.

After controlling for regional factors, the number of adult laborers in the household passes the test at a 1% level of significance, indicating that the number of adult laborers in the household has a significant positive impact on participation benefits. Specifically, for each additional adult laborer in the household, the participation benefits increase by 0.061%. The possible reason is that the more adult laborers a farming household has, the deeper their involvement in FCS projects tends to be, which in turn leads to greater participation benefits. This result is in line with those of numerous studies (e.g., Coulibaly-Lingani et al., 2011; Dolisca et al., 2006; Hua et al., 2017; Oli & Treueb, 2015; Shrestha & Shrestha, 2017).

The education of the workforce at home had a negative effect on earnings at the 5% level of significance; specifically, for each additional year in mean education level, earnings decreased by 0.092%. A possible reason for this is that the higher the education level of a household, the more likely it is to participate in non-farming livelihood activities with higher earnings. Similarly, the results of Wu et al. (2017) and Jiao et al. (2017) showed that increases in the education level of farm households will make their livelihood strategies less agriculture-oriented. However, Shresthaa and Shresthab (2017) investigated farmers undertaking forest management in Nepal and found a positive effect of education level on participation in forest management.

The coefficient of *the presence of a family member with major or chronic disease* is negative and significant at a 5% level, indicating that this variable had a significant negative effect on participation earnings; specifically, participation earnings were reduced by 0.034% if there was a family member with a major or chronic disease in the household. The possible reason for this is that if a household member is seriously or chronically ill, other household members may need to care for them, which leads to a reduction in the opportunity to participate in livelihood activities, and therefore, a reduction in participation earnings.

The coefficient of *whether a household labourer has non-farm production skills* is negative and significant at a 5% level, indicating that non-farm production skills had a significant negative effect on participation earnings. Specifically, if a household labourer had non-farm production skills, participation earnings were reduced by 6.6%. A possible reason for this is that if there is a household labourer with non-farm production skills, the labourer may be more inclined to choose non-farm livelihood activities and pay less attention to agricultural livelihood activities, so their participation in FCS projects will be relatively lower, and participation earnings will naturally be less.

The coefficient of *the number of times family members participated in labour skills training organised by the village (community) in a year* is negative and significant at the 5% level, meaning that the higher the number of labour skills training farmers participated in, the less their participation earnings. The possible reason is that the more actively farmers participate in labor skills training, the more proficient they may become in their labor skills, and the more diverse the range of labor projects they can choose to participate in. Therefore, for FCS sink projects that do not have a clear comparative advantage, their level of participation might be lower, and the benefits from participation may be fewer.

Fourth, after controlling for geographical factors, the coefficient of *the ease of borrowing of a household* is negative and significant at the 1% level, indicating that the ease of borrowing had a significant negative effect on participation earnings, meaning that the easier it was to borrow, the lower the participation earnings. The possible reason is that the easier it is for farmers to borrow money, the more likely they are to actively invest or participate in agricultural and non-agricultural projects with quicker results and higher earnings, thus reducing their attention to and participation in FCS project development, with lower participation earnings.

Fifth, after controlling for geographical factors, the relationships between farmers and their neighbours and between farmers and village cadres are positive and significant at the 5% level, indicating that the friendlier and closer the relationships between farmers and their neighbours, the lower their participation earnings in FCS projects, and the better the relationships with village cadres, the lower their participation earnings. The possible reason for this is that rural areas contain typical kinship and human societies, and the closer farmers are to their relatives and neighbours, the more likely they are to have access to various livelihood opportunities. Thus, FCS projects do not have comparative advantages, and therefore, may be relatively less attractive to farmers, resulting in fewer earnings from participation. Similarly, the better the relationship between farmers and village cadres, the more opportunities they have to access various resources, and when farmers enjoy many choices, FCS projects will lack comparative advantages and attractiveness, making them less motivated to participate.

The coefficient of *the distance between the home and the nearest market* is positive and significant at the 5% level, indicating that the distance had a significant positive effect on participation earnings. For every 1% increase in the distance between a farmer's home and the nearest market, participation earnings increased by 0.044%. This result is similar to a finding by Kemkes (2015), who showed that farmers whose homes are further away from the nearest marketplace are more motivated to participate in FCS projects. A possible reason for this is the longer the distance between farmers'home and the nearest market, i.e., the more remote they are, the fewer livelihood opportunities they have, the fewer their livelihood options, and the more attractive the FCS projects are to this group with their extremely limited options. Therefore, farmers living in more remote areas will generally participate more actively in FCS projects.

Sixth, farmers' earnings in Yunnan Province are lower than that in Sichuan Province at the 1% level. This indicates that all other things being equal, farmers in the Yunnan FCS project area received lower earnings from participation than those in Sichuan. Possible explanations are that the overall level of economic and social development of the project communities in Tengchong, Yunnan, is significantly higher than that of the project communities in Liangshan, Sichuan, and that farmers have a wider choice of livelihood activities and more opportunities to generate earnings. The project communities in Ganluo and Yuexi counties have a high incidence of poverty and a large number of poverty-stricken people. In other words, the opportunity cost for farmers to participate in the FCS project is higher, which leads to lower motivation and degree of participation in the FCS project, and naturally lower participation earnings.

In summary, after controlling for geographical differences, farmers' natural capital and physical capital had a significant positive effect on their participation in FCS projects, while human capital, financial capital, and social capital had a significant negative effect on their participation. That is, the more forestable land included in the FCS project and the more convenient the transportation near the household, the more earnings they obtained from participation. Farmers with relatively weak human capital, financial capital, and social capital received more labour earnings from the project instead. The main reason for this is that the CDM AR standard of *land eligibility* requires project plots to have been forest-free since 31st December 1989, and lands have to be adjacent, resulting in a large number of plots ineligible for the projects. On the other hand, FCS projects are unattractive to farmers with relatively good human capital, financial capital, and social capital, while the poverty-stricken people with fewer resources are more involved in short-term and arduous labour, and obtain relatively smaller labour earnings.

10.3 Summary

In this chapter, multiple linear regression models were used to analyse the effects of five forms of livelihood capital and their influence on farmer earnings in FCS projects. The results show that, after controlling for geographical differences, the natural and physical capital of farmers had a significant positive effect on their participation in the FCS projects, while human, financial, and social capital had a significant negative effect on participation. They indicate that in current carbon reforestation efforts, most of the poverty-stricken people are involved in short-term and arduous labour and obtain relatively meagre earnings; farmers with relatively greater natural capital get more asset earnings, and farmers with relatively rich natural and physical capital acquire more management and maintenance earnings.

To conclude, this study offers the following three insights. First, for poverty-stricken households with relatively disadvantaged human capital, financial capital, and social capital, it is necessary to increase technical training and assistance to ensure they obtain fairer and more reasonable participation earnings than before. Second, the improvement of community roads and communication infrastructure is essential for farmers to participate in FCS projects. Third, policies such as *Land Compliance* made to select sites for FCS projects should favour poverty-stricken communities, and land boundaries should be more flexible.

References

Awono, A., Somorin, O. A., Atyi, R. E., & Levang, P. (2014). Tenure and participation in local REDD+ projects: Insights from southern Cameroon. *Environmental Science & Policy, 35*, 76–86.

Cai, F., & Wang, M. Y. (2004). Changes in China's urban labour participation rate and it's policy implications. *Chinese Social Sciences (English Edition), 25*(4), 15–27.

Coulibaly-Lingani, P., Savadogo, P., Tigabu, M., & Per-Christer, O. (2011). Factors influencing people's participation in the forest management programme in Burkina Faso, West Africa. *Forest Policy and Economics, 13*(4), 292–302.

Ding, Y., & Ma, P. P. (2013). Research on forest carbon sequestration and economic development in ethnic minority areas in west Sichuan Province: A case study of Yuexi County, Liangshan Yi Autonomous prefecture, Sichuan Province. *Rural Economy, 5*, 38–41.

Dolisca, F., Carter, D. R., McDaniel, J. M., Shannon, D. A., & Jolly, C. M. (2006). Factors influencing farmers' participation in forestry management programmes: A case study from Haiti. *Forest Ecology and Management, 236*(2–3), 324–331.

Ellis, F. (2000). *Rural livelihoods and diversity in developing countries* (pp. 5–20). Oxford University Press.

Fang, Y., Fan, J., Shen, M., & Song, M.-Q. (2014). Sensitivity of livelihood strategy to livelihood capital in mountain areas: Empirical analysis based on different settlements in the upper reaches of the Minjiang River, China. *Ecological Indicators, 38*, 225–235.

Guo, X. L., Zhou, L. H., Chen, Y., & Yang, G. J. (2017). Impact of farmers' livelihood capital on livelihood strategies in typical desertification areas—An example from Hangjin County, inner Mongolia autonomous region. *Journal of Ecology, 37*(20), 6963–6972.

Hao, W. Y., Yang, D. S., Zhang, J., et al. (2014). A study on the relationship between sustainable livelihood capital and livelihood strategies of farmers and herders: An example from the Linzhi region of Tibet. *Arid Zone Resources and Environment, 28*(10), 37–41.

Hua, X., Yan, J., & Zhang, Y. (2017). Evaluating the role of livelihood assets in suitable livelihood strategies: Protocol for anti-poverty policy in the Eastern Tibetan Plateau, China. *Ecological Indicators, 78*, 62–74.

Ickowitz, A., Sills, E., & Sassi, D. C. (2017). Estimating smallholder opportunity costs of REDD+: A pantropical analysis from households to carbon and back. *World Development, 95*, 15–26.

Jiao, X., Pouliot, M., & Walelign, S. Z. (2017). Livelihood strategies and dynamics in rural Cambodia. *World Development, 97*, 266–278.

Kemkes, R. J. (2015). The role of natural capital in sustaining livelihoods in remote mountainous regions: The case of Upper Svaneti, Republic of Georgia. *Ecological Economics, 117*, 22–31.

Li, X. P., & Wang, M. (2014). Livelihood vulnerability in the framework of multidimensional poverty "Action-Structure" analysis: An empirical investigation and theoretical analysis based on the Wuling mountain region. *Journal of Centre China Normal University (Humanities and Social Sciences Edition), 5*, 1–9.

Li, Z. B., & Li, J. M. (2015). Research on barriers and countermeasures for participation of poor people in ethnic areas in tourism poverty alleviation. *Ethnic Forum, 6*, 15–21.

Liang, Y. C., Li, S. Z., & Li, C. (2011). The impact of non-farm participation on agricultural technical efficiency: A new explanation at the farmer level. *Soft Science, 25*(5), 102–107.

Meng, J. J., Aimu, Y. L., Liu, Y., et al. (2013). Research on the relationship between sustainable livelihood assets and livelihood strategies of farming and herding households: The case of Wuzheng County in Erdos City. *Journal of Peking University (Natural Science Edition), 49*(2), 321–328.

Moser, C. (1998). The asset vulnerability framework: Reassessing urban poverty reduction strategies. *World Development, 26*(1), 1–19.

Ndjondo, M., Gourlet-Fleury, S., Manlay, R. J., et al. (2014). Opportunity costs of carbon sequestration in a forest concession in central Africa. *Carbon Balance and Management, 9*(1), 4.

Oli, B. N., & Treueb, T. (2015). Determinants of participation in community forestry in Nepal. *International Forestry Review, 17*(3), 311–325.

Qiu, L. L., Yan, F., Paudel, K. P., & Koondhar, M. A. (2018). Influence of rural households' livelihood capital on income derived from participation in the forest carbon sequestration project: A case from the Sichuan and Yunnan provinces of China. *International Forestry Review, 20*(4), 538–558.

Rakatama, A., Pandit, R., Ma, C., & Iftekhar, S. (2017). The costs and benefits of REDD+: A review of the literature. *Forest Policy and Economics, 75*, 103–111.

Shi, Z. L., & Yang, Y. Y. (2012). Household endowment, household decision making and rural migrant labour return. *Sociological Research, 3*, 157–181.

Shrestha, S., & Shrestha, U. B. (2017). Beyond money: Does REDD+ payment enhance household's participation in forest governance and management in Nepal's community forests? *Forest Policy and Economics, 80*, 63–70.

Su, F., Pu, X. D., Xu, Z. M., & Wang, L. A. (2009). A study on the relationship between livelihood capital and livelihood strategy: A case study of Ganzhou District, Zhangye City. *China Population, Resources and Environment, 19*(6), 119–125.

Su, F., & Shang, H. Y. (2012). The influence of farmers' livelihood capital on their risk coping strategies-an example in Zhangye City. *Heihe River Basin. China Rural Economy, 8*, 79–96.

Sunderlin, W. D., de Sassi, C., Sills, E. O., et al. (2018). Creating an appropriate tenure foundation for REDD+: The record to date and prospects for the future. *World Development, 106*, 376–392.

Wan, D. S. (2005). Changes in the distribution of urban residents' income in China: 1988–1999. *Economics (Quarterly), 4*, 45–66.

Wang, Y. I. (2007). Research on the benefit mechanism of the poor in tourism poverty alleviation: A case study of ethnic areas in Sichuan. *Economic System Reform, 4*, 92–96.

Wu, Y. (2016). Research on the influence of livelihood capital on livelihood strategies of poor mountain farmers–based on survey data from Pingwu and Nanjiang counties in Sichuan province. *Agricultural Economic Issues, 3*, 88–94.

Wu, Z., Li, B., & Hou, Y. (2017). Adaptive choice of livelihood patterns in rural households in a farm-pastoral zone: A case study in Jungar, Inner Mongolia. *Land Use Policy, 62*, 361–375.

Xing, C. J., & Ge, Z. J. (2013). Concentrated contiguous poverty alleviation and development: Macro situation, theoretical basis and realistic choice: Analysis and reflection based on China's rural poverty monitoring and related results. *Guizhou Social Science, 5*, 123–128.

Yang, F., Paudel, P. K., Cheng, R. Z., et al. (2018). Acculturation of rural households participating in a clean development mechanism forest carbon sequestration programme: A survey of Yi Ethnic areas in Liangshan, China. *Journal of Forest Economics, 32*, 135–145.

Zeng, W. Z., Zhang, J. Y., & Yang, F. (2016). Forest carbon Sequestration for poverty alleviation: Theoretical discussion and realistic reflection. *Rural Economy, 5*, 17–22.

Zhang, L., Zhao, X. Y., Hou, C. C., et al. (2012). The impact of ecological compensation on the livelihood capital of farming households - an example of the Yellow River water recharge area in Gannan. *Glacial Permafrost, 34*(1), 186–195.

Zhao, W. J., Yang, S. L., & Wang, X. (2016). A study on livelihood capital and livelihood strategies based on logistic regression model—an example of dry hot river valley Dai ethnic areas in Xinping County, Yunnan. *Resource Science, 38*(1), 136–143.

Zhao, W. J., Yang, S. L., & Xu, R. (2015). The influence of livelihood capital on farmers' livelihood strategy choice in Yuanjiang dry and hot river valley area: an example from Xinping County *China Population Resources and Environment* (S2), 162–165.

Chapter 11
Analysis on the Functions of the Community in FCS Projects and the Factors Influencing These Functions

As a basic organisational unit, the community plays a key role in FCS projects. As most farmers live and farm in households in scattered plots, the transaction costs between carbon sink buyers and farmers are very high; it is necessary to bring together scattered farmers through the community, which is to organise them to participate in carbon sink transactions. In the process, the community plays a role in ensuring a project that benefits the poverty-stricken people. The community has a direct or indirect influence on the execution, organisation, post-maintenance, and revenue distribution of the project, and is an indispensable unit in carrying out PAFCS. The role of communities is determined mainly by their functions. This chapter defines both community and community functions and explores the mechanisms of community functions. According to the A (Adaption) G (Goal attainment) I (Integration) L (Latency pattern maintenance) model (Parsons, 1949), the community's function is explicated, while communities involved in the Novartis Sichuan FCS Project are selected as the study sample. Based on a field survey and case interviews conducted in the selected communities, the functions of the community are empirically analysed, and based on the results, relevant mechanisms are proposed, with a view to providing a reference for relevant policymakers and future researchers.

11.1 Theoretical Analysis of Community Functions

Communities play an important role in the operation of FCS projects, and a theoretical analysis of community functions is necessary. These functions are defined and categorised in this section. This part of the research is a prerequisite for the evaluation of community functions and the study of influencing factors.

W. Zeng and F. Yang, *Poverty Alleviation Via Forest Carbon Sequestration*, International Research on Poverty Reduction, https://doi.org/10.1007/978-981-99-8191-5_11

11.1.1 Community Functions

Generally, community service and political propaganda are the two community functions in their traditional sense. The community service function includes the provision of free services as public goods (e.g., community care), the provision of convenience services for community residents, and the provision of service platforms among community members. The political propaganda functions of the community concern political stability, policy publicity, and political participation. In addition to these two basic functions, the community functions can be divided into economic, political, cultural, and social functions according to the areas they are servicing (Zhang, 2010).

According to the characteristics of FCS projects and the communities involved, the community functions include promoting the execution of the projects and poverty alleviation; specifically, the functions of publicity, organisation, and execution to carry out an FCS project, as well as the functions of poverty alleviation that arise in these executions. Under the dual objectives of carbon sink generation/trading and poverty alleviation, communities also have a special function to benefit the poverty-stricken people. There is an intrinsic link between community function, community participation, and community role. However, the community functions are influenced by the natural intrinsic and extrinsic attributes of a community due to its economic, social, and cultural resource endowments; community participation is the main way in which community functions are revealed and exercised; community function is the nature of the community itself, and community roles are the result of community function.

In PAFCS projects, community functions are crucial to the achievement of the project's predetermined goals. In the current context of collective ownership of rural land and the fragmentation of land plots, targeting poverty alleviation depends on both land and the understanding of community farmers, and the execution and maintenance of the projects also depend on the organisation and assistance of the community. The PAFCS projects are organised by village cadres and by capable people (leaders) of the communities, thus reducing the transaction cost of FCS buyers and sellers, which plays a crucial role in the promotion and stable execution of the overall project.

Community characteristics and functions affect the development of PAFCS projects. The original cultural endowment of a community, especially the ecological culture and concept, will influence the execution of a project. Due to the poor transportation, relatively low information accessibility, and relatively low degree of interpersonal interaction in poverty-stricken areas, the community often forms a more closed circle within itself. The ecological and development cognitive concepts within the community affect the recognition of the FCS project by community members, which in turn affects whether the project can proceed successfully. The threshold conditions for FCS projects include an original natural resource endowment, adequate climate and soil conditions, sizable plots for FCS, and low cost of forestable land. Whether the community has an effective formal or informal organisation system, whether it has harmonious interpersonal relationships, and whether a consistent attitude can be formed are key to the smooth maintenance of an FCS project.

Another essential community function is to make sure FCS projects are to benefit the poverty-stricken people. Wise and Cacho (2005) found that the execution of carbon sink projects may be accompanied by unequal political games, in which women, landless people, poverty-stricken people, and indigenous people are often marginalised and excluded from projects; even if they are included, their legitimate interests are often embezzled, and their motivation is easily discouraged. The community, as the representative of local organisations, is to coordinate the interests of different parties so as to avoid *elite capture* and ensure that the poverty-stricken people can directly benefit from the improvement of labour skills and from labour income. In the long run, the infrastructure and organisational capacity of the community will be improved, which will help the poverty-stricken people escape poverty sustainably.

Through their combination, the community and the project benefit each other. The project will strengthen the community's functions, improve its ecological culture, organisational culture, and coordination skills, and strengthen its function of benefitting the poverty-stricken people. Through the improvement of the community function and others mentioned above, project leaders will no longer choose to negotiate with landowners in large areas while excluding small-scale landowners or landless foresters due to the increase in transaction costs (e.g., project initiation and execution and monitoring costs), and the project will be more oriented towards benefitting the poor and improving ecological environments both than before.

11.1.2 Composition of Community Functions

The AGIL functional analysis diagram was proposed by Parsons (1949), a professor at Harvard University. In Parsons' theoretical analysis, there are four basic functional elements that an agent's action system must satisfy. The first is the adaptation functional element, in which the system and the environment are closely related, and a certain influence relationship is bound to occur. The second is the goal attainment functional element. Any action system should have a certain goal, and the system must have the ability to determine the level and order of its goals and mobilise the system's internal dynamics to focus on achieving its goals. The third is the integration functional element. Any action is composed of various sub-parts, and in order to make the system function effectively as a whole, it is necessary to link these sub-parts so that the various parts can coordinate to achieve the goals. The fourth is the latency pattern maintenance functional element. If the system's operation is temporarily terminated, the original operation pattern must be preserved to ensure that the original system interaction can be restored when the system recovers. The coordination and stability of these four functional elements determine the effective functioning of the action system, and each subsystem can be infinitely divided into more detailed AGIL structural functions.

Parsons argued that from a functional perspective, the action system contains four functional subsystems, namely the behavioural organism, the personality system, the social system, and the cultural system. Among them, the social system, as a subsystem of the total action system, serves an integrative function in the action system, and its components simultaneously need to meet the functional conditions of AGIL. For example, a socioeconomic system serves an adaptive function, a political system serves a goal-achieving function, a legal system serves an integrative function (e.g. deterrence, regulation, and organisation), and family, education, and religion serve a potential pattern-maintaining function. There are two controlling relations within the action system, concerning information and energy. One is the relationship in which the higher-level information subsystem provides information guidance and teleological regulation for the lower-level information subsystem. For example, the cultural subsystem prescribes the formation of the social subsystem, the social subsystem serves the role of facilitating the adjustment of the personality subsystem, and the personality subsystem serves the role of facilitating the adjustment of the organic subsystem. Another is the relationship in which the higher-level energy control system provides the lower-level energy control system with the force of action and means of expression. For example, the organic subsystem provides energy for the personality subsystem, and the personality subsystem provides energy or power for the social subsystem.

From the perspective of Parsons' system theory, the community, as an important unit in PAFCS, will meet the four functional requirements of AGIL and play a corresponding functional role. The community adjusts and adapts to the FCS project through personnel organisation, institutional facilities, and revenue distribution, thus serving the A function (adaptive function) of the community. The G function (goal-achieving function) of the community is performed through community rules and regulations, commonly accepted codes of conduct, and community leaders or community cadres. The I function (integration function) of the community is achieved through the community's cultural environment and organisational system, which enable the community to reduce transaction costs and promote the project. As the community reaches consensus and coordinates interests via its common cultural customs and values, it is serving the L function (model maintenance function).

To serve its adaptation function, the community in an FCS project needs to be composed of four elements. One is a certain number of community members, with a certain ability to participate in the project. The second is the availability of specific natural resources, mainly land suitable for the project. Third, there is a certain infrastructure, such as basic water irrigation and transportation infrastructure for the project. Fourth, there is a basic community identity. It is essential to make full use of community organisation resources, social network resources, human resources, and community culture and customs in project design, execution, maintenance, and monitoring to strengthen the sustainability of participation. Project goals are to be clarified at the early stage to ensure social, economic, and environmental benefits (Fig. 11.1).

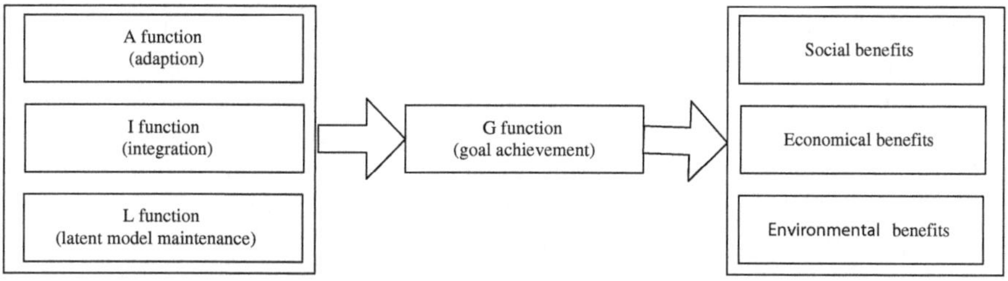

Fig. 11.1 Composition of community functions

11.2 Evaluation of Community Functions

Whether FCS projects can be developed is closely related to whether they can alleviate poverty (Zhang & Huang, 2019), while the pro-poor function of FCS projects is, in turn, inextricably linked to their performance in community participation, creation, and governance. Community participation plays an important role in this process, as it enhances the willingness and motivation of farmers to participate, targets poverty-stricken people and vulnerable groups, and gives them encouragement and assistance without harming the interests of non-poor people, thus enhancing the sustainability of FCS projects while protecting the environment and improving the investment efficiency of the FCS projects. This study selected communities involved in the Novartis Sichuan FCS Project as the study sample and adopted a standardised approach to quantitatively assess their functions.

11.2.1 Overview of Communities

Communities involved in the Novartis Sichuan FCS Project were selected as the study area. This FCS project was launched in 2010, and the project owner is the Sichuan Dadu River Afforestation Bureau. The FCS purchaser is the Novartis Group of Switzerland, and the project is registered with the UN CDM-EB, which is the first afforestation and carbon reduction project to directly cooperate with domestic and foreign enterprises in China. The data in this section were obtained from a field questionnaire and case interviews conducted in 26 communities. It should be noted that 27 AR communities were originally selected, but the Bashi village in Wanchang Township was excluded in consideration of various factors such as geography, climate, and soil.

The researchers selected five counties and three nature reserves in the state of Liangshan Yi Autonomous Prefecture in southwest Sichuan Province, which include 16 townships and 26 rural communities in Zhaojue County, Yuexi County, Yuexi Shenguozhuang Nature Reserve, Meigu County, Leibo County, Leibo Mamize Nature Reserve, Ganluo County, and Ganluo Maanshan Nature Reserve. These 26 communities are located in the southwestern part of Sichuan Province in the upper reaches of the Yangtze River, in the transverse mountain range from the Qinghai-Tibet Plateau to the Yunnan-Guizhou Plateau, with complex topography, annual average precipitation of 800–1350 mm, annual average temperature of 7.5–15.5 °C, and elevation of 305–5958 m. The communities have good neighbourly relations and high cohesion and rely on and trust the local government. There is an abundant labour force in the communities and relatively few migrant workers. Details for these 26 communities in 2016 are summarised in Table 11.1.

11.2.2 Evaluation of Community Functions

In order to comprehensively evaluate the functions and roles undertaken by the sample communities in the Novartis Sichuan FCS Project, the researchers decomposed the community functions into organisation and publicity function, revenue distribution function, and project maintenance function and quantified the functions.

11.2.2.1 Selection of Indicators

Although the main purpose of FCS projects is to increase carbon sinks and improve the ability of forestry to mitigate climate change in order to build an ecological environment, the fundamental reason they are sought after by various communities is that they contribute to the establishment of a mechanism to benefit the poverty-stricken people. Therefore, with reference to relevant literature on carbon sink poverty alleviation, green poverty alleviation, and sustainable development theory, and taking into account the characteristics of communities, the researchers divided the functions of FCS projects into three dimensions: organisation and publicity function, revenue distribution function, and project maintenance function.

First, the organisation and publicity function is one of the most basic functions of communities in FCS projects, referring to the overall cohesion and mobilisation ability and the propaganda ability of participating communities after the implementation of an FCS project. Considering the feasibility of field research and the specific requirements of FCS, this function consists of three secondary indicators: community cohesion, community awareness of FCS, and the number of community participants in the project.

Second, the revenue distribution function concerns whether the FCS project can generate economic benefits. This concerns the number of poverty-stricken people benefitting from the project and the way and extent they benefit. The secondary indicators are the number of poverty-stricken people being benefitted, temporary work opportunities, net labour income, and net income from timber.

Third, the project maintenance function refers to the community's ability to maintain the project after its implementation, specifically the ability to ensure environmental protection of the region and protection of the interests of farmers and herdsmen, especially in the context of the conflict between the grazing of farmers

Table 11.1 Data of 26 communities (2016)

Community	Groups	Ownership	Forest (ha)	Number of labourers	Average annual temperature (°C)	Average annual precipitation (mm)	Annual income per capita (RMB)	Temporary job opportunities
Heshamu	0	State	147.1	415	10.9	1020.7	3840.1	27,860
Kumo	0	State	75.2	48	10.9	1020.7	3327.1	14,351
Aerbagu	0	State	212.8	347	10.9	1020.7	3643.7	40,977
Mugejuemu	3	Collective	181.3	200	13.3	1113.2	3846.4	35,132
Yierjuemu	2	Collective	300.8	160	13.3	1113.2	3582.5	58,871
Hongqi	4	Collective	104.7	390	13.3	1113.2	3292.2	18,982
Zhuluo	3	Collective	151.3	210	13.3	1113.2	3132.9	29,372
Pingqiao	2	Collective	121.0	205	13.3	1113.2	3181.6	25,781
Shaku	3	Collective	81.2	340	13.3	1113.2	3969.6	18,003
Habu	2	Collective	80.0	270	13.3	1113.2	3476.8	17,463
Laiwu	4	Collective	162.5	305	13.3	1113.2	3292.2	35,475
Magu	3	Collective	62.2	210	13.3	1113.2	3122.6	11,878
Weilejue	5	Collective	219.4	340	11.4	814.6	3013.2	39,774
Huowo	6	Collective	290.0	230	11.4	814.6	3210.1	52,575
Nilejue	6	Collective	222.2	400	11.4	814.6	3476.8	44,263
Dagudui	5	State	537.9	800	13.8	850.1	3530.4	97,512
Minzu	2	State	316.2	98	13.8	850.1	3486.4	62,329
Hamu	2	Collective	96.1	600	14.1	1268.7	2950.4	26,673
Meisan	4	Collective	47.0	980	14.1	1268.7	2861.6	12,239
Ashamo	7	Collective	43.9	1000	14.1	1268.7	3076.8	18,830
Qialuo	4	Collective	147.5	600	14.1	1268.7	2092.8	44,244

(continued)

Table 11.1 (continued)

Community	Groups	Ownership	Forest (ha)	Number of labourers	Average annual temperature (°C)	Average annual precipitation (mm)	Annual income per capita (RMB)	Temporary job opportunities
Maqianmen	2	State	110.3	124	14.1	1268.7	3692.7	23,246
Sanpiyan	2	Collective	101.4	300	14.1	1268.7	3230.4	21,370
Shishier	3	Collective	236.0	90	14.1	1268.7	2849.6	102,360
Shuangmuc	4	Collective	100.4	121	14.1	1268.7	2956.6	43,292
Laer	4	Collective	41.7	700	14.1	1268.7	2708.9	9854

Table 11.2 Indicators of community functions in FCS projects

	Primary indicator	Secondary indicator	Code	Indicator explanation
Community functions	Organisation and publicity	Community cohesion	S11	Situation of community cohesion (0–5)
		Awareness of FCS	S12	Community awareness of FCS projects (0–5)
		Participants	S13	Number of people in the community participating in the FCS project (people)
	Revenue distribution	Beneficiaries	J11	Number of poor people benefitting from FCS projects in the community (people)
		Job opportunities	J12	Number of temporary job opportunities created by FCS projects for the community (person)
		Net income	J13	Total annual net income of labour force involved in FCS projects in the community (RMB)
		Income from timber	J14	Total annual net income from inter-harvested and main logging timber from FCS projects (RMB)
	Project maintenance	Degree of soil erosion	H11	Area of soil erosion/total land area of the community * 100%
		Conflict resolution of grazing and project	H12	Number of grazing/major livestock number in project area * 100% (cattle, sheep, horses)
		Net absorption of greenhouse gas	H13	Net absorption of greenhouse gas by FCS projects (tons CO_2-e)

Note Community cohesion and community awareness of FCS were measured using the Likert-5 scale, with increasing degrees from 1 to 5; degree of soil erosion is an inverse indicator, and all other indicators are positive

and herdsmen and FCS. It is necessary to focus on the community's ability to resolve conflicts. Therefore, the secondary indicators of the project maintenance function consist of conflict resolution between grazing and the project (obtained by dividing the number of grazing animals in the project area by the number of livestock there), soil erosion, and net absorption of greenhouse gas (Table 11.2).

11.2.2.2 Data and Methodology

The entropy value method was adopted to assign weights to the indicators in order to calculate the comprehensive function of the community involved in FCS projects.

Considering that there are 10 secondary indicators in the evaluation indexes, with different metrics respectively, the Max–Min (i.e., maximum–minimum) method was adopted to process the data in a dimensionless manner so that the accuracy of the results was not affected if the indicators were measured directly. The processing methods are as follows.

$$P_{ij} = \frac{x_{ij} - \min x_j}{\max x_j - \min x_j} \quad P_{ij} = \frac{\max x_j - x_{ij}}{\max x_j - \min x_j} \tag{11.1}$$

In formula 11.1, x_{ij} and P_{ij} denote the original value of the *j*-th index of the *i*-th community and its standardised value, while $\min x_j$ and $\max x_j$ denote the minimum and maximum values of the *j*-th index, respectively.

On this basis, the weights were further determined by the entropy value method, and the main calculation method is as follows:

$$H_j = -k \sum_{j}^{n} (f_{ij} \ln f_{ij}) \tag{11.2}$$

$$W_j = \frac{(1 - H_j)}{\sum_{j=1}^{m} (1 - H_j)} \tag{11.3}$$

In the above two formulas, H_j is the information entropy, W_j is the entropy value, $k = 1n\, n^{-1}$, $f_{ij} = \frac{Y_{ij}}{\sum_{j=1}^{n} Y_{ij}}$, Y_{ij} is the standardised value, and when $f_{ij} = 0$, $f_{ij}\ 1n\ f_{ij} = 0, 0 \leq W_j \leq 1 \Delta \sum_{j=1}^{m} W_j = 1$.

11.2.2.3 Function Evaluation

The observed data for each indicator of the community function were calculated using the entropy method, resulting in the overall evaluation score of functions and sub-functions of 26 communities in southwest Sichuan in 2013 and 2016 (see Tables 11.3 and 11.4 for details).

The results in Table 11.3 show that the functions of 26 communities were efficient, while it can be seen that after the execution of the FCS projects, the total function score in 2016 was 8.58, which is significantly higher than the 7.81 in 2013. The highest total function score in 2016 was 0.55 and the lowest was 0.14, while the highest score in 2013 was 0.57 and the lowest was 0.12. The function score of communities such as Dagudui, Shuangmachao, and Yierjuemu increased and showed an upward trend in ranking, with total function scores above 0.3; communities such as Shaku and Hamu also exhibited increased scores but had slight decreases in ranking; communities such as Huowo and Weilejue had relatively stable total function scores.

Table 11.3 The evaluation of functions of 26 communities

Community	2013		2016	
	Score	Ranking	Score	Ranking
Sum	7.81	–	8.58	–
Heshamu	0.25	19	0.29	17
Kumo	0.26	17	0.23	22
Aerbagu	0.35	9	0.42	6
Mugejuemu	0.33	11	0.38	8
Yierjuemu	0.40	5	0.46	4
Hongqi	0.23	20	0.34	11
Zhuluo	0.25	18	0.20	24
Pingqiao	0.16	24	0.25	21
Shaku	0.19	22	0.23	23
Habu	0.12	26	0.25	20
Laiwu	0.37	7	0.20	25
Magu	0.20	21	0.14	26
Weilejue	0.32	12	0.34	12
Huowo	0.27	14	0.32	14
Nilejue	0.37	6	0.32	13
Dagudui	0.53	2	0.55	1
Minzu	0.34	10	0.39	7
Hamu	0.28	13	0.30	15
Meisan	0.35	8	0.43	5
Ashamo	0.26	16	0.35	10
Qialuo	0.42	3	0.47	3
Maqianmen	0.14	25	0.30	16
Sanpiyan	0.18	23	0.27	18
Shishier	0.57	1	0.35	9
Shuangmuchao	0.41	4	0.52	2
Laer	0.26	15	0.26	19

As Table 11.4 indicates, the scores for the organisation and publicity function, revenue distribution function, and project maintenance function of each community fluctuated from 2013 to 2016 but showed an overall upward trend, especially the project maintenance function, which changed very significantly, indicating that communities' ability to maintain their FCS project became prominent. In terms of the organisation and publicity function, the total scores in 2016 decreased slightly compared with 2013, and the scores of each community fluctuated slightly (with little ups or downs), which may be due to the fact that after the execution of the project, the preliminary role of publicity weakened with the progress of the project. In terms

Table 11.4 The evaluation of sub-functions of 26 communities

Community	Organisation and publicity				Revenue distribution				Project maintenance			
	2013		2016		2013		2016		2013		2016	
	Score	Ranking	Score	Ranking	Score	Ranking	Score	Ranking	Score	Ranking	Score	Ranking
Sum	2.58	–	2.23	–	3.11	–	3.54	–	2.13	–	2.76	–
Heshamu	0.08	18	0.06	18	0.09	15	0.12	15	0.08	11	0.11	10
Kumo	0.12	8	0.03	26	0.10	13	0.12	15	0.04	24	0.08	20
Aerbagu	0.11	10	0.07	15	0.16	6	0.20	5	0.08	11	0.15	3
Mugejuemu	0.10	13	0.14	1	0.15	8	0.10	18	0.08	11	0.13	6
Yierjuemu	0.11	10	0.09	12	0.18	3	0.21	2	0.11	5	0.16	2
Hongqi	0.07	19	0.07	15	0.12	11	0.16	8	0.04	24	0.11	14
Zhuluo	0.09	15	0.08	14	0.08	17	0.08	22	0.08	14	0.04	25
Pingqiao	0.01	26	0.13	5	0.06	23	0.06	23	0.09	7	0.07	22
Shaku	0.06	20	0.05	21	0.07	21	0.09	21	0.06	21	0.09	15
Habu	0.03	24	0.10	11	0.05	24	0.04	26	0.03	26	0.08	19
Laiwu	0.18	2	0.05	20	0.11	12	0.10	18	0.08	15	0.05	24
Magu	0.10	14	0.05	24	0.04	26	0.06	23	0.06	20	0.03	26
Weilejue	0.14	6	0.11	8	0.10	14	0.12	15	0.08	9	0.11	10
Huowo	0.09	16	0.07	15	0.07	21	0.10	18	0.11	4	0.15	3
Nilejue	0.19	1	0.14	3	0.09	15	0.06	23	0.09	6	0.13	8
Dagudui	0.16	4	0.13	4	0.16	6	0.16	8	0.21	1	0.26	1
Minzu	0.05	22	0.11	7	0.17	4	0.19	6	0.12	2	0.09	18
Hamu	0.06	21	0.05	21	0.15	8	0.18	7	0.07	16	0.07	21
Meisan	0.13	7	0.11	8	0.15	8	0.21	2	0.06	19	0.11	10

(continued)

Table 11.4 (continued)

Community	Organisation and publicity				Revenue distribution				Project maintenance			
	2013		2016		2013		2016		2013		2016	
	Score	Ranking	Score	Ranking	Score	Ranking	Score	Ranking	Score	Ranking	Score	Ranking
Ashamo	0.12	9	0.11	8	0.08	17	0.13	14	0.06	22	0.11	10
Qialuo	0.15	5	0.12	6	0.19	2	0.21	2	0.08	10	0.14	5
Maqianmen	0.02	25	0.05	21	0.05	24	0.16	8	0.07	17	0.09	15
Sanpiyan	0.05	23	0.04	25	0.08	17	0.14	12	0.06	23	0.09	15
Shishier	0.08	17	0.08	13	0.38	1	0.15	11	0.12	3	0.12	9
Shuangmuchao	0.17	3	0.14	2	0.17	4	0.25	1	0.07	18	0.13	7
Laer	0.10	12	0.06	19	0.07	20	0.14	12	0.09	8	0.07	23

of the revenue distribution function, the total score of its corresponding community function increased from 3.11 in 2013 to 3.54 in 2016, and the revenue distribution function of each community exhibited an increasing trend, indicating that with the execution of the FCS project, its influence on the distribution of community revenue increased. In terms of the project maintenance function, the total score of its function increased from 2.13 in 2013 to 2.76 in 2016, with an obvious rapid growth rate, which shows that the carbon sink project played a more important role.

11.3 Analysis on the Factors Influencing Community Functions

To reveal the influence of community resource endowment and the socioeconomic resources of FCS projects on community functions, the researchers conducted a quantitative study using regression analysis and found a U-shaped relationship between a community's resource endowment and its functions, while the degree of trust in government, the degree of community integration, and the degree of government support for the project each had a positive impact on community functions.

11.3.1 Indicators and Methods

As was seen in the previous section, the various functions of communities in FCS projects are conducive to the smooth execution of the projects. How can we promote FCS projects more effectively to maximise their benefits for communities? In previous studies and community governance theories, the factors affecting community functions are often divided into two categories. One is community resource endowment (Liu et al., 2019). These 26 communities are all located in the southwestern part of Sichuan Province, with complex climate types and ecological environments, and numerous community resource endowments can have an impact on a community's functions. Variables such as temperature, precipitation, and labour force size were selected to measure the impact of resource endowments. The second category is community socioeconomic status (Fang & Gao, 2012). The community's trust in the government and the community's dependence on the livestock industry will affect its function. The external environment will also play a role in the execution of the project; thus, indicators such as the degree of integration a community, the government support for FCS, and a community's dependence on the livestock industry were selected to measure the influence of community socioeconomic status on the community functions.

A multiple linear regression model was adopted to explore the factors influencing a community's functions. Considering that there are many influencing factors, some of which may have a non-linear relationship with a community's functions, all variables were log-transformed and standardised in order to eliminate the influence of differences in magnitude. Considering also that some variables, such as temperature and precipitation, may have an 'inverted U'-shaped curve relationship with explained variables, the quadratic terms of these variables were included in the model to eliminate the influence of the difference. The specific model settings are shown in the following equation:

$$LnY_i = \beta_0 + \beta_{1n} LnX_{1ni} + \beta_{2n} LnX_{2ni}^2 + \alpha LnZ_{it} + \varepsilon_{it} \quad (11.4)$$

In formula 11.4, LnY_{i} is the explained variable, representing the function of 26 communities, and the subscript i indicates the i-th community; $LnX_{1\mathrm{ni}}$ represents the explanatory variables, including the resource endowment (temperature, precipitation, etc.) and socioeconomic status (i.e., degree of integration of a community, degree of government support for FCS, etc.) of community I; $LnX_{2\mathrm{ni}}^2$ represents the squared term of the explanatory variables, which include temperature, precipitation, and sunshine hours; Z_{i} represents the control variables, which include the number of community groups, land ownership, land use rights, etc., and ε_{it} is a random disturbance term. The specific variable selection is shown in Table 11.5 (the squared terms are not listed).

11.3.2 Empirical Results

This section focuses on the empirical analysis of the data in 2016 with the help of Stata 15.0. The regression results are shown in Table 11.6.

According to the regression results, arable land area, temperature, sunshine hours, labour force (in the variable of community resource endowment), trust in government, intra-community togetherness, government support for the project, and livestock reliance (in the variable of community socio-economic status) had significant effects on community function. The three variables of temperature, sunshine hours, and government support for the project had an especially significant effect on overall community function. The effects of their benefits were very significant, indicating that suitable temperature, sunshine, and government support can greatly influence a community's function in FCS projects.

Among community resource endowments, temperature and sunshine hours displayed an obvious U-shaped relationship with community function in FCS projects, with coefficients of 0.122 and 1.234, respectively, indicating that certain degrees of temperature and sunshine are beneficial to the overall function of FCS projects, though too high a temperature and too much sunshine are not conducive to the growth of trees, and projects may fail. The coefficient between the size of the labour force and community function was 0.178, which shows a significant positive correlation, indicating that the larger the labour force, the greater its professional knowledge and skills, and the easier it is for a community to mobilise farmers to

Table 11.5 Indicators in the regression

Variables		Definition and value	Specific indicator
Explained variable	Community function	Function (Y)	Value of function
Explanatory variable	Resource endowment	Arable land (X_{11})	Community arable land area (ha)
		Temperature (X_{12})	Average annual temperature (°C)
		Precipitation (X_{13})	Average annual precipitation (mm)
		Sunshine hours (X_{14})	Average annual sunshine hours (hours)
		Frost-free period (X_{15})	Frost-free period (days)
		Number of labourers (X_{16})	Total labour force in the community (persons)
		Grazing land (X_{17})	Area of land suitable for grazing (ha)
	Socio-economic status	Population distribution (X_{21})	Degree of aggregation of community population distribution (1–5)
		Trust in government (X_{22})	Degree of trust in government by community (1–5)
		Community rapport (X_{23})	Degree of rapport within community (1–5)
		Government support (X_{24})	Degree of government support for project (1–5)
		Livestock dependence (X_{25})	Degree of community dependence on livestock (1–5)
Control variable		Group numbers (Z_{11})	Number of community village neighbourhoods (number)
		Land ownership (Z_{12})	State = 1; Collective = 2; Individual = 3
		Land-use rights (Z_{13})	State = 1; Collective = 2; Individual = 3

Note Population distribution, degree of trust in the government, degree of integration of a community, degree of government support for the project, and degree of livestock dependence were measured using the Likert-5 scale, in increasing order of degree from 1 to 5

Table 11.6 Regression results

	B	Std.error	Sig.
G	0.453**	1.389	0.042
$\mathrm{Ln}X_{11}$	−0.078*	0.298	0.099
$\mathrm{Ln}X_{12}$	−0.237	1.238	0.167
$\mathrm{Ln}(X_{12})^2$	0.122***	2.943	0.002
$\mathrm{Ln}X_{13}$	0.637	1.716	0.165
$\mathrm{Ln}\ (X_{13})^2$	0.765	2.348	0.234
$\mathrm{Ln}X_{14}$	−0.415	1.145	0.289
$\mathrm{Ln}\ (X_{14})^2$	1.235***	2.145	0.000
$\mathrm{Ln}X_{15}$	0.768	1.916	0.698
$\mathrm{Ln}X_{16}$	0.178*	0.303	0.072
$\mathrm{Ln}X_{17}$	0.115	0.388	0.773
$\mathrm{Ln}X_{21}$	-0.175	0.426	0.691
$\mathrm{Ln}X_{22}$	0.546**	0.545	0.043
$\mathrm{Ln}X_{23}$	0.738**	0.396	0.035
$\mathrm{Ln}X_{24}$	0.174***	0.326	0.007
$\mathrm{Ln}X_{25}$	−0.754**	0.474	0.046
$\mathrm{Ln}Z_{11}$	0.017*	0.487	0.073
$\mathrm{Ln}Z_{12}$	1.175	1.087	0.308
$\mathrm{Ln}Z_{13}$	1.425	1.110	0.231
R^2	0.851	F value	32.216

Note ***, **, and * indicate 1%, 5%, and 10% significance levels, respectively

participate in the project. The coefficient of arable land with community function was –0.078, which shows a negative correlation, probably because cultivated land squeezed out afforestation areas, making the number of forest trees insufficient to highlight the advantages of the project. The coefficient of precipitation, frost-free period, and community suitable grazing land with dependent variable were not significant, probably because precipitation, as well as frost period conditions in these 26 communities, are relatively stable, and the grazing land area accounts for a small proportion of the total community area, thus having little impact on an FCS project.

In the community socioeconomic status variable, the coefficients of trust in the government, degree of integration of a community, and degree of government support for the project with the dependent variable were 0.546, 0.738, and 0.174, respectively, showing significant positive correlations indicating that communities with higher trust in the government, more integration, and higher government support for the project were more likely to believe that FCS projects would bring benefits. This is also related to the current grassroots governance structure in China. The coefficient of the degree of livestock reliance with a dependent variable was −0.754, probably because these 26 communities have more dependence on livestock farming than elsewhere and are prone to forest-pastoral conflicts. Communities that are more dependent on livestock farming will experience low project benefits. The relationship between the degree of community aggregation and community function was not significant, probably because the overall distribution of these 26 communities is relatively uniform.

11.4 Summary

Communities in FCS projects depend on human and material resources at the village level. The village community, village cadres, and capable people or leaders help reduce the transaction costs of FCS buyers and sellers, and they play a crucial role in the promotion and smooth execution of the project. The community requires the composition of corresponding elements to serve its functions. The first is a certain number of community members with a particular level of participation capacity. The second is the availability of specific natural resources, mainly land suitable for the project. Third, there is a certain infrastructure, such as basic water irrigation and transportation infrastructure, for the project. Fourth, there is a basic community identity.

The community function guarantees the execution of projects. In the sample analysis, the overall level of community function rose; the total level of function in 2016 was significantly higher than that in 2013, which indicates that the community functions were brought into play, and they helped improve the social, economic, and environmental levels of the communities. The community function is consistent with the goal of the overall FCS project and positive for the execution of the project. From 2013 to 2016, the values for the organisation and publicity function, revenue distribution function, and project maintenance function of each community fluctuated, but the overall trend was up, especially for the revenue distribution function and project maintenance function, which increased very significantly, indicating that the function of the communities gradually changed from organisation and publicity to project maintenance as FCS projects advanced. This also indicates that communities have a direct or indirect influence on project execution, organisation, post-project maintenance, and revenue distribution and that they are indispensable units in the process of PAFCS.

The positive function of the community depends on the trust of the government, the interaction between the government and the community, and the resource endowment of the community itself. The study results show that communities with suitable forest land, temperature, and sunshine are conducive to the execution of FCS projects, and the larger the community workforce, the easier it is to drive projects to achieve the corresponding goals. In addition, communities with a higher level of trust and a higher degree of integration of a community and a higher level of government support for the project will have a long-term relationship of trust, and be more likely to believe that FCS projects will bring benefits and accept carbon sink projects.

References

Fang, S. L., & Gao, L. (2012). Research on the influencing factors and countermeasures of forest carbon sequestration projects in Guangdong. *Jiangsu Agricultural Science, 40*(11), 6–8.

Liu, Z., Wang, L. G., Cheng, S. Y., & Huang, J. X. (2019). Analysis of tourists' willingness to compensate for carbon and its influencing factors in forest tourism sites based on BRT–Taking Mingyue Mountain National Forest Park as an example. *Journal of Northwest Forestry College, 34*(02), 267–272.

Parsons, T. (1949). *The structure of social action[M]*. Free press.

Wise, R., & Cacho, O. (2005). A bioeconomic analysis of carbon sequestration in farm forestry: A simulation study of Gliricidia sepium. *Agroforestry Systems, 64*(3), 237–250.

Zhang, J. Z. (2010). The community participation in the perspective of community function. *Journal of Zhengzhou Institute of Aviation Industry Management (Social Science Edition), 29*(5), 125–130.

Zhang, Y., & Huang, Y. L. (2019). Do forest carbon sequestration projects help reduce poverty? *Forestry Economic Issues, 1*, 71–76.

Chapter 12
Analysis of Community Elite Willingness to Organise FCS Projects

The participation of community elites is often indispensable to FCS projects, from the selection of project plots, background survey, execution and promotion, verification and certification, as well as conflict resolution and dispute mediation. Their efforts are needed to maximise community interests. Community elites are not only important participants and promoters of FCS projects but also effective coordinators and supervisors who talk to government and forestation entities externally and communicate with community residents internally. To reduce the development cost of FCS projects and promote sustainability and poverty alleviation, efforts should be made to mobilise community elites. To this end, this chapter draws on elite theory and social exchange theory to analyse the willingness of rural elites to organise community farmers to participate in FCS projects and the factors influencing this process, with a view to providing useful insights to promote the transformation of community elites from participants in projects to promoters of poverty alleviation.

12.1 Identification of and Theoretical Hypotheses for Factors Influencing the Willingness of Community Elites to Organise

The term *elite* originated from France in the seventeenth century, meaning the *selected few* or *excellent people* (Wang, 2009). Pareto, a pioneer of modern elite theory, referred to the most powerful, capable, or prestigious part of a class or hierarchy as the elite (Pareto, 2010). He believed that human history is a history of continuous elite turnover, with declining elites gradually retreating from the stage and emerging elites coming to the stage. This theory is built on social heterogeneity; that is, unequal distribution of social resources in a society leads to the phenomenon of a minority ruling, organising, leading, and coordinating the majority in political, economic, and social fields. The social basis for the existence of elite governance is said to include

W. Zeng and F. Yang, *Poverty Alleviation Via Forest Carbon Sequestration*, International Research on Poverty Reduction, https://doi.org/10.1007/978-981-99-8191-5_12

psychological needs, organisational requirements, and the social division of labour (Wang, 2014). Its rationality lies in advantages in cost, efficiency, and order (Wang, 2013). Since ancient times, rural elites have been an important force in the economic and social development of rural China (Huang & Liu, 2013). Their strengths are reflected in three ways: first, they are more capable in their own development; second, they possess more resources; and third, they can exert greater influence on other individuals within a group.

In recent years, there has been increasingly widespread attention regarding the positive role of rural elites in poverty alleviation, especially in industrial poverty alleviation. As some Chinese researchers have argued, rural elites have superior resource endowments, more human, economic, social, and political capital and market experience than others do as well as prestige and social status in the community, and they can influence industrial poverty alleviation effectively (Fan, 2016; Su & Guo, 2018). Zeng and Zeng (2017) found through a case study that rural elites can have a good understanding of local intrinsic resources and cultural heritage and play a unique role in enhancing the self-development capacity of poverty-stricken individuals through industrial cooperation, social capital sharing, and changing the perceptions of the poverty-stricken people. Si (2017) examined the poverty alleviation model in Wuzhou of Guangxi from the perspective of elitism and found that poverty alleviation there centred on an approach in which the elites mobilised ordinary villagers, added value to social capital, and effectively integrated poverty alleviation forces to form a synergy. Ma (2012), in a study on FCS and the development of poverty-stricken minority areas in Western Sichuan, concluded that community elites build a community's self-development capacity and get involved in project promotion, participation, management, and supervision, which can reduce development costs and guarantee the smooth execution of FCS projects.

This study draws on social exchange theory, which is based on the core idea that the basic mode of human interaction is the exchange of material and social resources (Cook et al,. 2013). According to contemporary exchange theory, rural elites are willing to organise villagers to participate in FCS projects because, through organisational activities, rural elites can obtain *optimal value*, including both monetary benefits and non-monetary ones, e.g., social capital accumulation, network expansion, and interpersonal relationship optimisation. Only when rural elites realise that organisational behaviour can bring *optimal value* to them are they willing to engage in organisational behaviour; otherwise, they are not. This is also consistent with the hypothesis of the rational economic man. By inference, demographic, human capital, family, project, social capital, institutional, and elite characteristics will influence the value judgements and decision-making behaviours of rural elites. Based on the above analysis, a number of research hypotheses are proposed.

Hypothesis 1: Demographic characteristics affect rural elite willingness to organise FCS projects

Numerous theoretical and empirical studies have shown that individual behavioural decisions are closely related to demographic characteristics. Rural elites too are influenced by gender and age when making decisions on whether they are willing to organise villagers to participate in FCS projects. Generally speaking, men are physically stronger than women and would respond to the new in a more objective and stable way than women (Tong, 2014). Generally, in rural areas of China, men dominate the external affairs of the family, while women's main job is to take care of the lives of family members. Therefore, men may be more willing to organise than women. In addition, as people get older, their physical abilities gradually decline, their willingness to accept new things and their desire to pioneer and innovate diminishes, and their risk aversion increases. Accordingly, the older the rural elite, the weaker their willingness to organise will be.

Hypothesis 2: Human capital characteristics affect rural elite willingness to organise FCS projects

Human capital generally refers to knowledge, experience, information, and various skills and abilities. The human capital of rural elites refers to the sum of various competencies resulting from their education, training, health, and migration mobility. Specifically, the human capital of rural elites consists of their education level, occupational skills, health status, and work experience. For this reason, three variables, namely education level, forestry-related work experience, and health status, were identified as forms of human capital. In general, rural elites with higher education level, forestry-related work experience, and good health will be more willing to organise.

Hypothesis 3: Family characteristics affect rural elite willingness to organise FCS projects

It has been shown that family characteristics had a significant effect on the willingness of rural elites to participate in the Chinese New Rural Areas project, in which they led villagers to transform the farming way and become rich (Tang & Zhao, 2009; Zheng, 2011). In this study, two variables, source of family income and strength of family support, were selected to analyse the organisational intentions of rural elites. The fact that family income is derived mainly from agriculture indicates that rural elite families are more dependent on agricultural income, including earnings from FCS projects, and thus, their willingness to organise and participate may also be stronger. Rural elites cannot organise public affairs without family support, so the more supportive their family, the more likely they will be to organise.

Hypothesis 4: Project characteristics affect rural elite willingness to organise FCS projects

Project characteristics refer to the benefits and costs that rural elites perceive in their decision on whether to organise. Two variables were selected for analysis: perceived benefits and perceived feasibility. Rural elites are willing to encourage villagers to participate in FCS projects because the projects can benefit both villagers and rural elites; villagers can gain monetary and material benefits; and rural elites can expand their social relationship with forestation entities (enterprises). Other considerations include that they can enhance mutual trust with villagers and improve their social capital in addition to gaining monetary and other material benefits. Therefore, the perception of project benefits will influence rural elite willingness to organise: the higher the perceived project benefits, the higher the willingness to organise. FCS projects are more difficult than general afforestation projects, and rural elite perceptions of the difficulty of project execution will also influence their decisions. Whether villagers can complete the project according to the standard is an important factor influencing rural elite decisions on whether to help organise the project. If rural elites believe that villagers can complete the project according to the standard, they will be inclined to organise; otherwise, they will not be willing to help.

Hypothesis 5: Social capital characteristics influence rural elite willingness to organise FCS projects

Social capital is a resource that people possess in their position in a social structure. A person's social capital can be measured in three dimensions: social network, reciprocity norms, and trust. In this chapter, two variables, the relationship with local cadres and the relationship with villagers, were selected to examine the social capital of rural elites. It is well-known that China is a country that places great importance on human relationships. Therefore, if rural elites have close ties with local cadres and local villagers, they will trust each other; thus, they will organise actively the FCS projects that can help both the villagers and themselves.

Hypothesis 6: Institutional characteristics affect rural elite willingness to organise FCS projects

The institutional environment is the behavioural context that has an impact on the willingness of rural elite to organise an FCS project. The organisational decisions of rural elites are often embedded in the current institutional environment rather than made independently. Elite knowledge of policies concerning national FCS, poverty alleviation, and targeted poverty alleviation were identified as variables to measure the institutional environment related to FCS. The higher the rural elites' awareness of relevant policies, the more they will perceive support for FCS projects, and the higher will be their willingness to organise.

Hypothesis 7: Elite type affects the willingness to organise FCS projects

Rural elites are also called *village elites* and *rural capable people*. They are generally classified into three types: political, economic, and social elites, and include rural cadres, highly respected senior members of rural communities, farmers with higher education, clan leaders, and wealthy leaders. The study divided rural elites into institutional elites and extra-institutional elites. Intra-institutional elites generally refer to rural cadres, while extra-institutional elites include highly respected elderly people in rural communities, farmers with higher education, clan leaders, and wealthy leaders. Different categories of rural elites have different types of human and social capital; their willingness to organise may also differ.

Table 12.1 Distribution of sample

County	Village	Valid sample/person
Qingchuan	Huajian, Laixi, Luoyigou, Fangshi, Yaolin	28
Pingwu	Wuyi, Xinhua, Tongjiang, Tielong, Xiaohe	27
Beicuan	Yongxing, Tongbao, Huangdimiao, Anmian, Zhenghe	28
Mao	Bieli, Heping, Dapeng, Hunsuigou, Shaba	29
Li	Rierzu, Ganbao, Xionger, Lianhe, Shajin	26
Meigu	Weiliejue, Huowo, Nilejue, Jimu, Waguo	27
Yuexi	Habu, Magu, Hongqi, Shaku, Mugejuemu	27
Ganluo	Heshalin, Aerbagu, Kumo, Dashitou, Jini	28
Zhaojue	Ashamo, Maqianmen, Shisier, Qialuo, Hamu	25
Leibo	Minzu, Dagudui, Yierwo, Dashipan, Nuilong	28

12.2 Investigation of Community Elite Willingness to Organise

The data were derived from field surveys conducted by the research team in two projects, i.e., the Northwest Sichuan FCS Project and the Novartis Sichuan FCS Project, covering 10 counties of Qingchuan, Pingwu, Beichuan, Mao, Li, Meigu, Yuexi, Ganluo, Zhaojue, and Leibo. The survey was conducted in their counties and districts. The formal survey used a combination of stratified and random sampling. A total of 300 questionnaires were distributed, and 273 valid samples were collected for a sampling efficiency of 91%. The method employed to identify and select rural elites was a combination of recommendations by others and self-confirmation. First, the investigator described the attributes and characteristics of rural elites to local villagers (i.e., power, ability, and prestige), and then asked villagers to recommend rural elites based on the descriptions. Then, recommended candidates were randomly selected. If they confirmed their identity as rural elite, the survey was conducted; otherwise, the survey was cancelled. The effective sample distribution and descriptive statistics of the sample are shown in Tables 12.1 and 12.2, respectively.

12.3 Econometric Analysis of Community Elite's Willingness to Organise FCS Projects

To explore the factors influencing the willingness of community elites to organise FCS programmes, an econometric model was constructed, and the rationality of the model was confirmed using data derived from previous surveys. In this section, the model's construction, methodology, and analysis are presented, and the proposed theoretical hypotheses are verified.

Table 12.2 Descriptive statistics of the sample

Indicator		Number of people willing to organise/person	Frequency/%
Gender	Male	122	54.71
	Female	29	58.00
Education	Primary school or below	28	41.79
	Junior high school	86	59.72
	High school (secondary school)	30	61.22
	College or above	7	53.85
Forestry-related work experience	Yes	132	75.43
	No	19	19.39
Health status	Poor	31	41.33
	Fair	72	60.00
	Good	48	61.54
Source of family income	Agriculture	112	63.28
	Other	39	40.63
Family support	Family supports participation in public affairs	96	78.05
	Family doesn't support participation in public affairs	55	36.67
Perceived benefit	FCS project is to bring benefits	115	82.14
	FCS project is not to bring benefits	36	27.07
Perceived feasibility	Feasible	42	29.37
	Infeasible	109	83.85
Relationship with local cadres	Trust each other	90	51.14
	Don't trust each other	61	62.89
Relationship with villagers	Trust each other	128	78.53
	Don't trust each other	23	20.91
Awareness of policies	Aware of national support for FCS projects	103	83.74
	Not aware of	48	32.00
Type of elite	Institutional elite	104	83.87
	Non-institutional elite	47	31.54

12.3.1 Model Construction

According to the research hypotheses, demographic, human capital, household, project, social capital, institutional characteristics, and elite type were selected to construct an empirical model of rural elite willingness to organise FCS projects. The functional form of the model is set as follows:

$$Will = \beta_0 + \sum_{i=1}^{2} \beta_i dem_i + \sum_{j=3}^{5} \beta_j huc_j + \sum_{k=6}^{7} \beta_k fam_k + \sum_{l=8}^{9} \beta_l pro_l + \sum_{m=10}^{11} \beta_m soc_m + \beta_{12} sys_{12} + \beta_{13} cae_{13} + \varepsilon \quad (12.1)$$

In Formula 12.1, *Will* is the willingness of rural elites to organise villagers to participate in FCS projects. When they are willing to organise, Will = 1; otherwise, Will = 0; dem_i, huc_j, fam_k, pro_l, soc_m, sys_{12}, and cae_{13} represent demographic characteristics, human capital characteristics, family characteristics, project characteristics, social capital characteristics, institutional characteristics, and rural elite type, respectively.

12.3.2 Methodology

Whether rural elites are willing to lead the organisation of villagers to participate in FCS projects is a binary choice variable, and it is appropriate to choose a logistic regression model for the analysis, which is as follows.

$$Ln\left(\frac{p_i}{1-p_i}\right) = \alpha + \sum_{j=1}^{m} \beta_j X_j \quad (12.2)$$

In formula 12.2, p_i is the explained variable, namely the probability of occurrence of rural elite willingness to organise villagers to participate in FCS projects; i represents the *i-th* rural elite; X_j is the explanatory variable, including 13 variables such as dem_i, huc_j, fam_k, pro_l, soc_m, sys_{12}, and cae_{13}; and β_j is the coefficient to be estimated.

12.3.3 Variables

The definitions and values of the explained and explanatory variables are shown in Table 12.3.

Table 12.3 Definitions and values of variables

Variable	Definition	Mean	S.D	Correlation
Explained variables				
Rural elite willingness to organise FCS projects s (Y)	Willing = 1;Unwilling = 0	0.553	0.498	
Explanatory variables				
Demographic characteristics				
Gender (X_1)	Male = 1; Female = 0	0.817	0.387	+
Age (X_2)	Age (years)	45.795	6.058	–
Human capital characteristics				
Education (X_3)	Elementary school or lower = 1; Middle school = 2; High school = 3; College or higher = 4	2.029	0.785	+
Forestry-related work experience (X_4)	Yes = 1; No = 0	0.641	0.481	
Health status (X_5)	Poor = 1; Fair = 2; Good = 3	2.011	0.750	+
Family characteristics				
Source of family income (X_6)	Agriculture = 1; Other = 0	0.648	0.478	+
Family support (X_7)	Family support for participation in public affairs = 1; Otherwise = 0	0.451	0.498	+
Project characteristics				
Perceived benefits of the FCS project (X_8)	FCS project is to bring benefits = 1; Otherwise = 0	0.513	0.501	+
Perceived feasibility (X_9)	Perceived difficulty of project execution = 1; otherwise = 0	0.524	0.500	–
Social capital characteristics				
Relationship with local cadres (X_{10})	Trust each other = 1; Otherwise = 0	0.645	0.479	+
Relationship with villagers (X_{11})	Trust each other = 1; otherwise = 0	0.597	0.491	+
Institutional characteristics				
Awareness of policies (X_{12})	Aware of national support for FCS projects = 1; Otherwise = 0	0.451	0.498	+

(continued)

Table 12.3 (continued)

Variable	Definition	Mean	S.D	Correlation
Elite characteristics				
Elite type (X_{13})	Institutional elite = 1; Non-institutional elite = 0	0.454	0.499	?

Note +, −, and ? indicate positive, negative, and uncertain correlation of the explanatory variables with the explained variables, respectively

12.4 Logistic Regression Analysis

The data were analysed using the forward stepwise regression method for the logistic model, with a significance level of 10% as the threshold. Insignificant variables were gradually eliminated. The results are shown in Table 12.4.

In the regression results, the significant variables were age, forestry-related work experience, source of household income, perceived benefits, perceived feasibility, relationship with villagers, awareness of policies, and elite type.

As shown in Table 12.4, only the coefficient of *age,* one variable of the demographic characteristics, is positive and significant, resulting in Hypothesis 1 being partially supported. This result indicates that age had a significant effect on rural elite willingness to organise FCS projects; however, unlike the hypothesis, the empirical tests showed that the willingness of rural elites to organise increased with age. The reason for this may be that farmers always have very complex emotions about the land, showing multiple entanglements of leaving the farm, aversion to the farm, and

Table 12.4 Results of logistic regression

Variable	B	S.E.	Wald	Sig.
X_2	0.086**	0.034	6.357	0.012
X_4	1.408***	0.469	9.025	0.003
X_6	1.112**	0.451	6.069	0.014
X_8	0.745*	0.448	2.768	0.096
X_9	−0.777*	0.447	3.020	0.082
X_{11}	2.504***	0.471	28.237	0.000
X_{12}	1.776***	0.492	13.019	0.000
X_{13}	0.931**	0.469	3.928	0.047
Constant	−7.855***	1.858	17.883	0.000
−2 Log likelihood		101427		
Cox and Snell R^2		0.543		
Nagelkerke R^2		0.727		
Prediction accuracy		87.2%		

Note *, **, and *** indicate significance at the 10%, 5%, and 1% levels, respectively

love for the farm (Luo et al., 2012). The younger rural elites are, the more they intend to leave the farm, and the older they are, the more they tend to love the farm. This is not just an economic issue but a problem of local sentiment and socio-psychological differences among rural elites. In addition, younger rural elites have more human capital advantages such as knowledge and skills and are more likely to run non-farm businesses, so they are less willing to organise FCS projects.

As shown in Table 12.4, only the coefficient of *forestry-related work experience,* only one variable of the human capital characteristics, is positive and significant, resulting in Hypothesis 2 being partially supported. This result indicates that rural elites who have (had) forestry-related jobs were more willing to organise FCS projects; while education and health status were not significant. The reason for this may be that education is only a relative advantage for rural elites, as the mean value is only at the junior high school level. The reason for health status being insignificant may be that the surveyed sample was relatively healthy, and there was no statistically significant difference.

As shown in Table 12.4, only the coefficient of *household income source,* one variable of the family characteristics, is positive and significant, resulting in Hypothesis 3 being partially supported. This result indicates that the rural elites whose household income was derived mainly from agriculture were more willing to organise FCS projects; lack of family support was not a significant factor, indicating that these rural elites had relatively strong decision-making abilities and that their decision-making behaviour excluded interference from family. This is also more consistent with the group characteristics of rural elites.

The coefficient of project characteristics is positive and significant at 5% level, resulting in Hypothesis 4 being supported. This result indicates that the more rural elites perceived the benefits of FCS projects, the more willing they were to organise project execution; the more they perceived the project execution as complicated and difficult, the lower their willingness to organise.

As shown in Table 12.4, only the coefficient of *the relationship with villagers,* one variable of the social capital characteristics, is positive and significant, resulting in Hypothesis 5 being partly supported. This result indicates that the stronger the relationship with villagers and mutual trust, the more willing the rural elites were to organise the project. The relationship with local cadres was not significant. The reason may be that, on one hand, a large proportion of the sample were local village cadres themselves; on the other hand, the role of rural elites in FCS projects is the link between enterprises and farmers, and they tend to prioritise their relationship with these two.

The coefficient of institutional characteristics is positive and significant at 5% level, resulting in Hypothesis 6 being supported, which indicates that the more rural elites perceived the state's supportive policies for FCS, the more willing they were to organise villagers to participate in the project.

In addition, the coefficient of elite type is positive and significant at 5% level, resulting in Hypothesis 7 being supported, which indicates that different categories of rural elite had different degrees of willingness to organise. The willingness to organise was significantly stronger among the institutional elites than the non-institutional elites. The reason may be that, on one hand, the institutional elites had a higher perception of national FCS policy and a stronger sense of responsibility to both promote FCS projects and alleviate poverty; on the other hand, institutional elites are more likely to use institutional authority to organise public affairs, which non-institutional elites do not have.

12.5 Summary

Based on elite theory and social exchange theory, this chapter investigated data on 273 rural elites in two regions in Sichuan Province and used a binary logistic regression model to analyse their willingness to organise FCS projects. The results show that age, forestry-related work experience, main source of household income, perceived benefits, perceived feasibility, relationship with villagers, awareness of national FCS policy, and elite type were significant factors influencing the willingness of rural elites. The willingness to organise community farmers to participate in FCS projects increased with age, institutional (cadre) status, forestry-related work experience, household agriculture income, the perception that FCS projects could provide benefits, perceived ease of project execution, trust of village members, and awareness of national FCS policy.

The findings have obvious policy implications. First, it is important to mobilise non-institutional elite groups of *outside the system* and to promote synergistic development of diverse rural elites. The success of poverty alleviation through FCS projects depends also on the non-institutional elites, such as respected community elders, farmers with higher education, clan leaders, and wealthy leaders. It is necessary to expand elite power in rural governance by incorporating non-institutional elites in village governance in order to enhance the effectiveness of FCS for poverty alleviation. Second, technical guidance should be provided to rural elites to mitigate the impact of psychological and behavioural barriers caused by the lack of technical knowledge so as to enhance their willingness to organise and to improve their organisational capacity. Finally, it is vital to transform village economic elites from participating subjects to poverty alleviation subjects. On the one hand, the development of PAFCS cannot exclude the reasonable personal interests of rural elites; on the other hand, through publicity, recognition, and encouragement, rural economic elites are to be urged to take social responsibility for poverty alleviation by mobilising and organising farmers to participate in the projects, particularly disadvantaged and poverty-stricken farmers.

References

Cook, K. S., Cheshire, C., Rice, E. R. W, et al. (2013). *Social exchange theory//Handbook of social psychology* (pp. 61–88). Springer.

Fan, J. Y. (2016). The "What to do " and "what not to do" of rural capable people in poverty alleviation. *People's Forum, 16*, 84–85.

Huang, B., & Liu, Z. Y. (2013). An analysis of the phenomenon of rural elite governance in the context of villager autonomy. *Economic System Reform, 3*, 86–90.

Luo, B. L., He, Y. L., Wang, S., et al. (2012). Land contracting rights: analysis of farmers' willingness to withdraw and its influencing factors—based on farmers' questionnaire in Guangdong Province. *China Rural Economy, 6*, 4–19.

Ma, P.P. (2012). Research on forest carbon sequestration and development of poor minority areas in West Sichuan—A case study of poverty alleviation based on carbon sequestration in Yuexi, Liangshan. Master's Thesis, Sichuan Provincial Academy of Social Sciences.

Pareto. (2010). *The rise and fall of elites,* trans by Gong Weiming. Beijing Publishing.

Si, Y. T. (2017). Study on the targeted poverty alleviation model in Wuzhou City, Guangxi Zhuang Autonomous Region—based on the perspective of elitism. *Journal of Sichuan Administrative College, 5*, 93–99.

Su, J., & Guo, Z. F. (2018). A case study on the influence of rural economic elites on village development. *Journal of Academic Theory, 2*, 101–102.

Tang, X. Y., & Zhao, L. (2009). A study on the willingness of rural elites to participate in the construction of new rural areas: an example of migrant workers in urban areas in Northern Jiangsu Province. *Anhui Agricultural Science, 21*, 10220–10222.

Tong, R. (2014). *Feminist thought: a more comprehensive introduction.* Westview Press.

Wang, T. (2009). Elite Governance in Community Integration - A Paradigm Choice in the Perspective of Chinesized Social Capital. *Academia, 5*(1), 175–180.

Wang, Y. (2013). An analysis of the rationality of elite governance: cost, efficiency and the advantages of order. *Theoretical World, 8*, 15–18.

Wang, Y. (2014). The social basis of elite governance: Psychological needs, organisational requirements and social division of labour. *Theoretical World, 4*, 22–26.

Yang, F., Zhuang, T. H., & Zeng, W. Z. (2016). Analysis of rural elite's willingness to organise FCS projects and its influencing factors. *Technology Management Research., 36*(4), 201–206.

Zeng, M., & Zeng, W. (2017). Rural elite participation in endogenous poverty alleviation: An example of relevant practices in Wuzhou City, Guangxi Autonomous Region. *Journal of Theory, 1*, 92–95.

Zheng, M. H. (2011). Exploring the ability and willingness of rural economic elites to lead villagers to wealth. *Agricultural Archaeology, 6*, 39–41.

Chapter 13
Analysis of Elite Capture in FCS Projects

Accurately targeting poverty alleviation resources to poverty-stricken farmers is a focal issue in anti-poverty research (Gong & Zeng, 2019), and avoiding the deviation of poverty alleviation targets due to elite capture is an urgent issue in FCS projects. Literature has shown that elite capture exists in FCS project development (Cacho et al., 2005; Benessaiah, 2012; Hu et al., 2015; Chen et al., 2017; Xing and Li, 2013), but it generally describes the marginalisation of the disadvantaged poverty-stricken people in a qualitative way, and there is a lack of scientific quantification of the extent of poverty bias. Based on FCS project data for villages in Sichuan and Yunnan provinces, this chapter quantifies the extent of elite capture in FCS projects and reveals the key factors that lead to the bias, with the aim of mobilising the positive role of community elites in the development of FCS projects and avoiding elite capture. The purpose of this chapter is to provide a scientific basis and a reference for decision-making to avoid elite capture and to better achieve creating benefits and development opportunities for the poverty-stricken people as emphasised in FCS projects.

13.1 Theoretical Analysis

Elite capture is a common phenomenon in attempts at poverty alleviation (Gong & Zeng, 2019). This section provides a typical theoretical basis for the study of the elite capture phenomenon. The study starts with an analysis of the process of poverty alleviation. It then defines the concept and measurement of elite capture and discusses factors that may lead to elite capture.

W. Zeng and F. Yang, *Poverty Alleviation Via Forest Carbon Sequestration*, International Research on Poverty Reduction, https://doi.org/10.1007/978-981-99-8191-5_13

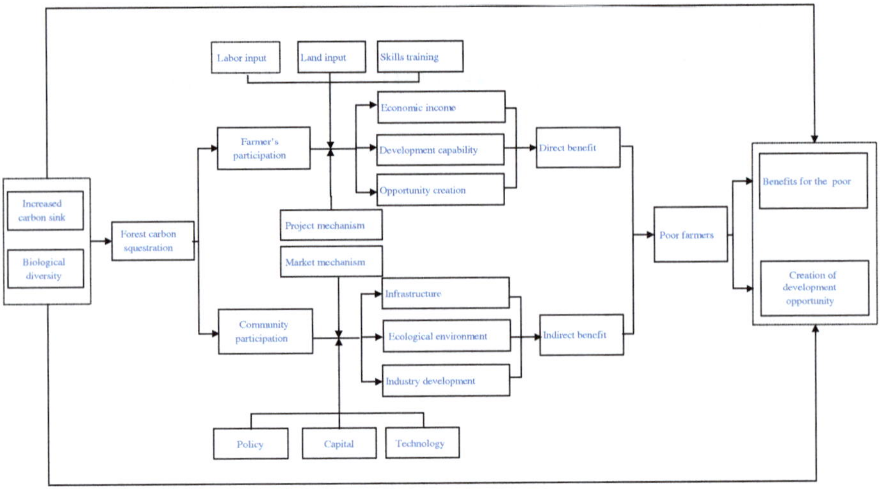

Fig. 13.1 FCS poverty alleviation roadmap

13.1.1 PAFCS Process

With the aim of creating benefits and development opportunities for the poverty-stricken people, there are two possible paths to poverty alleviation through FCS: direct benefits realised by farmer participation, and indirect benefits realised by community participation. In the former case, farmers obtain income, development capacity, and opportunities through labour input, land input, and participation in skills training, thus obtaining direct benefits. These benefits are subject to the operation mechanism of the FCS project, such as whether community residents are given priority to participate and whether land is collectively owned forest land. In the latter case, with the support of government policy, financial investment, and technological resources, FCS projects are intended to promote community infrastructure, improve the ecological environment, and upgrade industry, which residents are able to indirectly benefit from (Fig. 13.1).

13.1.2 Measurement of Elite Capture

In elite capture, resources that are supposed to be transferred for the majority are appropriated by a few people, who are usually politically or economically powerful (Dutta, 2009). In this chapter and in the context of PAFCS, elite capture refers to institutional, economic, and traditional elites in villages or communities controlling the participation channels of FCS projects and FCS resources by virtue of their social capital, natural capital, and other resource endowments, with poverty-stricken farmers marginalised or even excluded from project participation. As a result, while

FCS projects favour anti-poverty, it is difficult to achieve the goals of poverty alleviation. Generally, elite capture of PAFCS is the elite group's capture of the *participation opportunities* and *participation channels* of FCS projects, and thus, the degree of elite capture can be measured by the proportion of elite participation. Based on the results of Pan and Christiaensen (2012), Wang et al. (2007), and Hu et al. (2015), a formula was constructed to measure elite capture of FCS projects.

$$RE_1 = \frac{\sum_{i=1}^{N} I_{i(IE)}(IE = 1)}{\sum_{i=1}^{N} P_i}$$

$$RE_2 = \frac{\sum_{i=1}^{N} I_{i(EE)}(EE = 1)}{\sum_{i=1}^{N} P_i} \tag{13.1}$$

$$RE_3 = \frac{\sum_{i=1}^{N} I_{i(SE)}(SE = 1)}{\sum_{i=1}^{N} P_i}$$

In formula 13.1, *RE* denotes the extent of elite capture, RE_1 denotes the extent of institutional elite capture, RE_2 denotes the extent of economic elite capture, and RE_3 denotes the extent of traditional elite capture.

IE denotes institutional elites, i.e., farmers (themselves, or a member of their household) who are cadres of a township, village group, etc. or public officials in a government agency.

EE denotes economic elites, who have relatively prominent economic power in the community, i.e., large village/community farmers, private owners, etc. In this chapter, farmers with per capita income in the top third are considered economic elites.

SE indicates traditional elites possessing local prestige or family power, usually representing the attitudes and opinions of most community members, i.e., clan chiefs or people with prestige in the village/community.

13.1.3 Factors Affecting Elite Capture

There are many reasons for the emergence of anti-poor bias, which can be broadly classified into four categories. The first category is the degree of farm household heterogeneity, which includes the differentiation of the household in terms of economic level, social capital, and cultural strength, as well as the differences in household voices in community governance caused by these differentiations. Those villages/communities with a greater degree of heterogeneity are more likely to experience elite capture. Hu et al. (2015) identified differences in literacy, especially in high school-level literacy, as an important factor affecting the degree of elite capture. The second category is the degree of democratisation of governance in the village/community. Some scholars argue that the higher the degree of democratisation, the more transparent the governance, the fairer the rights of farmers in terms of access to resources, etc., and the less significant the elite capture phenomenon (Bardhan & Mookherjee, 2005). Other scholars argue that the current pan-democratisation problem is prominent, and the logic of *what is approved by the majority must be right* makes it difficult to promote FCS projects. At the same time, elite groups themselves have a certain degree of influence and can influence or even manipulate democratic intentions. Thus, the higher the degree of democratisation, the more prominent the phenomenon of elite capture (Wang et al., 2007). The third category is the individual characteristics of local government leaders, especially those of village group cadres. Hu and Wang (2017) found that the tenure of village cadres is an important factor affecting elite capture. Bardhan and Mookherjee (2005) found that the political accountability of local government leaders is highly correlated with elite capture; as accountability diminishes, elite capture becomes likely. The fourth category is the resource endowment of the village/community. Some scholars have focused on the fact that poverty alleviation bias is related to natural factors such as the distribution of resources and the climate of the village/community itself, which are associated with migration.

Drawing on the existing literature and the status quo of poverty alleviation, the natural resource endowment of villages was used as the control variable, and degrees of heterogeneity and democratisation as well as the cadre characteristics of farmers were used as independent variables to analyse the reasons for the existence of poverty alleviation bias in FCS. The degree of heterogeneity included three dimensions, i.e., income level, variability of occupation, and the clustering of the residence; the degree of democratisation had two dimensions, i.e., frequency of village general meetings and the availability of village rules and regulations from both formal and informal systems; the characteristics of cadres had three dimensions, i.e., household registration, education level and democratic evaluation; and resource endowment included land, population distribution, natural conditions, etc. The theoretical framework is shown in Fig. 13.2.

The study involved five major categories of variables. The first category was poverty alleviation bias, i.e., the degree of elite capture. The second category was heterogeneity. The heterogeneity of income was measured by the variance of per capita household income; occupational heterogeneity was measured by assigning values to different occupations and finding the variance; and the clustering of residences was measured by the variance of household distance from the village committee. The larger the variance, the higher the degree of heterogeneity. The third category was democratisation. Formal democratic governance was measured by the number of village assemblies held each year, and informal democratic governance was measured by the availability of village rules and regulations. The fourth category

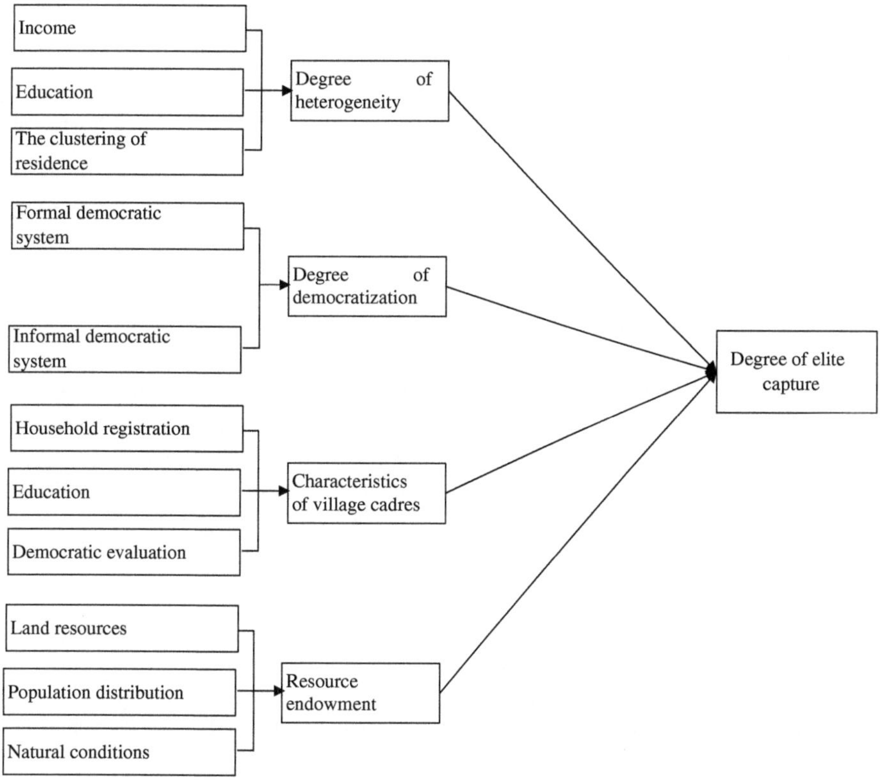

Fig. 13.2 Analytical framework for extent of elite capture and factors affecting it

was the characteristics of village cadres. The household registration was measured by whether the cadres were registered in the village, the education level of the village group cadres was measured by the number of years of education, and the democratic evaluation of the cadres was measured as the mean value of farmer evaluations of them in the village. The fifth category was village resource endowment. The village land resource was measured by the area of cultivated land; population distribution was measured by the number of ethnic groups in the village; and the natural condition of the village was measured by the frequency of disasters (see Table 13.1 for details).

13.2 Empirical Results and Discussion

The researchers collected data from 928 farm households over several months. Based on these sample data, the study assessed the extent of elite capture and ascertained the influence of factors such as heterogeneity, democratisation, village cadre characteristics, and resource endowment.

Table 13.1 Variable selection and interpretation

Primary indicator	Secondary indicator	Interpretation	Type
E: Degree of elite capture	Degree of institutional elite capture	RE_1	Numerical
	Degree of economic elite capture	RE_2	Numerical
	Degree of capture by traditional elites	RE_3	Numerical
V: Degree of heterogeneity	Income	V_1: variance of per capita household income	Numerical
	Occupation	V_2: agriculture = 0, labour or other = 1, as per the variance value of the whole village	Numerical
	Clustering of residence	V_3: variance of the distance of the household from the village committee	Numerical
De: Degree of democratisation	Formal democratic system	De_1: frequency of village assembly	Numerical
	Informal democratic system	De_2: availability of village rules and regulations, yes = 1, no = 0	Nominal
L: Characteristics of village cadres	Household registration	L_1: registered locally = 1, not = 0	Nominal
	Education	L_2: actual years of education of cadres	Numerical
	Democratic evaluation	L_3: farmer evaluations	Numerical
E: Resource endowment	Land resources	$E1$: area of arable land	Numerical
	Population distribution	E_2: number of ethnicities	Numerical
	Natural conditions	E_3: frequency of natural disasters	Numerical

13.2.1 Data Description

The data were collected from the Northwest Sichuan FCS Project and the Novartis FCS Project. The survey involved 27 administrative villages in five counties (Ganluo, Yuexi, Meigu, Zhaojue, and Leibo) in Southwest Sichuan and 28 villages in five counties (Beichuan, Li, Mao, Pingwu, and Qingchuan) in Northwest Sichuan. A total of 928 samples were collected from the 55 villages. The data included information on farmers participating in the FCS projects, information on village cadres, and village characteristics.

13.2.2 Descriptive Analysis

The area of the Novartis Sichuan FCS project is dominated by the Yi ethnic group, while in the case of the Northwest Sichuan FCS project, although some villages are located in ethnic minority areas, the proportion of ethnic minorities is low. Farmers in the area for the Novartis Sichuan FCS project earn income mainly from agriculture, and most of them farm full-time, while farmers in the area in the Northwest Sichuan FCS project earn income mainly from labour, and most are part-time farmers. So, in terms of absolute income level, that of Northwest Sichuan FCS project is relatively higher than that of the Novartis FCS project, but the differences in income heterogeneity and occupational heterogeneity between the two project areas were small, indicating that the heterogeneity of employment and income within the regions was low. The communities in the Novartis Sichuan FCS project are more isolated, and thus, most village cadres are local farmers with relatively low education levels. Yi people in the project areas tend to establish village rules and regulations under the traditional governance systems and leadership of Bimo and clan chiefs and thus have better informal democratic institutions than in the Northwest Sichuan FCS project. While the population in the Northwest Sichuan FCS project lives in a relatively concentrated area with small differences in distance from the village committee, farm households in the Novartis Sichuan FCS project live in a relatively dispersed area with high variance in residential distance (Table 13.2).

Table 13.2 Means and variances of independent variables

Indicator	Overall		Northwest Sichuan FCS project		Novartis Sichuan FCS project	
	Mean	Variance	Mean	Variance	Mean	Variance
V_1	0.106	0.002	0.099	0.003	0.114	0.002
V_2	0.063	0.002	0.039	0.001	0.089	0.003
V_3	15.371	259.983	0.567	0.206	30.723	58.966
De_1	3.164	0.769	3.214	0.767	3.111	0.795
De_2	0.491	0.255	0.179	0.152	0.815	0.157
L_1	0.455	0.253	0.179	0.152	0.741	0.199
L_2	1.618	0.537	1.643	0.534	1.593	0.558
L_3	3.164	0.769	3.214	0.767	3.111	0.795
E_1	8.762	34.509	7.643	52.121	9.922	14.799
E_2	0.693	0.178	0.441	0.202	0.955	0.019
E_3	2.673	1.002	3.036	0.925	2.296	1.832

Table 13.3 Deviations in the pro-poor benefit of FCS projects

Indicator	Northwest Sichuan FCS project		Novartis Sichuan FCS project		Overall	
	Mean	Variance	Mean	Variance	Mean	Variance
RE_1	0.094	0.000	0.097	0.000	0.095	0.000
RE_2	0.204	0.001	0.198	0.002	0.201	0.001
RE_3	0.059	0.002	0.125	0.001	0.092	0.002
RE	0.258	0.001	0.251	0.003	0.254	0.002

13.2.3 Measure of Elite Capture

The empirical results showed that the overall degree of elite capture was 0.254, among which the degree of institutional elite capture was 0.095, the degree of economic elite capture was 0.201, and the degree of traditional elite capture was 0.092, with economic elite capture the highest. The variance of the degree of elite capture was only 0.002, indicating that there was little difference in the extent of elite capture among the sample villages. The most obvious difference was in the degree of institutional elite capture, which was 0.125 in the Novartis Sichuan FCS Project and only 0.059 in the Northwest Sichuan FCS Project. The main reason is that the Novartis Sichuan FCS Project area is a minority one, and traditional informal governance institutions such as Bimo and clan leaders are better preserved, while the informal governance system is gradually dying out in the Northwest Sichuan FCS Project as the villages modernise. Even if a small number of them do have informal governance, it is no longer prominent (Table 13.3).

13.2.4 Factors Affecting Elite Capture

Four factors affecting the extent of elite capture were analysed: the degree of heterogeneity, the degree of democratisation, the characteristics of village cadres, and resource endowment. The results are shown in Table 13.4.

There was little difference in the degree of elite capture among the 55 villages or between the Northwest Sichuan Project and the Novartis FCS Project, with distribution at the 25% scale line, indicating that elite capture is common in FCS projects. This is closely related to the fact that the FCS projects rely on elite groups, especially projects with new challenges such as long project period, difficult management and care of young forests, and intensified conflicts among foresters and farmers. The success of the projects cannot be achieved without elite groups.

Farmer heterogeneity and informal democracy were important aspects of elite capture. In villages with greater income and occupational heterogeneity, there are often obvious differences in farmers' voices in decision-making, resulting in poverty alleviation resources in the hands of a small number of elite groups. The traditional

Table 13.4 Results of analysis of factors affecting pro-poor benefits of FCS projects

Variable	RE_1	RE_2	RE_3	RE
V_1	0.121*	0.132**	0.043*	0.132*
V_2	0.603**	0.107*	0.133*	0.302
V_3	−0.305	−0.384*	−0.036	−0.361*
De_1	—	—	—	—
De_2	0.228*	0.248***	0.280**	0.239**
L_1	0.016***	0.017*	0.025*	0.020*
L_2	0.062	0.067	0.148	0.073
L_3	—	—	—	—
E_1	0.131	0.144	0.116	0.143
E_2	0.286*	0.312*	0.445***	0.295**
E_3	−0.013	−0.014**	−0.132	−0.031*

Note ***, **, and * indicate significance at the 1, 5, and 10% levels, respectively

governance system in ethnic minority areas is relatively well preserved, and informal democracy is still an important auxiliary to village grassroots governance; thus, elite capture is relatively prominent in ethnic minority areas and villages with better retention of informal democracy.

At present, FCS projects are supported by grassroots cadres, and almost all village and group cadres are involved in their execution. The village cadres who belong to their own village often have a deeper and more comprehensive understanding of the village situation and public opinion, and they (particularly the institutional elite) can drive other farmers to participate. Village cadres from other villages in FCS projects often have limited functions, such as providing guidance, and it is difficult for them to lead the local farmers; thus, the extent of institutional elite capture is smaller.

The presence of informal democracy positively influences capture by economic elites, mainly because institutional and traditional elites in villages tend to be aligned with economic elites; in other words, economic elites tend to be village cadres. The frequency of natural disasters was highly correlated with capture by economic elites because the proportion of migrant workers is often higher in villages with more frequent natural disasters, and migrant farmers are often economic elites in villages. Traditional elite capture was highly correlated with the size of the minority population. Elements of the traditional governance system, such as clan chiefs, basically no longer exist except in minority areas; thus, the percentage of traditional elites was relatively small.

13.3 Summary

Based on survey data of 55 villages, this chapter quantified and analysed the bias of FCS projects and explored the factors affecting this bias in four aspects, namely the degree of heterogeneity of farmers, the degree of democratisation of villages, the characteristics of village cadres, and the resource endowment of villages. The results showed that elite capture exists in FCS projects at this stage, and the overall degree of elite capture is 0.254, among which, the degree of institutional elite capture is 0.095, the degree of economic elite capture is 0.201, and the degree of traditional elite capture is 0.092, with the economic elite capturing the largest proportion. This indicates that FCS depends on the role of institutional elites, and it is closely related to non-institutional traditional elites, such as respected community elders, highly educated farmers, and clan leaders. FCS poverty alleviation focuses on the organisation of community elites, and their lawful interests should be respected. But, most importantly, their capturing of resources should be prevented.

References

Bardhan, P., Mookherjee, D. (2005). Decentralizing antipoverty programme delivery in developing countries. *Journal of Public Economic 89*(4), s675–704.

Benessaiah, K. (2012). Carbon and livelihoods in Post-Kyoto: Assessing voluntary carbon markets. *Ecological Economics 77*, 1–6.

Cacho. O.J., Marshall, G.R., Milne, M. (2005) Transaction and abatement costs of carbon-sink projects in developing countries. *Environment and Development Economics 10*(5), 597–614.

Cheng, M., Zheng, Y.F., Xu, J. X. et al. (2017). A study of elite capture dilemma and countermeasures in participatory poverty alleviation governance. *Rural Economy* (9), 56–62.

Dutta, D. (2009). Elite capture and corruption: concepts and definition. *National Council of Applied Economic, 4*, 20–25.

Gong, R. F., & Zeng, W. Z. (2019). Does elite capture and mass capture exist - empirical evidence from PAFCS. *Journal of Guangdong University of Finance and Economics, 1*, 60–68.

Hu, L. W., Wang, S. G.,Wang, N. (2015). Does elite capture exist in poor village mutual funds - empirical evidence based on 30 poor village mutual fund pilot villages in 5 provinces. *The Economist* (9). 78–85.

Hu, L., & Wang, S. G. (2017). Does China's file-building for poverty-stricken family face the challenge of elite capture? *Management World, 1*, 89–98.

Pan, L., & Christiaensen, L. (2012). Who is vouching for the input voucher? decentralized targeting and elite capture in Tanzania. *World Development, 40*(8), 1619–1633.

Wang, S. G., Park, A., Chaudhuri, S., & Datt, G. (2007). Rural poverty alleviation and village-level poverty targeting in china's new era. *Management World, 1*, 56–64.

Xing. C. J. (2104). *Elite capture in rural poverty alleviation resource allocation—perspectives on institutions, power and social structure.* Ph.D., Dissertation of China Agricultural University.

Xing, C. J., & Li, X. Y. (2013). A study of elite capture and deviation from the objectives of financial poverty alleviation programs. *Chinese Administration, 9*, 109–113.

Chapter 14
Evaluation of the Trickle-Down Effect of PAFCS

Studies have shown that FCS plays an important role in improving regional infrastructure, increasing the income of residents in poverty-stricken areas, and expanding the degree of openness to the outside world (Ding & Ma, 2013; Ehrenstein, 2018; Grassi et al., 2017; Røttereng, 2018; Wu et al., 2015). However, during the pilot period, a project may have negative impacts. For example, in order to avoid carbon leakage, farmer income from mushroom and herbal medicine harvesting is reduced for a period (Li et al., 2015), and farmer livelihood activities such as forest planting and grazing are often restricted (Chen, 2010; Jindal et al., 2012), preventing the development of the agricultural industry. So, does FCS promote local economic development, and does it have a trickle-down effect on poverty alleviation? It is important to accurately assess the impact of carbon afforestation on local economic development.

This study uses the propensity score matching-double difference (PSM-DID) method to ascertain the net impact of FCS on the county's economic development at the overall and dynamic level. Specifically, the propensity score matching method was adopted to match counties with FCS projects (treatment group) with counties without FCS projects (control group) in order to solve the endogeneity problem caused by the fact that the treatment and control groups do not satisfy the parallel trend assumption before the project. Then, a multi-period double difference method was employed to estimate the net effect of FCS on the county's economic development in order to reduce the estimation error. Finally, the impact of FCS on the county's economic development was ascertained.

W. Zeng and F. Yang, *Poverty Alleviation Via Forest Carbon Sequestration*, International Research on Poverty Reduction, https://doi.org/10.1007/978-981-99-8191-5_14

14.1 Hypotheses for the Trickle-Down Effect of FCS

The trickle-down effect is an economic concept first proposed by Hirschman (1958). In the process of economic development, special preferential treatment is initially granted to priority groups or regions, who will later provide benefits to the poverty-stricken class, disadvantaged groups, or poverty-stricken regions through consumption, employment, etc. (Aghion & Bolton, 1997).

Over the past 40 years or so, the trickle-down effect, inclusive growth, and special government-led poverty alleviation have become important ways to achieve poverty reduction in China. In the early stage of economic development, developed regions could benefit poverty-stricken regions through the trickle-down effect and drive their economic development. However, with further economic growth, due to the characteristics of market mechanisms, relying solely on these mechanisms cannot achieve the trickle-down effect and narrow the gap between the rich and the poor, but will instead lead to the Matthew effect, creating increasingly serious imbalances in income and wealth distribution. In addition, due to remote locations and lack of human capital, it is hard to realise the trickle-down effect in all poverty-stricken regions and groups, and inclusive growth is also difficult to achieve. Therefore, it is necessary to rely on government-led poverty alleviation to help poverty-stricken groups increase their income and get rid of absolute poverty, but this does not mean that the trickle-down effect does not exist.

This chapter explores the trickle-down effect of carbon sink afforestation in promoting county economies through regional FCS projects, thereby boosting employment, increasing the number of livelihood strategies available to the poverty-stricken people, reducing poverty vulnerability, and ultimately helping the poverty-stricken people increase their incomes for sustainable development.

Global climate change has received widespread attention from governments and scientific communities. Measures to promote low-carbon economic development must meet the conditions of high potential, easy execution, quick results, and low cost (Huang, 2008). In recent years, with the advent of the post-Kyoto era and the signing of international agreements such as the Paris Agreement (2015), the role of afforestation in addressing global climate change has become more prominent (Zeng et al., 2017). Therefore, carrying out pilot FCS projects is not only an international requirement for China to address climate change, but also an intrinsic requirement both to promote domestic energy conservation and emission reduction and to overcome resource and environmental constraints. According to incomplete statistics, as of the end of 2017, there were 110 active FCS projects in China, among which 75 were AR projects, accounting for 68.18% of the total.

Sichuan Province is rich in forestable lands, and its climate and soil conditions are extremely suitable for tree growth. It is an important natural carbon pool for the world and is among the first batch of afforestation pilot provinces. For example, among the five CDM FCS projects registered with the UN CDM-EB, two are in Sichuan. Here, the province was selected as the research object to investigate the impact of FCS projects on regional economic development.

Four AR projects are underway in Sichuan Province: two CDM projects, one CCER project, and one VCS project. These have been officially implemented and registered with CDRC. There are 13 counties involved in the projects, including six national-level poverty-stricken counties and nine minority counties. The Northwest Sichuan FCS Project was officially launched in 2004, covering an area of 2251.8 ha with 28 villages of 21 towns in five counties (Li, Mao, Beichuan, Qiangchuan, and Pingwu). The Novartis Sichuan FCS Project was officially launched in 2010. The project is located in the Liangshan Ethnic Region, which is in the upper reaches of the Yangtze River and contains precious and endangered species such as the giant panda. This project covers an area of 4196.8 ha and 27 villages of 17 towns in five counties (Ganluo, Yuexi, Meigu, Zhaojue, and Leibo). Furthermore, the Sichuan Yingjing County Afforestation Project ('Yingjing Project') was launched in Yingjing County in 2011, and the Audi-Panda Habitat Multiple Benefit Forest Restoration and Afforestation Carbon Sink Project ('Audi Project') was launched in Jinyang and Mianning counties in 2012. The implementation of carbon sink afforestation projects in Sichuan is shown in Table 14.1.

One of the purposes of afforestation is to increase the income of residents and alleviate rural poverty. Therefore, the projects also have a strong role in promoting the development of the county's economy. Combining with the existing literature,

Table 14.1 Basic information of carbon sink afforestation projects in Sichuan Province

County	Type	Name	Launched (year)	National poverty-stricken county	Ethnic county	Crediting period (year)	Area (ha)
Mianning	CCER	Audi project	2012	No	Yes	30	153.4
Jinyang	CCER		2012	Yes	Yes	30	181.7
Lixian	CDM	Northwest Sichuan FCS project	2004	No	Yes	20	747.8
Maoxian	CDM		2004	No	Yes	20	234.9
Beichuan	CDM		2004	No	No	20	200.2
Qingchuan	CDM		2004	No	No	20	878.3
Pingwu	CDM		2004	No	No	20	190.6
Ganluo	CDM	Novartis Sichuan FCS project	2010	Yes	Yes	30	924.3
Yuexi	CDM		2010	Yes	Yes	30	1245
Meigu	CDM		2010	Yes	Yes	30	731.6
Zhaojue	CDM		2010	Yes	Yes	30	441.8
Liebo	CDM		2010	Yes	Yes	30	854.1
Yingjing	VCS	Yingjing project	2011	No	No	30	159.2

Source Collected and compiled by the researchers

this chapter discusses the influence of afforestation on the county's economic development from four aspects: industrial structure, capital accumulation, income, and fiscal revenue and expenditure.

First, FCS projects are conducive to upgrading local agriculture-based industries. On the one hand, the continuation of a project will reduce the use of agricultural arable land (Chong et al., 2017; Fu et al., 2017; Wu et al., 2018) and enhance non-farm employment (Carton & Anderson, 2017); on the other hand, it, to a certain extent, offsets industrial energy emissions, becoming a low-cost option for industrial and energy enterprises and thus indirectly promoting the development of secondary industries. Therefore, FCS can promote the optimisation and upgrading of regional industrial structures.

Second, FCS projects can expand multiple financing channels and improve regional financing (Ji & Wang, 2012). Enterprises with strong social and environmental responsibility, NGOs, individuals, and some investors participate in FCS projects, which provide diverse financing channels for local areas (Gaast et al., 2018). Innovative forms of carbon finance such as CDM project financing, enterprise low-carbon transition financing, and emission right financing can enhance local financing (Sun et al., 2018; Wood et al., 2016) and promote regional economic development.

Third, FCS projects provide local employment opportunities for residents and increase their income. Farmers receive economic benefits from the project in terms of labour, carbon sink, timber, and forest by-products, as well as short-term or long-term job opportunities, thus increasing the overall income and savings of residents (Zeng et al., 2017). In addition, FCS can bring advanced afforestation technologies to the project area and develop the afforestation skills of residents (Ke et al., 2010), improving local human capital and thus indirectly promoting regional economic development.

Fourth, carbon sink trading can increase government revenues, which in turn can prompt local governments to increase capital investment in the local economy. On the one hand, the inclusion of afforestation into the carbon trading market can effectively transform high-quality ecological resources into income and improve the financial revenue of the local government (Pandit et al., 2017). On the other hand, along with the increase in the local government's fiscal revenue, its expenditure on regional economic construction will also increase accordingly, thus promoting the sustainable development of the regional economy and society.

In summary, FCS projects have an important impact on local economic development by promoting the optimisation and upgrading of industrial structures, enhancing channels for financing, increasing the income and savings of local residents, and improving fiscal revenues. However, carbon afforestation is a long-term and continuous economic project, and its effect on local economic development is difficult to see in the short term. Accordingly, a theoretical framework for the impact of afforestation projects on local economic development was constructed (see Fig. 14.1), with the following research hypotheses proposed:

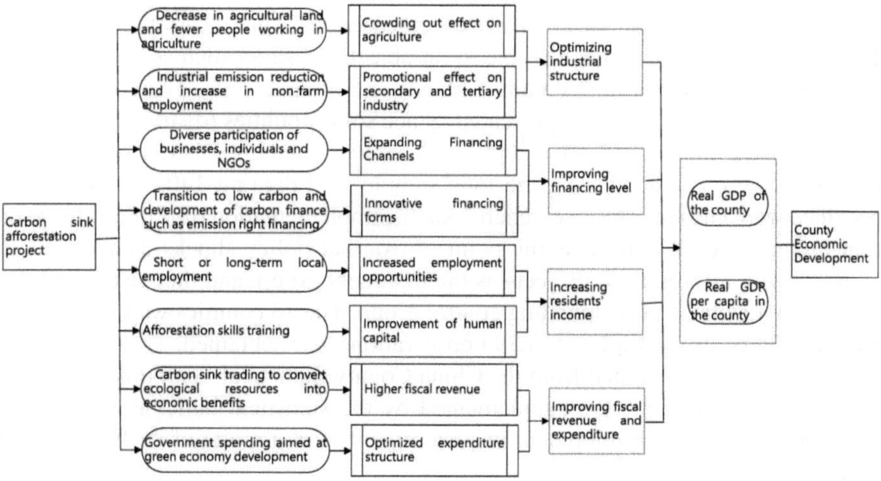

Fig. 14.1 Theoretical framework

Hypothesis 1: Afforestation projects can promote local economic development.

Hypothesis 2: The economic benefits of an afforestation project are not obvious in the short term but will gradually become apparent in the long term in promoting local economic development.

Hypothesis 3: Possible effects of afforestation projects on local economic development are optimising local industrial structure, increasing local capital accumulation, and improving local income and expenditure.

14.2 Data Sources, Model Construction, and Variable Selection

The researchers need to construct a quantitative assessment model of the trickle-down effect and select an appropriate regional sample to collect data for its quantitative assessment. In this section, the researchers introduce the econometric model of the trickle-down effect of afforestation, the data sources used to ascertain the quantitative model, and the variables.

14.2.1 Data Sources

Considering the availability of research data, the period from 2000 to 2016 was selected as the time range of the sample. By the end of 2018, a total of 13 counties in Sichuan Province had successively executed carbon sink afforestation projects. The selection of project sites was based on objective factors such as natural conditions, e.g., the land did not contain wetlands and organic soils. Therefore, a good *quasi-natural experiment* was presented. Since some counties (districts) in Sichuan Province were established after 2000, such as the Qianfeng District of Guang'an City, they were excluded from the total sample, and data on 140 counties in the province were chosen. The researchers took the implementation time of an afforestation project in each county as the point of external policy shocks, considered 13 counties with afforestation projects as the treatment group, and then used the PSM method to match the corresponding control group. The 13 counties were matched to a total of 34 control samples, and 799 observations were obtained.

Most data were obtained from the China County Statistical Yearbook from 2000 to 2016. Missing data were supplemented by the statistical bulletins on national economic and social development or government work reports of each county for each calendar year.

14.2.2 Model Construction

The researchers adopted the double difference method to estimate the effect. However, due to the different levels of economic development between counties (districts) in Sichuan Province, the treatment and control groups may have significant differences, which may lead to bias in the estimation results. Therefore, PSM was adopted to match the treatment group with a similar control group and then used DID to ascertain the net effect of the project on the county's economic development within the matched sample. The model was set as follows:

$$Y_{it} = \beta_0 + \beta_1 \cdot fcs \cdot t + \sum \beta_X \cdot control + r_i + y_t + \varepsilon_{it} \tag{14.1}$$

In formula 14.1, Y_{it} is a proxy variable to measure the level of economic development of the county. β_0 is a constant term to measure the level of the county's economic development. *fcs* is used to distinguish the treatment group from the control group, and t is a dummy variable to distinguish before and after the project. The cross-product term $fcs \cdot t$ is the core explanatory variable to measure whether a carbon afforestation project was executed, β_x indicates the net impact of carbon afforestation on the county's economic development, *control* is the control variable, and β_X is the coefficient of each control variable. And r_i denotes individual fixed effects that do not change over time, y_t is a time fixed effect, and finally ε_{it} is the random disturbance term.

To estimate the dynamic effects of an afforestation project affecting the county's economic development, the model was set as follows:

$$Y_{it} = \beta_0 + \sum \beta_k \cdot after_k + \sum \beta_X \cdot control + r_i + \varepsilon_{it} \tag{14.2}$$

In formula 14.2, $after_k$ is the cross-multiplier term of $fcs \cdot t^k$, which is the dummy variable of a county in the year k since the start of the project. For example, if County A started an afforestation project, the variable $after_k$ was assigned the value of 1 in the k-th year and 0 in the other years. β_k was used to measure the policy effect of the project on the county's economic development in the k-th year after the project started.

To test the mechanism of the afforestation project affecting local economic development, the model was set as follows:

$$control = \beta_0 + \beta_j \cdot fcs \cdot t + \varepsilon_{it} \tag{14.3}$$

Formula 14.3 is an ordinary least squares regression where the above control variables are explained variables and $fcs \cdot t$ is an explanatory variable, aiming to examine the effect of the project on various economic development drivers.

14.2.3 Variable Selection

The focus of this study was the impact of an afforestation project on a county's economic development. Considering that other economic factors may also affect the development of a county's economy, control variables were introduced, as shown in Table 14.2.

Explained variables. Regarding the explained variables, Y_{it} is used to measure the level of economic development of counties. Referring to Liu and Zhao (2015), the researchers used the logarithm of the real gross product of the county ($\ln gdp_{it}$) and the logarithm of the real gross product per capita ($\ln pgdp_{it}$) as the explained variables. The real GDP of each county is derived from the nominal GDP divided by the GDP deflator index as 2000 being the base period, and the real GDP per capita is calculated in the same way.

Core explanatory variables. The cross-multiplier term $fcs \cdot t$ is the core explanatory variable, representing whether or not the project is executed, in which fcs is a dummy variable reflecting whether this county was chosen as a pilot county for an afforestation project; it was assigned 1 if yes, otherwise 0. $\boldsymbol{t}$ stands for whether the project was executed in the county; it was assigned 1 if yes, otherwise 0. β_1 is a double difference estimator reflecting the net impact of the carbon afforestation project on the county's economic development.

Control variables. The first category reflects the industrial structure of the county. China's economic development is often accompanied by changes in regional industrial structure (Shi & Gao, 2006; Wang & Fan, 2004), and the industrial structure plays a crucial role in the sustainable development of county economies. The researchers use two indicators of the value added of primary and secondary industries as a share of nominal regional GDP (*pfir* and *psec*), respectively, to measure the industrial structure of county economies.

Table 14.2 Definitions and descriptive statistics for main variables

Type	Name	Definition	Sample size	Mean	S.D.	Min.	Max.
Dependent variable	*lngdp*	Log value of real GDP of the county	799	11.947	1.233	8.878	15.327
	lnpgdp	Log value of real GDP per capita in the county	799	9.017	0.837	7.295	11.022
Explanatory variable	*fcs.t*	Dummy variable with project = 1; otherwise = 0	799	0.277	0.448	0	1
Control variable	*pfir*	Value added of primary industry/county GDP	799	0.328	0.134	0.032	0.631
	psec	Value added of secondary industry/county GDP	799	0.368	0.190	0.054	1.096
	pfix	Fixed asset investment/county GDP	799	0.829	0.700	0.079	5.897
	psav	Resident deposit balance/county GDP	799	0.611	0.289	0.103	1.836
	pfin	Year-end balance of loans from financial institutions/county GDP	799	0.551	0.455	0.028	3.013
	pinc	General budget expenditure/county GDP	799	0.048	0.07	0.011	1.867
	pexp	Fiscal general budget revenue/county GDP	799	0.47	0.464	0.058	3.018
	dens	Log value of population density	799	4.019	1.347	0.878	6.175
	mino	Minority county	799	0.617	0.486	0	1
	pove	National poverty-stricken county	799	0.553	0.497	0	1

Note The sample is a panel of 47 counties for 2017

The second category reflects the capital accumulation capacity of the region. County capital stock can have an important impact on the local economy, especially in the economic development of rural areas (Li et al., 2014; Wang, 2018). On the one hand, as a typical feature of China's transitional economy (Li & Yin, 2005), a high savings rate can directly or indirectly increase the level of social investment and consumption, which in turn promotes economic growth (Attanasio et al., 2000; Wang, 2008). Therefore, the share of residents' savings deposit balances to regional GDP (*psav*) was selected to measure the level of savings in the county. On the other hand, regional fixed asset investment and economic growth have a mutually reinforcing effect (Song, 2011) and are long-term drivers of regional economic development; therefore, an indicator of the share of fixed asset investment in regional GDP (*pfix*) was introduced.

The third category reflects the level of fiscal revenue and expenditure of the county. Reasonable fiscal revenues and expenditures help ensure the normal operation of local public services and provide a good environment for local economic development (Yan & Gong, 2009; Jin, 2007). The researchers used two variables, the share of local general budget revenue and expenditure in regional GDP (*pinc, pexp*), respectively, to measure the levels of government fiscal revenues and expenditures. In addition, the researchers selected three control variables (population density, status as ethnic minority county, and status as national-level poverty-stricken county) for propensity score matching in order to obtain a more similar control group.

14.3 Empirical Analysis

According to the research methodology described in the previous section, a quantitative assessment of the trickle-down effect of afforestation projects was conducted, including generalised and dynamic evaluations. This was followed by a robustness test on the stability of the findings and a discussion of the mechanism underlying the trickle-down effect.

14.3.1 Average Effect of FCS Projects on the County's Economic Development

Based on the model in formula 14.1, *lngdp* and *lnpgdp* are used as the explained variables to estimate the net effect of implementing an afforestation project on the county's economic development, i.e., the average change of the selected indicators of the treatment group compared to that of the control group. Table 14.3 also reports the regression results when not controlling for other economic factors. Columns (1) and (2) are the estimated results without the control variables, and columns (3) and (4) are the estimated results when the control variables are introduced. It can be seen that the coefficients of the cross-products are significant and positive at the 1% level, regardless of whether the control variables are introduced. This indicates that the afforestation project made a significant contribution to the county's economic development.

Table 14.3 Regression results of the average effect of FCS project on the county's economic development

	(1)	(2)	(3)	(4)
	Lngdp	*Lnpgdp*	*lngdp*	*lnpgdp*
fcs·t	1.236***	1.178***	0.204***	0.166***
	(12.39)	(12.15)	(3.27)	(2.81)
pfir			−5.536***	−4.872***
			(−20.93)	(−19.47)
psec			0.856***	1.439***
			(4.28)	(7.61)
pfix			−0.039	−0.023
			(−1.13)	(−0.72)
psav			0.515***	0.601***
			(4.90)	(6.04)
pfin			−0.249***	−0.208***
			(−4.09)	(−3.61)
pinc			0.948***	0.785***
			(4.45)	(3.89)
pexp			0.509***	0.363***
			(7.27)	(5.48)
Time effect	*Control*	*Control*	*Control*	*Control*
Region effect	*Control*	*Control*	*Control*	*Control*
_cons	11.768	8.846	12.989	9.621
	(417.78)	(323.21)	(89.33)	(69.91)
N	799	799	799	799
R^2	0.170	0.164	0.738	0.750

Note Figures in parentheses are t-values calculated by robust standard errors for county clustering, and 1, 5, and 10% significance levels are denoted by ***, **, and *, respectively

When considering other factors affecting regional economic development, the coefficient of *pfir* is significantly negative and the coefficient of *psec* is significantly positive, indicating that optimisation of the three industrial structures promoted regional economic development, which is consistent with the findings of existing studies (Yang & Zhang, 2018; Zhao & Tang, 2018). The coefficient of *pfix* is negative but non-significant, which may be explained by the possible crowding-out effect of fixed asset investment on other short-term investment projects in the region and the lagged economic benefits of fixed asset investment. *psav* has a significantly positive coefficient, indicating that the higher the level of residents' savings, the stronger their ability to convert them into an effective investment, which has a boosting effect on regional economic development. *pfin* has a significantly negative coefficient, which indicates that financial loans have a suppressive effect on local economic development, possibly because loans are not effectively converted into investment in the county. The findings of Wang et al. (2014) help explain this result. In line with the researchers' expectations, the financial revenue and expenditure of regional governments made significant contributions to local economic development.

14.3.2 Dynamic Effects on the County's Economic Development

The above results reflect the average effect of the project on the county's economic development but do not examine the sustainability effect from a dynamic perspective nor allow a conclusion regarding whether a project has a short- or long-term effect. Therefore, the researchers estimated the dynamic effects (Table 14.4), i.e., the researchers further analysed whether the effect of the carbon afforestation on the county's economic development was sustainable and whether there was a difference between its short-term effect and long-term effect over time, on a yearly basis. According to the estimation results, the coefficient of $after_1$–$after_3$ increased gradually and became significant, indicating that the project had a gradually increasing effect on the county's economic development. However, it is worth noting that, firstly, after controlling for other influencing factors, although the coefficient of the cross-product term gradually increased over time, the coefficients of $after_1$ and $after_2$ were not significant, and the coefficient of $after_3$ was significant at the 10% level, which indicates that the economic effect of afforestation was not significant in the first two years. Second, the coefficient of column (4) was negative, although it was not statistically significant. It can be inferred that in the first year, a project may have had a negative effect on per capita GDP. One possible explanation for this is that afforestation projects do not produce significant economic results in the short term and have a crowding-out effect on agricultural production, while part-time farmers with reduced arable land cannot immediately engage in non-agricultural employment in the short term, which then has a dampening effect on local economic development.

In summary, afforestation projects do not promote regional economic development in the short term and exhibit an obvious lag effect. On average, the economic impact of carbon afforestation begins to appear only in the third year.

14.3.3 Robustness Tests

14.3.3.1 Single Difference Method

To test whether the DID method was reliable in estimating the effect of afforestation projects on a county's economic development; the researchers also applied the traditional single difference method to the estimation. The researchers used the counties

Table 14.4 The dynamic effects of FCS projects on the economic development of counties

	(1)	(2)	(3)	(4)
	lngdp	*Lnpgdp*	*lngdp*	*lnpgdp*
after$_1$	0.343	0.306	0.016	−0.052
	(1.61)	(1.48)	(0.14)	(−0.48)
after$_2$	0.481**	0.456**	0.093	0.071
	(2.25)	(2.20)	(0.82)	(0.67)
after$_3$	0.573***	0.550***	0.200*	0.179*
	(2.68)	(2.66)	(1.77)	(1.68)
Control variable	*No*	*No*	*Yes*	*Yes*
Time effect	*Control*	*Control*	*Control*	*Control*
Region effect	*Control*	*Control*	*Control*	*Control*
_cons	11.925	8.996	13.033	9.654
	(440.31)	(342.82)	(89.60)	(70.25)
N	799	799	799	799
R^2	0.017	0.017	0.736	0.748

Note Numbers in parentheses are t-values calculated from the robust standard errors of county clustering; 1, 5, and 10% significance levels are denoted by ***, **, and *, respectively

with an afforestation project as the sample for the single difference method to compare their economic development changes before and after the project (Table 14.5).

Compared to Table 14.3, it can be seen that the estimation result of *fcs* is still significantly positive; therefore, there is no significant difference between the estimations of the single difference method and the DID method, without controlling for other influencing factors. However, after controlling for other factors, the coefficients

Table 14.5 Estimation results based on the single difference method

	(1)	(2)	(3)	(4)
	lngdp	*lnpgdp*	*lngdp*	*lnpgdp*
Fcs	1.236***	1.178***	0.556***	0.514***
	(17.27)	(17.23)	(7.65)	(7.51)
Control variable	*No*	*No*	*Yes*	*Yes*
Time effect	*Control*	*Control*	*Control*	*Control*
Region effect	*Control*	*Control*	*Control*	*Control*
_cons	11.175***	8.348***	12.891***	9.623***
	(223.58)	(174.96)	(47.29)	(37.48)
N	221	221	221	221
R^2	0.590	0.589	0.808	0.812

Note Figures in parentheses are t-values calculated for robust standard errors of county clustering, and 1, 5, and 10% significance levels are denoted by ***, **, and *, respectively

Table 14.6 Robustness test results for replacement matching methods

	(1)	(2)	(3)	(4)
	lngdp	*lnpgdp*	*lngdp*	*lnpgdp*
fcs·t	1.213***	1.163***	0.259***	0.205***
	(12.21)	(12.06)	(3.97)	(3.25)
Control variable	*No*	*No*	*Yes*	*Yes*
Time effect	*Control*	*Control*	*Control*	*Control*
Region effect	*Control*	*Control*	*Control*	*Control*
_cons	11.386	8.637	13.088***	10.086***
	(269.91)	(210.97)	(60.84)	(48.39)
N	408	408	408	408
R^2	0.280	0.275	0.794	0.793

Note Figures in parentheses are t-values calculated from robust standard errors of county clustering, and 1, 5, and 10% significance levels are denoted by ***, **, and *, respectively

of the single difference method were higher than those of DID, in terms of both real GDP and real GDP per capita. The DID method was more accurate.

14.3.3.2 Replacement of Matching Methods

Before conducting the DID empirical analysis, the researchers used the kernel density PSM method to match the control group for the treatment group based on the reflective variables of industrial structure, capital accumulation, and fiscal revenue and expenditure. To test the robustness of the empirical results, they decided to use the one-to-one nearest neighbour PSM method for matching. Matched samples of 24 counties were obtained, including 12 each for the treatment and control groups, and the estimation results remained robust (see Table 14.6).

14.3.3.3 Exclusion of Counties Starting Project Later

In the research sample, the 13 counties in the treatment group started projects in 2004 or from 2010–2012 and did not have a uniform starting time point. To exclude potential policy factors that may have (also) affected economic development at different times, the researchers excluded the treatment group from 2010–2012 and retained only the five counties that started the project in 2004. With the one-to-one nearest neighbour PSM method to match the control group, 170 samples in 10 counties were obtained.

According to the estimation results (see Table 14.7), after other factors affecting economic development were controlled for, the effect of the afforestation project on the county's economic development remained significant: the longer the project period, the larger the coefficient of $after_k$, i.e., the dynamic effect was more significant. However, it is worth noting that while the dynamic effect of $after_1 \sim after_6$ was not significant, it was positive and increasing from year to year, which supports the conclusion that the effect of an afforestation project on economic development is not short-term but has a lagged effect. Therefore, our conclusion is robust after excluding the sample of counties that started the project later.

14.3.4 *Mechanism of Afforestation Project Effect on Economic Development*

According to the empirical research, a project can effectively promote the economic development of a county. What is the mechanism of this effect? To further test Hypothesis 3, the researchers used model 14.3 to estimate the effect, and the regression results are shown in Table 14.8. The coefficients of $fcs \cdot t$ represent the effects of the project on various economic growth drivers in each county. The coefficients in column (1) are significantly negative and in column (2) are significantly positive, indicating that the project was conducive to promoting the development of secondary industries and thus improving the regional industrial structure. By comparing with Table 14.3, it can be found that the optimised industrial structure promoted the development of the county's economy; therefore, a project can promote the growth of the regional economy by improving the local industrial structure. The coefficients of columns (3)–(5) are significantly positive, indicating that it effectively enhanced the regional capital stock. But, with Table 14.3, the loan balance of financial institutions did not promote regional economic development but rather had a negative impact; therefore, inefficient financial loans can inhibit the positive effect. The coefficients of columns (6) and (7) are significantly positive, and combined with Table 14.3, it can be seen that a project promoted regional economic growth by improving local government financial revenues and expenditures. In summary, projects promote local economic development mainly by optimising the local industrial structure, increasing the savings rate of residents, and elevating the financial revenue and expenditure of the local government.

14.4 Summary

Based on panel data of 140 counties in Sichuan Province from 2000 to 2016, this chapter explored the impact of afforestation projects on economic development by using the PSM-DID method and analysed the former's trickle-down effect. The results show the following: (1) An afforestation project has a significant effect on the economic development of counties, and this conclusion still holds after robustness testing. (2) The longer the implementation time, the greater the effect on local

Table 14.7 Robustness test results after excluding counties that started the project later

	(1)	(2)	(3)	(4)
	lngdp	*lnpgdp*	*lngdp*	*lnpgdp*
fcs·t	0.327***	0.233**	—	—
	(2.85)	(2.13)	—	—
$after_1$	—	—	0.102	−0.012
	—	—	(0.62)	(−0.08)
$after_2$	—	—	0.029	−0.018
	—	—	(0.18)	(−0.12)
$after_3$	—	—	0.224	0.188
	—	—	(1.33)	(1.17)
$after_4$	—	—	−0.130	−0.133
	—	—	(−0.42)	(−0.45)
$after_5$	—	—	0.481	0.555*
	—	—	(1.56)	(1.89)
$after_6$	—	—	0.102	0.075
	—	—	(0.32)	(0.25)
$after_7$	—	—	0.466**	0.415**
	—	—	(2.38)	(2.24)
$after_8$	—	—	0.517***	0.445**
	—	—	(2.80)	(2.54)
$after_9$	—	—	0.611***	0.513***
	—	—	(3.36)	(2.97)
$after_{10}$	—	—	0.768***	0.660***
	—	—	(4.20)	(3.81)
$after_{11}$	—	—	0.843***	0.736***
	—	—	(4.68)	(4.31)
$after_{12}$	—	—	1.031***	0.930***
	—	—	(5.67)	(5.40)
_cons	13.604***	10.598***	13.486***	10.501***
	(63.39)	(51.81)	(63.62)	(52.29)
N	170	170	170	170
R^2	0.788	0.814	0.835	0.858

Note Figures in parentheses are t-values calculated for robust standard errors for county clustering, and 1, 5, and 10% significance levels are denoted by ***, **, and *, respectively

Table 14.8 Test results of the mechanism for FCS projects promoting economic development

	(1)	(2)	(3)	(4)	(5)	(6)	(7)
	pfir	*psec*	*Pfix*	*Psav*	*pfin*	*pinc*	*pexp*
fcs·t	−0.126***	0.132***	0.594***	0.210***	0.137***	0.018*	0.300***
	(−11.53)	(8.75)	(6.88)	(7.54)	(3.62)	(1.83)	(6.83)
_cons	0.346	0.348	0.742***	0.580	0.531***	0.045***	0.426***
	(112.29)	(81.50)	(30.46)	(73.75)	(49.67)	(16.17)	(34.32)
N	799	799	799	799	799	799	799
R^2	0.150	0.092	0.059	0.070	0.017	0.004	0.058

Note Figures in parentheses are t-values calculated for robust standard errors of county clustering, and 1, 5, and 10% significance levels are denoted by ***, **, and *, respectively

economic development. (3) An afforestation project promotes local economic development by optimising the local industrial structure, increasing the savings rate of residents, and expanding the revenues and expenditures of the regional government.

Combining the above findings, the researchers suggest the following policy implications: (1) Continue to expand the coverage area of afforestation, increase special investment, and guide it towards ecologically fragile and poverty-stricken areas to enhance local economic development. (2) Establish and improve the long-term stability mechanism of project operation, prevent potential natural and market risks, and thereby guarantee the long-term driving force of the project on regional economic development. On the one hand, evaluations of their effectiveness should not be limited to the short term but pay attention to their long-term effects. On the other hand, measures should be undertaken to guide participating farmers to pay more attention to their long-term benefits, which may increase their willingness to participate.

References

Aghion, P., & Bolton, P. (1997). A theory of trickle-down growth and development. *The Review of Economic Studies, 64*(2), 151–172.

Attanasio, O. P., Picci, L., & Scorcu, A. E. (2000). Saving, growth, and investment: A macroeconomic analysis using a panel of counties. *Review of Economics & Statistics, 82*(2), 182–211.

Carton, W., & Andersson, E. (2017). Where forest carbon meets its maker: Forestry-based offsetting as the subsumption of nature. *Society & Natural Resources, 30*(7), 829–843.

Chen, C. Y. (2010). Forestry carbon sinks and farmers' livelihoods: The example of the first global forestry carbon sequestration project. *World Forestry Research, 23*(5), 15–19.

Chong, J., Zhang, H., Tang, Z., et al. (2017). Evaluating the coupling effects of climate variability and vegetation restoration on ecosystems of the Loess Plateau, China. *Land Use Policy, 69*(1), 134–148.

Ding, Y., & Ma, P. P. (2013). Research on forest carbon sequestration and economic development in ethnic minority areas in West Sichuan Province: A case study of Yuexi County, Liangshan Yi Autonomous Prefecture, Sichuan Province. *Rural Economy, 5*, 38–41.

Ehrenstein, V. (2018). Carbon sink geopolitics. *Economy and Society, 47*(3), 1–25.
Fu, B., Meng, W., Yue, C., et al. (2017). Effects of land-use changes on city-level net carbon emissions based on a coupled model. *Carbon Management, 8*(3), 245–262.
Gaast, W. V. D., Sikkema, R., & Vohrer, M. (2018). The contribution of forest carbon credit projects to addressing the climate change challenge. *Climate Policy, 18*(1), 1–7.
Grassi, G., House, J., Dentener, F., et al. (2017). The key role of forests in meeting climate targets requires science for credible mitigation. *Nature Climate Change, 7*(3), 220–226.
Huang, D. (2008). Forest carbon sequestration: An important way to reduce emissions in the Post-Kyoto Era. *Forestry Economics, 10*, 12–15.
Jindal, R., Kerr, J. M., & Carter, S. (2012). Reducing poverty through carbon forestry? Impacts of the N'hambita community carbon project in Mozambique. *World Development, 40*(10), 2123–2135.
Ji, X., & Wang, X. L. (2012). Carbon financial innovation and low carbon poverty alleviation. *Agricultural Economic Issues, 1*, 79–87.
Jin, C. P. (2007). Spatial variability of fiscal policy effects and regional economic growth. *Management World, 7*, 47–56.
Ke, S. F., Li, Z., Zheng, Y., et al. (2010). Analysis of employment effects of afforestation initiatives in China. *Agricultural Economic Issues, 31*(3), 98–103.
Li, G. C., Fan, L. X., & Feng, Z. Z. (2014). Capital accumulation, institutional change and agricultural growth- an empirical estimation of agricultural growth and capital stock in China from 1978 to 2011. *Management World, 5*, 67–79.
Li, J. H., Ming, H., & Yu, W. Y. (2015). Comparative analysis of forestry carbon sequestration project implementation in Sichuan Province. *Journal of Sichuan Agricultural University, 3*, 332–337.
Li, Y., & Yin, J. F. (2005). High savings, high investment and China's economic growth in the process of labour transfer. *Economic Research, 40*(2), 4–15.
Liu, R. M., & Zhao, R. J. (2015). Do national high-tech zones promote regional economic development?–A validation based on double difference method. *Management World, 8*, 30–38.
Pandit, R., Neupane, P. R., & Wagle, B. H. (2017). Economics of carbon sequestration in community forests: Evidence from REDD+ piloting in Nepal. *Journal of Forest Economics, 26*, 9–29.
Røttereng, J. K. S. (2018). The comparative politics of climate change mitigation measures: Who promotes carbon sinks and why? *Global Environmental Politics, 18*(1), 52–75.
Sun, M. J., Peng, H. J., & Cong, J. (2018). A review of carbon finance and forestry carbon sequestration project financing. *Forestry Economic Issues, 38*(5), 90–98.
Shi, L., & Gao, F. (2006). Regional economic disparity: An empirical study based on economic structural transformation. *Management World, 5*, 35–44.
Song, L. Z. (2011). Re-examining the relationship between fixed asset investment and economic growth in China: 1980–2010. *Macroeconomic Research, 11*(1), 17–21.
Wang, J. Y. (2018). Rural financial development, capital stock enhancement and rural economic growth. *Quantitative Economics and Technical Economics Research, 35*(2), 64–81.
Wang, X. L., & Fan, G. (2004). Trends and influencing factors of regional disparities in China. *Economic Research, 1*, 33–44.
Wang, X. H., Wen, T., & Wang, D. X. (2014). County rural financial inhibition and internal inequality of farmers' income. *Economic Science, 2*, 44–54.
Wang, W. (2008). A study on the dynamic correlation between savings, investment and economic growth–analysis based on data from 1952–2006 in China. *Nankai Economic Research, 2*, 105–125.
Wood, B. T., Sallu, S. M., & Paavola, J. (2016). Can CDM finance energy access in Least Developed Countries? Evidence from Tanzania? *Climate Policy, 16*(4), 456–473.
Wu, G. Z., Wang, H. P., Zhou, N. D., et al. (2015). Analysis of social benefits of forest carbon sink industry development in Hunan Province. *Forestry Economics, 11*, 90–93.
Wu, X., Wang, S., Fu, B., et al. (2018). Land use optimization based on ecosystem service assessment: A case study in the Yanhe watershed. *Land Use Policy, 72*, 03–312.
Yan, C. L., & Gong, L. T. (2009). Fiscal spending, taxation and long-run economic growth. *Economic Research, 44*(6), 4–15.

Yang, Z. R., & Zhang, P. Y. (2018). Financial structure, industrial structure and economic growth-An empirical test from the perspective of new structural finance. *Economics (quarterly), 17*(2), 847–872.
Zhao, J., & Tang, J. (2018). Industrial structure change and economic growth: A China-Russia comparison. *China Economic Review, 47*, 219–233.
Zeng, W. Z., Liu, S., Yang, F., et al. (2017). A review of research on forest carbon sequestration from the perspective of poverty alleviation. *Agricultural Economic Issues, 2*, 102–109.

Chapter 15
Comprehensive Evaluation of PAFCS Performance

Many studies have shown that PAFCS projects have macro-scale impacts on sustainable development in poverty-stricken areas; meso-scale impacts on the collective economy of poverty-stricken communities, their organisational and development capacity, and the integration of FCS industries and local specialised industries; and micro-scale impacts on the income, development capacity, and development opportunities of poverty-stricken residents. The performance of poverty alleviation of FCS includes both economic and non-economic benefits, as well as short- and long-term performance (Chen, 2010; Chen et al., 2011; Chia et al., 2013; Chu & Zhou, 2012; Ding & Ma, 2013; Lv, 2014; Ma, 2012; Milder et al., 2010; Xu & Jv, 2009). However, discussions on the performance of PAFCS are generally based on qualitative analysis, and quantitative evaluations are rarely seen (Zeng et al., 2018). Therefore, in this chapter, the researchers employ feasible capacity theory (Sen, 1999) to construct an evaluation index system for the performance of PAFCS, taking into account the typical characteristics of diversity, spatial heterogeneity, and temporal dynamics of the effect of PAFCS. This study adopted the hierarchical analysis–entropy value weighting method to investigate two FCS projects in Sichuan Province with the aim of better exploring the poverty alleviation potential of FCS and deepening the quantitative evaluation of PAFCS.

15.1 Indicator System

The poverty alleviation effect of PAFCS is complex and is reflected at different levels and in differing categories. Therefore, an index system for evaluating poverty alleviation performance was constructed in this chapter, taking into account the characteristics of PAFCS, i.e. diversity, spatial heterogeneity, and temporal dynamics.

W. Zeng and F. Yang, *Poverty Alleviation Via Forest Carbon Sequestration*, International Research on Poverty Reduction, https://doi.org/10.1007/978-981-99-8191-5_15

15.1.1 Principles

There are six principles in constructing a performance evaluation system for PAFCS.

Objectivity principle. The selection of performance evaluation indexes should be based on recognised scientific theories, avoiding subjective assumptions as much as possible; the data should be objective; and measurement and data processing should be standardised to ensure accuracy and reliability.

Principle of combining comprehensiveness and representativeness. The indicators should reflect the purpose, role, and function of the index system, and each indicator should have a certain logic; at the same time, as few indicators as possible should be used.

Principle of hierarchy. A clear hierarchy for indicators should be established when constructing the index system, with clear relationships and reasonable articulation between each level as well as relatively independent and reasonable weighting for indicators at all levels.

Principle of operability. Each indicator should be practical and operational; the specified content can be measured clearly using existing tools; and each indicator can obtain sufficient information about the reality.

Principle of comparability. The indicators can be compared to certain standards so as to determine which is better or worse, including the comparability of different time and space scopes.

Principle of win–win situation. The sustainable development of FCS projects is indispensable for their poverty alleviation function, while the support of the government and the public is a prerequisite. Therefore, the performance evaluation should adhere to the basic principle of sustainable development of the projects and their benefiting the poverty-stricken people.

15.1.2 Selection of Indicators

According to Amartya Sen's (1999) viable capacity theory, poverty is manifested not only by low income but also in the absence of basic functional activities and the deprivation of viable capacity to realise basic functional activities. Therefore, the means and measures to combat poverty are not limited to increasing income but include enhancing development capacity and creating development opportunities for the poverty-stricken people. Development capacity is the endogenous driving force for poverty alleviation; development opportunities are the external driving force for poverty alleviation; and economic benefits are the most direct benefit of poverty alleviation (see Fig. 15.1 for the relationship between the three).

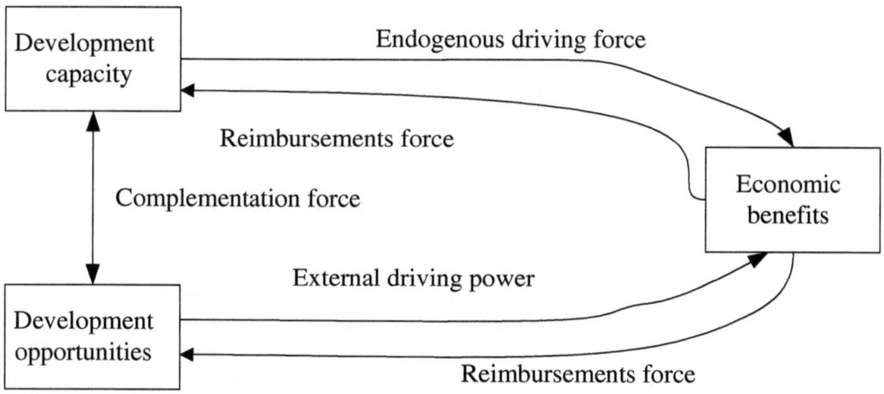

Fig. 15.1 Relationship between the criteria of the performance evaluation index system for PAFCS

FCS projects can both create wage and asset income for poverty-stricken farmers by directly providing jobs and land rental fees and enhance their development capacity through technical training. By improving the regional hard and soft environment, the accessibility of external capital and resources for a poverty-stricken area can be increased, creating opportunities for regional and individual development. Accordingly, drawing on research related to performance evaluation index systems for poverty alleviation (e.g. Chen et al., 2016; Ming et al., 2015; Xiang & Gao, 2013; Xu et al., 2013; You & Tian, 2009), a performance evaluation index system for PAFCS was constructed with three dimensions, i.e., economic benefits, development capacity, and development opportunities.

Based on theoretical and descriptive analyses of existing studies (e.g., Ming et al., 2015), this study accounted for typical characteristics of FCS projects and the principles of indicator selection to construct an evaluation system for PAFCS including these dimensions and 14 indicators (see Table 15.1 for details).

Criterion layer of economic benefits. Achieving multi-level and diverse employment and increasing the productivity and income of the poverty-stricken people is the most direct and primary goal of almost all poverty alleviation projects (Edwards, 2000). The criterion layer of economic benefits aims to evaluate the performance of promoting employment and income rise for poverty-stricken areas and people. Therefore, the indicators include employment and income. Among them, the percentage of local labourers reveals the proportion of local farmers employed by the project, which is an indication of the performance of poverty alleviation of regional farmers in general; the percentage of labourers from poverty-stricken households is an indication of the performance of targeted poverty alleviation of households. Project area refers to the area of FCS sites in each region, which, to a certain extent, can reflect the potential benefits that poverty-stricken areas and people can obtain from the project and is a kind of expectation for benefits. Income per capita is a homogeneous

Table 15.1 Performance evaluation index system for PAFCS

Target layer	Criteria layer	Indicator elements layer	Calculation formula, indicator, or value	Unit
PAFCS performance	Economic benefits	Percentage of local labourers	Number of local labourers employed by the project/ number of all labourers employed by the project * 100%	%
		Percentage of poverty-stricken labourers	Number of local poverty-stricken labourers employed by the project/ number of all labourers employed by the project * 100%	%
		Project areas	Size of project area	km^2
		Income per capita	Total income of collective economic organisations and farmers from project work, carbon sink trading, public welfare, compensation for grazing losses, etc./total number of people in the project community * 100%	RMB/ person
		Percentage of poverty-stricken population's income	Income of the poverty-stricken population participating in the project/ total income of collective economic organisations and community farmers participating in the project * 100%	%
	Development capabilities	Number of trainings	Number of relevant skills trainings	Instance
		Participation rate of training among the poverty-stricken people	$F_r = \frac{\sum_{i=1}^{n} F_i}{\sum_{I=1}^{n} X_i} \times 100\%$, where $i = 1, 2, \ldots, n$ denotes the i-th training; F_i denotes the number of poverty-stricken people participating in the i-th training; X_i denotes the number of all participants in the i-th training	%
		Community involved in project design, management, and monitoring	Yes = 1, or No = 0	

(continued)

Table 15.1 (continued)

Target layer	Criteria layer	Indicator elements layer	Calculation formula, indicator, or value	Unit
	Developmental opportunities	Road built during the project	Length	km
		Number of small water facilities built during the project	Number	
		Areas adopting advanced forestation techniques	Size of area	hm^2
		Job opportunities created	Number of jobs	
		Selection of tree species	Number of tree species planted in the project/number of local tree species totally	%
		Forest coverage rate	Forest coverage growth due to the project	%

value of the actual benefit for poverty alleviation; the income share of poverty-stricken populations reflects their share from the project, which is also an indication of household-to-person targeted poverty alleviation.

Criterion layer of development capacity. The improvement of economic conditions can lift the poverty-stricken people out of poverty temporarily, but it cannot guarantee their sustained development. The endogenous motivation to ensure effective poverty alleviation and not to return to poverty is to improve the self-development ability of the poverty-stricken people. Therefore, the number of skills training sessions or workshops, the participation rate of poverty-stricken populations in skills training, and the participation of communities in project design, management, and supervision are three indicators that can measure the extent of improvement of farmers' own capacity. Among them, the number of skills trainings refers to the total number of training sessions or workshops that the project organiser and its commissioners organise in the early, middle, and late stages of the FCS project. The trainings are related to the project's execution or aim to improve other feasible abilities of the trainees. Generally speaking, given other conditions unchanged, the higher the number of training sessions, the more types of skills the participants will master, the better they will employ their knowledge and skills, and the more human capital they will have. Under normal circumstances, the higher the participation rate of the poverty-stricken population, the easier it is for them to master skills and escape poverty, and the more effective the training is in benefiting them. The participation of communities in project design, management, and monitoring highlights the initiatives of communities in project execution, who serves as an intermediary between project organisers and farmers, playing the role of a link, coordinating interests, and resolving conflicts. Also, the democratisation of community management will be enhanced. Therefore, community participation is bound to give better play to the function of the FCS project to benefit the poverty-stricken people.

Criterion layer of development opportunities. According to social embedding theory, individual behaviour is always embedded in and influenced by the external environment. In the case of poverty alleviation, it is not enough to enhance the self-development ability of the poverty-stricken people to escape poverty; external obstacles that prevent them from doing so must be removed. Therefore, the creation of external development opportunities is also an important indicator of poverty alleviation performance (Ren & Yuan, 2017). Infrastructure is essential for poverty alleviation, as the repair of roads and irrigation facilities is conducive to improving the benefit and yield of agroforestry production and reducing transaction costs, thus increasing the income of poverty-stricken farmers (Ma, 2016). The indicator of areas adopting advanced forestation techniques refers to the application of agroforestry industry technology, such as ecological planting and intensive breeding, in the area of FCS projects. This expands FCS industries and extends industrial chains and is conducive to increasing farmers' income (Zhang, 2015). The development of FCS projects can cause farmers to lose forest revenue, so providing employment opportunities is an important part of the project process that can bring economic benefits to these farmers. Numerous empirical studies have shown that poverty-stricken areas in rural China are located in ecologically fragile regions, so the ecological improvement is at the root of poverty alleviation and development (Li et al., 2005; Zeng & Zhang, 2011). Some scholars have started to incorporate related indicators, such as regional forest cover, a biological abundance index, a vegetation cover index, and a land degradation index, into the performance assessment of poverty alleviation (Wang et al., 2017a, b; Zhuang et al., 2012). On the one hand, large-scale carbon sequestration AR is conducive to water conservation, soil and water conservation, and reducing the risk of natural disasters such as droughts and floods, mudslides, and landslides confronted by poverty-stricken people (Wang et al., 2017a, b). It is also conducive to the accumulation of ecological resources in the region, transforming ecological assets, and realising ecological benefits (Xu & Jv, 2009). On the other hand, it may also bring ecological risks to the region, such as plant pests and diseases, loss of native tree species, and reduction of biodiversity (Lin & Li, 2006). Given that forest cover is an important indicator of regional ecological environment, the selection of local native species to create mixed forests is an important way to avoid negative environmental impacts in practice, and for this reason, two indicators were selected: the proportion of local native species planted in the FCS project and the growth rate of forest cover brought by the project implementation.

15.2 Empirical Study on the Performance Assessment of PAFCS

This subsection clarifies the assessment objects (i.e., samples) and describes the sources of data and assessment methods, such as indicators and their weight.

15.2.1 Data Sources and Sample Analysis

15.2.1.1 Data Sources

Most data are from the following sources. The first source is field survey data. The researchers carried out on-site investigations in many FCS projects (areas) in Novartis Sichuan FCS Project and Northwest Sichuan FCS Project during the past 3 years. They also conducted discussions and in-depth interviews with relevant government departments, enterprises, and institutions and obtained first-hand survey data. The second source is internal information provided by the Dadu River Afforestation Bureau and Novartis Pharmaceuticals, including project contracts and information on the evaluation of the effectiveness of projects. The third source is public data, including the Sichuan Statistical Yearbook (2017), data from four provinces (cities), i.e., Aba Prefecture, Mianyang City, and other cities and states, as well as relevant government work reports, publicly published books such as *China's FCS Practice and Low Carbon Development*, and digital data from websites such as that of the China Green Carbon Sink Foundation.

The counties investigated in this study included Pingwu County and Beichuan Qiang Autonomous County in Mianyang; Mao County in Aba Prefecture; and Leibo County, Zhaojue County, and Ganluo County in Liangshan Prefecture. In the sampling process, the samples were concentrated in villages with large FCS projects, and all participating farmers in each village were included as survey objects. A total of 630 questionnaires were distributed; 630 were returned; 607 were valid; the sampling efficiency rate was 96.35% (see Table 15.2 for sample details).

15.2.2 Sample Characteristics

As can be seen in Table 15.3, 70.19% of respondents were above 45 years old, and most were male, accounting for 66.06% of the total sample. Only 7.09% of the surveyed farmers were educated in high school or above, while more than 50.00% were educated only in elementary school, which shows that the general education level of the surveyed areas was not high. Since the survey was focused on the Yi Autonomous Prefecture of Liangshan, the main ethnic groups were Yi, Tibetan, and Han. The fact that 68.04% of the households owned more than 10.00 mu of forest land

Table 15.2 Sample distribution

County	Township	Village	Households investigated
Pinwu	Gaocun	Wuyi	51
	Shuijing	Xinhua	33
	Pingtong	Xinyuan	27
Beichuan	Duba	Huangdimiao	49
	Qingpian	Zhenghe	34
	Xiaoba	Yongxin	29
Mao	Nanxin	Bieli	89
Leibo	Changhe	Mingzhu	33
	Gudui	Dagudui	71
Zhaojue	Zepu	Heshamu	35
		Aerbagu	38
		Kumo	30
Ganluo	Jimi	Galuo	37
	Pingba	Shuangmacao	35
		Shishier	39

indicates that the average household ownership of forest land area in this region is relatively large and that forest land resources are relatively abundant, which provides a basis for carrying out FCS projects. However, the main source of income for these households was from working outside the village, and more than 70.00% of the households had workers employed outside the village.

15.2.3 Description of Variables

In terms of the percentage of local labourers, the mean value was 60.25% and the standard deviation was 9.63%, with little difference between the maximum and minimum values. There were slight fluctuations in terms of the percentage of poverty-stricken labourers, the percentage of income of the poverty-stricken people, and the skill training rate of them, but as the process of poverty eradication continues to advance, there is variability in the number of the poverty-stricken people according to the poverty status of each region. The standard deviation of per capita income was 436.34 RMB (about 63 USD), which was due to differences among farmers as well as local industries. The mean value of tree species selection (65.22%) and standard deviation (18.56%) did not fluctuate much, which proves that projects had taken full account of the adaptability of trees. The descriptive statistics of each variable are detailed in Table 15.4.

Table 15.3 Basic information of the sample

Category	Content	Number	Percentage (%)
Age	Under 18 years old	22	3.62
	18–30 years old	47	7.74
	31–45 years old	112	18.45
	Above 45 years old	426	70.19
Gender	Male	401	66.06
	Female	206	33.94
Education level	Illiterate	41	6.75
	Elementary school	360	59.31
	Junior high school	163	26.85
	High school or above	43	7.09
Ethnicity	Yi	210	34.60
	Tibetan	138	22.73
	Han Chinese	198	32.62
	Other	61	10.05
Forest land area	Less than 10 mu	194	31.96
	10–30 mu	202	33.28
	30–50 mu	89	14.66
	50 mu or more	122	20.10
Number of people working outside the village	0–2 persons	442	72.81
	3–4 persons	95	15.65
	5 or more persons	70	11.51

15.2.4 Indicator Weights

Subjective empowerment and objective empowerment are two major methods for assigning weight to indicators. Subjective weighting methods rely on the subjective experience of experts to determine the importance of each indicator, including AHP, the least square method, expert survey method (Delphi method), binomial coefficient method, ring score method, etc. (Meng & Li, 2013). As AHP can hierarchise and simplify complex problems, it is more widely used. Objective assignment methods include principal component analysis and the entropy method, deviation and mean square difference method, multi-objective planning method, etc. (Li et al., 2014), among which the entropy method is more popular. Objective weighting methods have a strong theoretical basis and do not rely on subjective judgement, but the calculation is tedious and does not reflect the importance of decision makers to the indicators, so the calculated weights may be contrary to the actual impact of the indicators. According to the respective strengths and weaknesses of the subjective and objective empowerment methods, AHP was applied to initially determine the weights of the performance indicators for PAFCS, preserving the preferences of

Table 15.4 Indicator definitions and descriptive statistics

Indicator	Definition and assignment of values	Mean	S.D.	Max	Min
Percentage of local labourers	Number of local labourers employed by the project/ number of all labourers employed by the project * 100%	60.25	9.63	76.92	47.19
Percentage of poverty-stricken labourers	Number of local poverty-stricken labourers employed by the project/ number of all labourers employed by the project * 100%	23.73	24.06	68.68	0.71
Project area	Size of area (km^2)	826.67	809.07	2349	190.6
Income per capita	Total income of collective economic organisations and farmers from project work, carbon sink trading, public welfare and compensation, etc./total population of the project community (RMB)	569.44	436.34	1188.9	216.6
Percentage of poverty-stricken population income	Income of poverty-stricken population participating in the project/total income of collective economic organisations and community farmers participating in the project * 100%	25.09	20.95	54.16	0.71
Number of trainings	Number of relevant skills trainings	2.17	1.17	4	1
Participation rate of training among poverty-stricken	$F_r = \frac{\sum_{i=1}^{n} F_i}{\sum_{I=1}^{n} X_i} \times 100\%$, where $i = 1, 2, \ldots, n$ denotes the *i*-th training; F_i denotes the number of poverty-stricken people participating in the *i*-th training; X_i denotes total participants in the *i*-th training (%)	20.42	19.82	58.30	1.70
Community involved in project design, management, and monitoring	Yes = 1 or No = 0	0.5	0.55	1	0

(continued)

Table 15.4 (continued)

Indicator	Definition and assignment of values	Mean	S.D.	Max	Min
Road built during the project	Length (km)	6.87	1.91	9.3	4.2
Number of small water facilities built during the project	Number	2.83	0.75	4	2
Areas adopting advanced forestation techniques	Size of area (ha)	226.22	35.08	290	190.6
Job opportunities created	Number of jobs	16.67	8.30	27.55	6.90
Selection of tree species	Number of tree species planted in the project/ number of local tree species totally (%)	65.22	18.56	91.60	33.30
Forest coverage rate	Forest coverage growth due to the project (%)	17.44	15.70	42.97	3.19

decision makers for the indicators. The entropy value method was used to revise the initial weights, thus minimising subjectivity, making the assignment more reliable, and making performance evaluation results more accurate.

15.2.4.1 Hierarchical Analysis

AHP is a systematic and hierarchical method combining qualitative and quantitative analyses (Saaty, 1994), which provides strong practicality and effectiveness in dealing with complex decision-making problems. In this study, the method was applied to determine the preliminary weight of each indicator. The two steps were as follows:

Forming an expert team. Experts in this field were invited to form an expert team, and the importance of each index was compared and analysed one by one using the nine-digit scale method (Table 15.5). Ten experts were invited, in addition to scholars from the research fields of forestry, plantation forest ecology, forest carbon accounting, forestry economy, rural anti-poverty, etc., as well as forestry leaders from different counties. Each expert had their own focus in their research field, which effectively compensated for the bias of experts giving higher scores to indicators in their field.

Calculation of indicator weights at the criterion level. The judgement matrix was constructed by setting the three indicators of economic benefit (A), development capacity (B), and development opportunities (C) according to the 1–9 scale method (see Table 15.6).

Table 15.5 Nine-digit scale

a_{ij}	Indications
1	a_i and a_j have the same impact
3	a_i has a slightly stronger influence than a_j
5	a_i is more influential than a_j
7	a_i is significantly stronger than a_j
9	a_i is absolutely stronger than a_j
2,4,6,8	Middle values of adjacent judgements (above)
Reciprocal	Value of a_i compared with a_j, $a_{ij} = 1/a_{ij}$

Table 15.6 Judgement matrix

Criteria layer factors	Income (A)	Development capacity (B)	Development opportunities (C)	Weight
Income (A)	1	3	5	0.656
Development capacity (B)	1/3	1	1	0.187
Development opportunities (C)	1/5	1	1	0.158

According to the AHP calculation method, eigenvectors of the judgement matrix were calculated:

$\overline{\omega} = [0.656, 0.187\,0.158]$;

the maximum eigenvalues $\lambda_{\max} = 3.029$;

via a consistency test, the consistency index $CI = \frac{\lambda_{\max} - n}{n-1} = 0.015$, which was small enough to meet the requirements.

Checking the random consistency index table, when $n = 3$, the random consistency index $RI = 0.580$. According to the formula $CR = \frac{CI}{RI}$, the consistency ratio result is $CR = 0.025 < 0.1$, so the matrix has good consistency. The obtained weight values $\overline{\omega} = [0.656, 0.187, 0.158]$ can be used.

15.2.4.2 Entropy Value Method

Similar to the hierarchical analysis, the entropy value method of assigning a weight to indicators also requires two steps to complete.

Dimensionless of raw data. As different types of data for each indicator cannot be evaluated directly, to overcome the statistical bias, mathematical methods were used to eliminate the influence of the original magnitude (Zhan et al., 2016). According to the data characteristics, the extreme difference normalisation method (min–max normalisation) was used for processing, as follows.

Let there be m evaluation indicators and n ($n > 1$) evaluated units (regions); the original indicator data matrix can be formed as $X = \left(x_{ij}\right)_{m\times n}$ ($i = 1, 2, \ldots$..m; $j = 1, 2, \ldots$, n), *where* x_{ij} denotes the original value of the i th indicator of the j-th evaluated unit. Then, the dimensionless treatment of the forward indicator is $y_{ij} = \frac{x_{ij}-min(x_j)}{\max(x_j)-min(x_j)}$; the dimensionless treatment of the reverse indicator is $y_{ij} = \frac{\max(x_j)-x_{ij}}{\max(x_j)-min(x_j)}$, where y_{ij} denotes the data after dimensionless processing and $min(x_j)$ and $max(x_j)$ denote the minimum and maximum values of the original data of the i-th indicator in the n evaluated units. The forward indicator means that the larger the value, the stronger the performance of PAFCS, while the reverse indicator means the opposite.

Calculate the entropy value of the i-th indicator s_i. $s_i = -\text{k}\sum_{\text{j}=1}^{\text{n}} F_j \ln F_j$, $s_i = -k \sum_{j=1}^{n} F_j ln F_j$, where $k = \frac{1}{\ln n} > 0$, ln is natural logarithm, $s_i \geq 0$. If the values of this indicator in the n evaluated units under the j-th indicator are all equal, we obtain $F_j = \frac{y_j}{\sum_{j=1}^{n} y_j} = \frac{1}{n}$, then s_i takes the maximum value, that is, $s_i = -k\sum_{j=1}^{n} l_j \ln l_j = -k - k\sum_{j=1}^{n} \frac{1}{n} \ln \frac{1}{n} = k \ln n = 1$, so the value range of s_i is in the range from 0 to 1.

The entropy value method is adopted to calculate the weight of indicators. $F_{ij} = \frac{y_{ij}}{\sum_{i=1}^{n} y_{ij}}$ represents the contribution of the *i*-th scheme A_i, E_j is used to represent the total contribution of all schemes to the attribute y_j, i.e., $E_j = -k\sum_{i=1}^{n} F_{ij} \ln F_{ij}$, where the constant $k = \frac{1}{\ln m}$, so as to ensure $0 \leq E_j \leq 1$. It can be seen that the size of the weight coefficient is determined by the size of the variance of all indicators; thus, L_j is defined as the degree of consistency of the contribution of each indicator under the *j*-th attribute. Then the weight of each attribute w_j: $w_j = \frac{L_j}{\sum_{j=1}^{n} L_j}$, when $\text{L}_\text{j} = 0$, its weight is also 0. The final weights W_j are obtained by normalisation of w_j: $W_j = \frac{w_j}{\sum_{j=1}^{n} w_j}$. The results are shown in Table 15.7.

15.2.4.3 Analysis of Performance Evaluation Results

The evaluation unit is the administrative county, and the analysis of them is represented by codes X_1– X_6 with random order. The researchers estimated the dimensional scores of PAFCS performance, multiplying the weight determined by the hierarchical analysis–entropy weighting method by the dimensionless indexes, i.e. $M_j = \sum_{i=1}^{m} W_j * y_{ij}$, where M_j represents the score of the j-th evaluated area. Then the final result was the mean value of the weighted dimensional scores (see Table 15.8).

Result analysis

Based on its poverty alleviation performance in six counties, PAFCS is effective to an extent. The first reason is government attention. With the guidance of the

Table 15.7 Weights of performance evaluation indicators for PAFCS

Target layer	Criteria layer	Weight	Indicator layer	Weight
Performance evaluation indicators for PAFCS	Economic benefit (A)	0.656	Percentage of local labourers (A1)	0.136
			Percentage of poverty-stricken labourers (A2)	0.195
			Project areas (A3)	0.244
			Income per capita (A4)	0.253
			Percentage of poverty-stricken population income (A5)	0.172
	Development capability (B)	0.187	Number of training sessions (B1)	0.327
			Participation rate of trainings among the poverty-stricken population (B2)	0.295
			Community involved in project design, management, and monitoring (B3)	0.379
	Development opportunities (C)	0.158	Road built during the project (C1)	0.118
			Number of small water facilities built during the project (C2)	0.111
			Areas adopting advanced forestation techniques (C3)	0.080
			Job opportunities created (C4)	0.363
			Selection of tree species (C5)	0.145
			Forest coverage rate (C6)	0.182

Table 15.8 Performance scores of PAFCS

Target	M_A	M_B	M_C	M
X_1	0.675	0.538	0.669	0.648
X_2	0.749	0.682	0.603	0.713
X_3	0.669	0.709	0.561	0.659
X_4	0.731	0.552	0.640	0.683
X_5	0.719	0.639	0.567	0.679
X_6	0.698	0.694	0.680	0.694
Northwest Sichuan FCS project	0.680	0.647	0.637	0.667
Novartis Sichuan FCS project	0.733	0.624	0.604	0.692
Mean	0.707	0.636	0.620	0.680
Overall variance	0.029	0.068	0.046	0.021

ecological development philosophy of *green water and green mountains are golden, lucid water and lush mountains are invaluable assets*, government departments at all levels increasingly support FCS projects. Secondly, the natural environment is the basic guarantee. Two CDM FCS projects surveyed in this study are located in areas rich in natural resources in Sichuan Province, suitable for forestry development in terms of rainfall, climate, and altitude. Further, the average area of forest land for farmers is large, which is a favourable condition for the development of FCS projects. Also, frequent natural disasters in Sichuan Province in recent years have enhanced the awareness of all parties to protect the ecological environment and promote the development of such ecological projects as FCS projects. Finally, the recognition of farmers is the key reason. FCS projects cannot be carried out without the labour input of farmers. Most farmers have a low level of education. Therefore, in addition to bringing direct economic benefits to farmers by involving them in afforestation and management, training in afforestation skills has improved their knowledge and promoted their own development initiatives.

From the viewpoint of economic benefits, the input of labour power is dominant. The two projects have progressed into the operational stage with stable development. Meanwhile, the overall FCS project is a typical market-oriented project that pursues benefits, so economic benefits account for the highest percentage in most cases. Furthermore, in terms of development capacity, China's targeted poverty alleviation strategy encourages poverty-stricken households to participate in the execution and operation of projects, with the hope of increasing the farmers' own capacity through training and other means. However, the beneficiaries of FCS should not and cannot be limited to poverty-stricken people, and all farmers should obtain training and job opportunities. Some farmers will invest their skills in their own agricultural production in the process of reforestation practices.

From the perspective of development opportunities, FCS projects are usually located in remote and poverty-stricken mountainous areas in China, and these projects can drive the development of these areas by providing job opportunities. At the same time, in poverty-stricken areas that are ecologically fragile, afforestation projects can increase forest cover and thus improve the ecological environment of communities, improve infrastructure conditions, and break external barriers to regional development, thus highlighting the poverty alleviation function of FCS projects.

Finally, the differences among economic benefits (mean score 0.707), development capacity (mean score 0.653), and development opportunities(mean score 0.620) were not significant. The maximum difference (between economic benefits and development opportunities), only 0.053, is gradually narrowing compared with the early stage of the project operation; the total variance of the three dimensions are 0.029, 0.068, and 0.046, respectively, which shows that the main gap is still relatively concentrated in the latter two dimensions.

Comparative analysis

As a whole, the economic benefit of the Novartis Sichuan FCS Project (0.733) was slightly higher than that of the Northwest Sichuan FCS Project. The main reason is that the Novartis Sichuan FCS Project was carried out following the Northwest

Sichuan FCS Project and is more closely integrated with the goal of poverty alleviation in China, so the targeting of the projects was clearer. In contrast, although the Northwest Sichuan FCS Project signed a trade cooperation agreement with Hong Kong Low Carbon Asia Limited, the volume of carbon sink transactions was low after 2012, so farmers received less income from them. Second, in terms of development capacity, the Novartis Sichuan FCS Project (0.624) is populated mostly by the Yi ethnic group, and its own economic foundation is weak, so its urgency for development and momentum of participation was slightly lower than that of the Northwest Sichuan FCS Project (0.647). Furthermore, the development capacity score of the Northwest Sichuan FCS Project (0.637) was higher than that of the Novartis Sichuan FCS Project (0.604), because the former has certain advantages in terms of infrastructure improvement and ecological environment protection. In addition, the economic benefits of the Novartis Sichuan FCS Project were higher than the overall average value (0.707), while the Northwest Sichuan FCS Project is higher than the overall average in terms of development capacity and opportunities (0.636 and 0.620). Therefore, the two projects have their own outstanding features and have brought positive impacts to the development of their respective areas and the production and living standards of farmers.

Within the Northwest Sichuan FCS Project (see Fig. 15.2), County X_6 scored slightly higher (0.694) than County X_1 (0.648) and County X_3 (0.659). As seen in Fig. 15.3, County X_1 scored slightly lower than the overall average in terms of economic benefit (0.675) and development capacity (0.538), while the development opportunities score (0.669) was higher than the overall average. Overall, X_1's poverty alleviation performance score(0.648) was lower than the overall mean. County X_3 scored slightly lower (0.669) in terms of economic benefit and higher in terms of development capacity (0.709) and development opportunities (0.561) than the overall average. Then, county X_3's overall poverty alleviation performance score (0.659) was lower than the overall mean. County X_6's economic benefit score (0.698) was slightly lower than the overall mean, the development capacity score (0.694) was higher than the overall mean, and the development opportunities score (0.680) was higher than the overall mean. Then its poverty alleviation performance score (0.694) was higher than the overall mean.

Among the Novartis Sichuan FCS Project (see Fig. 15.4), County X_2 scored higher (0.713) than X_4 (0.683) and X_5 (0.679) in terms of poverty alleviation performance. Focusing on the counties (see Fig. 15.5), X_2 had a higher economic benefit score (0.749), a higher development capacity score (0.682), and a lower development opportunities score (0.603); it had a higher overall poverty alleviation performance score than the overall average. X_4 had a higher economic benefit score (0.731), a lower development capacity score (0.551), and a higher development opportunities score (0.640); it had a slightly higher overall poverty alleviation performance score than the overall average. X_5 economic benefit score (0.719) was higher, its development capacity score (0.639) was slightly higher, and its development opportunities score (0.567) was lower; X_5 overall poverty alleviation performance score was slightly higher than the mean.

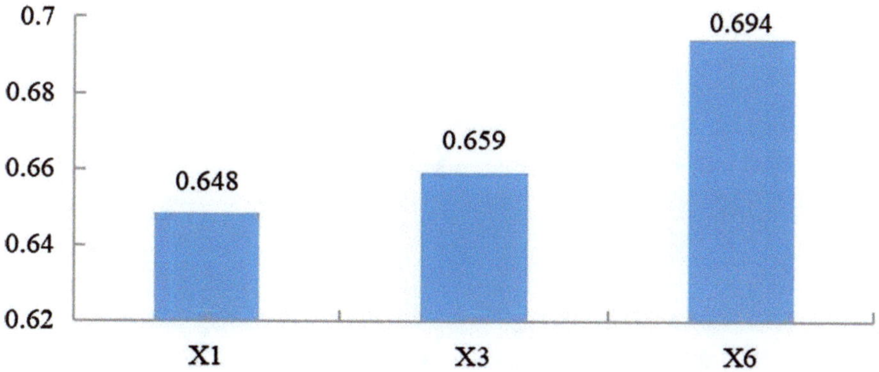

Fig. 15.2 Poverty alleviation performance scores of the Northwest Sichuan FCS Project

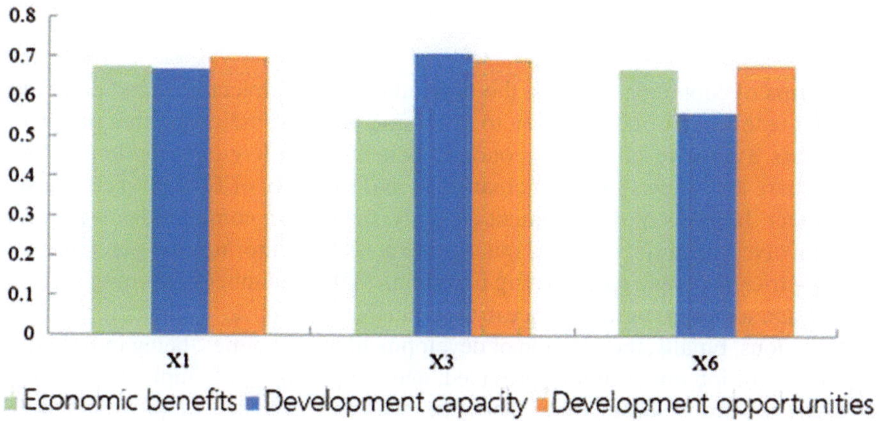

Fig. 15.3 Poverty reduction performance scores of the Northwest Sichuan FCS Project by dimension

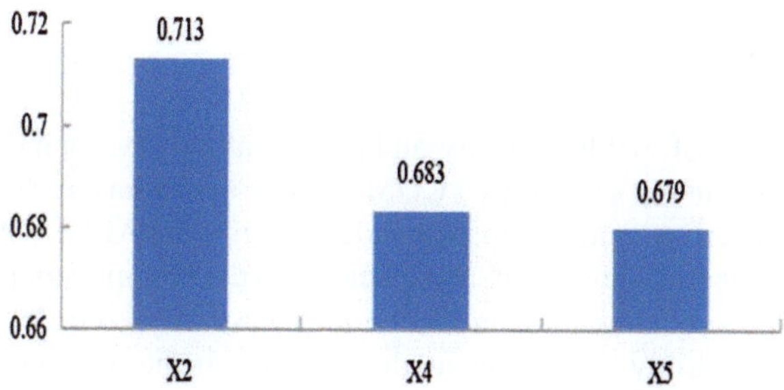

Fig. 15.4 Poverty alleviation performance scores of the Novartis Sichuan FCS Project in three counties

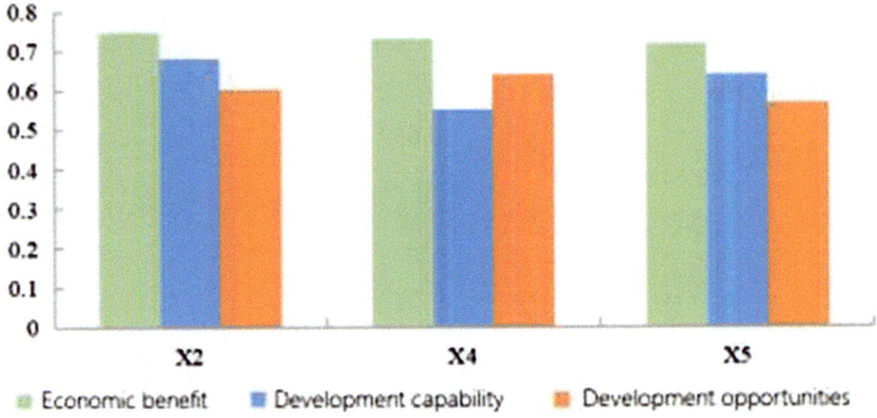

Fig. 15.5 Poverty reduction performance scores of the Novartis Sichuan FCS Project by dimension

In terms of economic benefits, the two indicators of project area and per capita income made a higher contribution to economic benefits, indicating that per capita capital has a positive effect on economic benefits. The low contribution of local labour share also shows that participation in and publicity of FCS projects are still insufficient. In terms of development capacity, the project communities were averagely involved in project design, management, and monitoring; the percentage of poverty-stricken people participating in training and the number of training sessions had average scores, so the difference in the contribution to poverty alleviation benefits is not obvious. Finally, the creation of development opportunities, being indicated by the number of job opportunities provided, and the provision of employment opportunities can address surplus rural labourers, who can learn new employment skills, which is important in improving the economic status of farm households.

15.3 Summary

Based on the theory of viable capacity and participatory development, this chapter leveraged the field survey data of two CDM FCS project areas in Sichuan Province and constructed a performance evaluation index system for PAFCS with three dimensions: economic benefits, development capacity, and development opportunities. The performance of poverty alleviation in six counties was analysed in depth using a hierarchical analysis–entropy weighting method. Through the analysis and empirical research, the following conclusions were reached.

The performance evaluation index system of PAFCS is feasible, universal, and reasonable. Based on the theory of feasibility, this study combined the subjective empowerment method with the objective empowerment method and established an index system in three dimensions: economic benefits, development capacity, and development opportunities, and it validated the index system in two CDM-AR project areas.

The FCS projects provide certain poverty alleviation benefits, and there is room for improvement. The evaluation of the six project counties showed that FCS projects offer multiple economic, social, and ecological benefits for the regions. In particular, they have been carried out in poverty-stricken areas, and their poverty alleviation function has been manifested, but there is still room for improvement, such as regarding the strict standards of FCS projects and imperfections in the carbon trading market.

The performance of PAFCS varies among projects and in different regions of the same project. Due to differences in regions, project capital investments, and social and cultural characteristics, the Northwest Sichuan FCS Project and the Novartis Sichuan FCS Project differ in terms of economic benefits, development capacity, development opportunities, and overall poverty alleviation performance. The higher investment in the Novartis Sichuan FCS Project is the main reason for the economic benefits higher than those in the Northwest Sichuan FCS Project. Because the Northwest Sichuan FCS Project has been executed for a longer period, related development capacity and development opportunities were slightly stronger than those in the Novartis Sichuan FCS Project.

PAFCS has both short-term and long-term performance. From the perspective of poverty alleviation performance, economic benefits still are dominant, but from the actual situation of the two FCS projects, their transaction amounts in the carbon sink market in China are still low; the carbon sink market in China is still in its initial stage and needs to develop and improve constantly. Although the economic benefits can quickly help the farmers and achieve poverty alleviation performance in the short term, the short-term performance cannot guarantee that farmers maintain poverty alleviation for a long period (i.e., without returning to poverty), so countermeasures in terms of creating development opportunities and improving development capabilities should be taken.

Although the two FCS projects exhibit certain poverty alleviation performances, they have not yet reached the most desirable results, and they are long-term projects that need continuing promotion. Therefore, the researchers conclude that several points need to be improved in the operation of future FCS projects, as follows.

Establish time-series monitoring and improve the dynamic mechanism

Due to the regularity of the growth of carbon sink forests, the strict monitoring of carbon sink measurement, and the special characteristics of the carbon trading market, the development of FCS projects has significant long-term and risk characteristics, which leads to the lagging and fluctuation of the poverty alleviation benefits of the projects. Therefore, it is necessary to establish time-series monitoring to realise the dynamic management of FCS projects and to use the performance index system to assess their performance at different times and spaces. Further, it is essential to carry out an objective and straightforward quantitative evaluation of project development, operation, and management; make timely adjustments to project system design and arrangement; and then tap the poverty alleviation potential of FCS projects.

First, the economic benefits of FCS projects should be quantified in a timely and accurate manner according to the actual carbon trading market, and the actual contribution of FCS projects to the income growth of farmers should be measured scientifically. The stability of the economic benefits of FCS projects should be maintained by relying on financial subsidies and financial instruments to ensure that farmers obtain stable and sustainable income. Second, according to the actual development stages of forests, the long- and short-term employment opportunities provided (or brought) by FCS projects should be measured accurately, avoiding substantial job losses in the middle of the project through employment transfer and job optimisation. Third, according to the dynamic development of the project, the needs of farmers for skills training should be identified, and according to their needs at different time points, the content of training should be adjusted and training methods optimised in a timely manner.

Strengthen management and enhance the effectiveness of poverty alleviation

First, the overall weighing of multiple benefits of FCS projects should be noticed, avoiding one-sided pursuit of ecological benefits. FCS projects have significant multiple benefits, not only increasing carbon stock for the purpose of market transactions but also providing long- or short-term employment and non-farm skills training as well as attracting external funds. However, FCS projects are market-oriented, and the carbon stock is intended for trading. Thus, project owners might pay more attention to the ecological benefits of the project, ignoring the economic and social benefits. Therefore, the government should act as an independent third party and intervene in the project owner's behaviour to balance ecological with economic and social benefits. Second, the exchange of experience should be strengthened to improve project system design. The natural and economic conditions of project regions differ, as do the organisation mechanism, operation mechanism, management mechanism, and other multiple links. Performance therefore varies significantly among different projects and among different regions of the same project. Thus, it is critical to rely on the performance index system and learn from the experience of different projects and different regions of a project in FCS poverty alleviation, summarising the dilemmas of poverty alleviation and optimising the system design of FCS projects.

Strengthen coordination and integrate multiple resources

On the one hand, the development and operation of FCS projects inevitably have a significant impact on the development of agriculture, pastoralism, and other industries, and there is a certain degree of conflict with the economic development planning of the project area. Therefore, it is necessary to strengthen coordination among forestry, agricultural, and other functional departments, especially in carrying out regional economic development planning in order to avoid conflicts in the same project area. Otherwise, local farmers might lack a sense of security. On the other hand, PAFCS cannot rely entirely on the forestry department; other functional departments, such as the poverty alleviation department and transportation department, should participate to form a synergy, and it is necessary to coordinate the operation of many functional departments. Specifically, the subjects and objects of projects should be coordinated in order to run well. Second, in addition to the government, it is necessary to make use of the leading enterprises, financial institutions, and other kinds of organisations to run the projects with their respective areas of expertise, to improve the carbon sink trading market, and to truly realise carbon sink trading so that farmers can convert carbon sink assets into income. Finally, actual local situations must be considered in order to create a long-term mechanism for PAFCS and to incorporate the economy, the environment, culture, and resources in planning. All the forces should actively participate to ensure long-term and sustainable poverty alleviation work.

Talent reserve and scientific and technological support

Recruiting and training talents is vital, and the key role of talents in project research, development, and management should not be overlooked. The development of carbon sink projects also needs to accelerate scientific research and resource innovation, which can be backed up by cooperation with universities and governments to form a system that attracts, cultivates, and manages talents for the development of FCS projects. First, rely on institutions of higher education, scientific research institutes, third-party institutions, etc., to carry out training to spread forestation knowledge in townships and administrative villages so as to gather a large number of grassroots talents with knowledge and skills for FCS project. Second, rely on experts and scholars in the field of FCS in order to continuously bring new ideas on the temperature, humidity, altitude, and other alpine planting dilemmas faced by FCS projects in practice. It would be of practical value to cultivate new tree varieties, innovate technically, and develop new models to expand the altitude of forestation lines upward. Third, continuously strengthen research on the methodological innovation of FCS projects. Carbon stock and carbon leakage should be calculated according to different tree species, climatic conditions, project areas, and economic and social conditions. Especially the carbon sink measurement model should be adjusted to include more tree species and plots into carbon trading.

Publicity and social influence

Given the initial stage of PAFCS, various theories and methods are being researched and developed, and social awareness of the poverty alleviation effect of FCS projects is not high. In this case, existing experience in the planning, design, and operation of FCS projects should be drawn on. At the same time, public awareness of the poverty alleviation performance of FCS projects should be enhanced so that the public can recognise and acknowledge the poverty alleviation performance of FCS projects and hence actively participate in carbon sink projects. First, radio, TV, brochures, newspapers, and village bulletin boards should be mobilised to continuously strengthen farmer awareness of FCS projects, reduce negative evaluations of projects, enhance recognition of the multiple benefits of projects, and strengthen support for projects. Second, through this summary and experience promotion, TV media and the Internet should be used to publicise information about the effectiveness and success of FCS projects so that more people and social institutions will pay attention to them and social resources will be attracted to promote the vertical extension of the FCS industry. Third, FCS project demonstration sites should be promoted to highlight their poverty alleviation benefits so that more people can discover the various benefits of FCS projects.

References

Chen, C. C., McCarl, B., Chang, C. C., et al. (2011). Evaluation on the potential economic impacts of Taiwanese biomass energy production. *Biomass and Bioenergy, 35*(5), 1693–1701.

Chen C. Y. (2010). Forestry carbon sequestration and farmers' livelihoods: The example of the world's first forestry carbon sequestration project.*World Forestry Research, 23*(5), 15–19.

Chen, S., Pan, H., & Lu, J. (2016). Targeted poverty alleviation performance and its influencing factors: a case study based on the east, middle and west China. *China Administration, 9*, 88–93.

Chia, E. L., Somorin, O. A., Sonwa, D. J., et al. (2013). Local vulnerability, Forest communities and Forest-carbon conservation: Case of southern Cameroon. *International Journal of Biodiversity and Conservation, 5*(8), 498–507.

Chu, R., & Zhou, F. (2012). Study of Kuznets inverted U-shaped relation between forest carbon sink and economic growth. *Journal of Central South University of Forestry and Technology, 32*(10), 94–99.

Ding, Y., & Ma, P. P. (2013). Research on forest carbon sequestration and economic development in ethnic minority areas in west Sichuan-taking Yuexi county, Liangshan yi autonomous prefecture, Sichuan province as an example.*Rural Economy, 5*, 38–41.

Edwards, A. C. (2000). Wage employment and rural poverty alleviation. In López, R., & Valdés, A. (Eds.), *Rural poverty in Latin America*. Palgrave Macmillan.

Hi, M., Qi, Y. B., Li, Y. M., et al. (2015). Will foresters participate in forestry carbon sequestration projects?–Taking the CDM forestry carbon sink pilot project as an example. *Agricultural Technology and Economics, 7*, 102–113.

Li, L., Jia, L., Zhao, X. X., et al. (2014). Application of the AHP and entropy weightmethod in evaluation on city waterenvironmental carrying capacity. *Resources and Environment in the Yangtze Basin, 23*(4), 456–460.

Lin, D. R., & Li, Z. Y. (2006). Policy studyofCDM forestry carbon-sink projects in China.*World Forestry Research, 19*(4), 52–56.

Li, S. C., Xu, Y. Q., & Fu, X. F. (2005). Spatial simulation using GIS and ANN for regional pauperization in China. *Resources Science, 27*(4), 76–81.

Lv, Z. (2014). *Forest carbon sequestration practices and low carbon development in China*. Peking University Press.

Ma, J. T. (2016). Conscientiously studying and implementing the important speech of General Secretary Xi Jinping and making concerted efforts to win the decisive battle against poverty. *Journal of the National School of Administration, 2*, 4–10.

Ma, P. P. (2012). Research on forest carbon sinks and development of poverty-stricken minority areas in Western Sichuan-A case study based on poverty alleviation through carbon sinks in Yuexi, Liangshan. Master's Thesis, Sichuan Academy of Social Sciences, Chengdu.

Meng, X., & Li, B. (2013). Research on multi-objective decision analysis model and application. *Modern Management Science, 7*, 42–44.

Milder, J. C., Scherr, S. J., & Bracer, C. (2010). Trends and future potential of payment for ecosystem services to alleviate rural poverty in developing countries. *Ecology and Society, 15*(2), 4.

Ren, C., & Yuan, M. B. (2017). Classification management: The problems in practice and focus direction in targeted poverty reduction policies. *Beijing Social Science, 1*, 100–108.

Saaty, T. L. (1994). How to make a decision: The analytic hierarchy process. *European Journal of Operational Research, 48*(1), 9–26.

Sen, A. (1999). *Development as freedom*. Oxford University Press.

Wang, H., Jiang, X. M., Chen, J. C., et al. (2017). A theoretical explanation for the relationships between "Green Hills" and "Golden Hills". *China Rural Economy, 4*, 2–12.

Wang, Y. H., Qian, L. Y., Chen, Y. F., et al. (2017), A Multidimensional and comprehensive poverty measurement of poverty-stricken counties from the perspective of ecological poverty. *Journal of Applied Ecology, 8*, 2677–2686.

Xiang, D. P., & Gao, F. (2013). On the impact of policyimplementation modes on the performance ofpoverty reduction. *Journal of Huazhong Normal University (Humanities and Social Sciences Edition), 52*(6), 12–17.

Xu, L. P., Ling, B., & Tan, T. Y. (2013). The research of China's rural pro-poverty-stricken community of interests comprehensive performanceevaluation model. *Issues in Agricultural Economics, 34*(12), 58–64.

Xu, Y. L., & Jv, H. (2009). Climate change and poverty: A case study of China (extract). *World Environment, 4*, 50–53.

You, X. C., & Tian, J. (2009). Comprehensivepoverty alleviation performance appraisalmethods and empirical research in national areas. *Science, Economy and Society, 27*(3), 7–13.

Zeng, W. Z., Cheng, Y., & Yang, F. (2018). A evaluation index system of forest carbon sequestration poverty alleviationperformance based on the CDM Forest Carbon Sequestration Project. *Journal of Nanjing Forestry University (Natural Science Edition), 42*(4), 9–17.

Zeng, Y. M., & Zhang, G. (2011). Spatial simulating in regional rural poverty based on GIS and BP neural network:A new appraisement method on regional rural poverty. *Geography and Geographic Information Science, 27*(2), 70–75.

Zhang, Y. B. (2015). Industrial convergence and interaction in the vision of agriculturalmodernization and its path. *Reform, 2*, 98–107.

Zhan, M., Liao, Z. G., & Xu, J. P. (2016). Character analysis of linear dimensionless methods. *Statistics & Information Forum, 31*(12), 17–22.

Zhuang, T. H., Zhang, H. X., & Yu, C. Y. (2012). Southwest minority poverty counties antipoverty comprehensive performance evaluation: Key counties for povertyrelief to 10 counties as an example. *Northwest Population, 33*(3), 89–93.

Chapter 16
Evaluation of Farmer Perceptions of the Effects of PAFCS

Farmers are important participants and stakeholders in FCS projects. Their positive evaluation of the multiple effects of a project reflects their understanding and support of these projects and directly determines their willingness to (continue to) participate. It helps consolidate the results of pre-forestation and reforestation and their long-term operation, and it is crucial to achieving the established goal of a win–win situation in combating climate change and reducing poverty. While FCS projects bring economic, social, and ecological benefits to farmers, they also inevitably have negative impacts that influence farmers' decisions. As pointed out earlier, these negative impacts include, but are not limited to, the externalisation of benefits, restricted participation of vulnerable groups, and restrictions on traditional livelihood activities such as grazing, understory planting, and fuel wood harvesting by farmers. In view of this, an in-depth analysis of the positive and negative poverty alleviation effects of FCS is important for improving the institutional design of FCS projects and optimising poverty alleviation. Therefore, the researchers used household questionnaire data from typical FCS projects and applied SEM to empirically test the relationship between farmers' perceived value of poverty alleviation and their willingness to support the later operation of the FCS project. The purpose of the study in this chapter was to provide a useful reference for poverty alleviation policy by targeting the interests of farmers.

16.1 Literature and Theoretical Hypotheses

In retrospect, perceived value has been well studied by academics. In this chapter, the researchers define the effect of FCS projects on poverty alleviation from the perspective of farmer perceptions and identify the main factors influencing these perceptions.

W. Zeng and F. Yang, *Poverty Alleviation Via Forest Carbon Sequestration*, International Research on Poverty Reduction, https://doi.org/10.1007/978-981-99-8191-5_16

16.1.1 A Definition of Farmer Perceived Value

Perceived value theory was introduced by Zeithaml (1988) in a study of consumer behaviour. Applied mainly in marketing, it refers to the comprehensive evaluation of consumer benefits from using products or services after comparing the perceived benefits with the perceived costs in the process of purchasing and consuming them, or the comparison between the gain and loss (from the products/services) by consumers (Flint et al., 2002). The core of the theory is the trade-off between perceived benefits and perceived payoffs. The dimensions revealing perceived value are becoming more diverse. Woodruff and Cardial (1996) proposed a basic model of the relationship between price, perceived quality, perceived benefits, perceived risks, and customer purchase intention. Li et al. (2009), Li et al. (2012), and Qin et al. (2016) studied the perception of residents in poverty-stricken areas regarding the positive and negative effects of tourism on poverty alleviation, including economic, social, and environmental effects, concluding that the perceived value is an important factor determining residents' behavioural attitudes. Zhang and Zhao (2014) conducted an empirical study on the perception of ecological compensation effects and the factors influencing farmers' willingness to participate in a farm retreat project. They concluded that improving farmer perceived value of ecological compensation is an important way to guarantee the sustainable development of ecological compensation projects. Based on perceived value theory, Han (2016) conducted a study of farmer behaviour in utilising agricultural waste and established a model with two dimensions, i.e., perceived benefits and perceived risks, with six sub-dimensions, namely perceived economic, social, and ecological benefits and perceived economic, psychological, and situational risks, showing that farmers' perceived value of resource utilisation indirectly influenced their utilisation behaviour through their willingness to utilise. Accordingly, in this chapter, the perceived value of an FCS is the subjective perception of farmers on whether the benefits of FCS projects in poverty-stricken areas outweigh the disadvantages. This is the subjective perception of farmers in the process of direct or indirect participation in project development. The theoretical framework of the study, shown in Fig. 16.1, draws on research results and combines the characteristics of FCS projects and their multiple effects.

16.1.2 Identification of Farmer Perceived Value

In usual economics methodology, farmer perceived value follows the principle of maximising utility, i.e., maximising perceived benefits and minimising perceived risks. However, farmer participation and benefits in FCS projects are diverse. The effects of PAFCS are reflected at both the macroscopic scale (such as in regional economic development and the organisational management capacity of poverty-stricken communities) and at the microscopic scale (such as in income, viability, and the creation of development opportunities). Both economic and non-economic

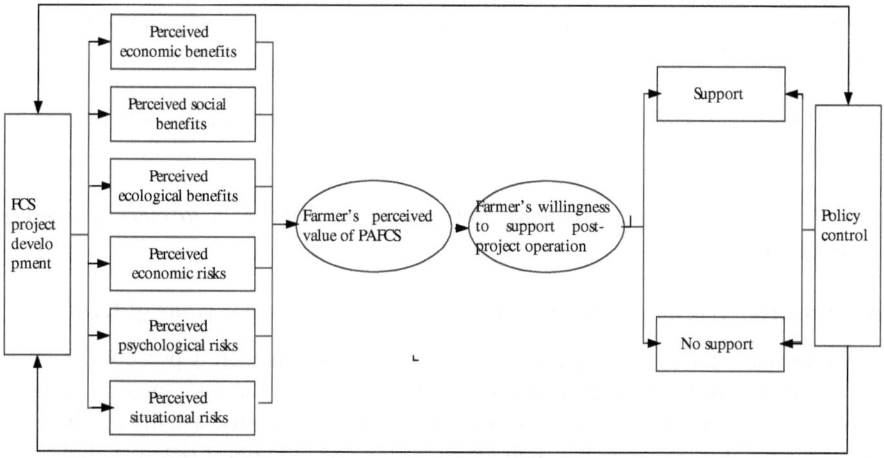

Fig. 16.1 Theoretical model of the perceived value of PAFCS

impacts, as well as short-term and long-term, positive and negative impacts, must be considered. A farmer's perceived value of PAFCS is not only limited to the maximisation of utility but also the farmer's comprehensive judgement, as a socialist, of the project development. This consideration is conducive to avoiding the bias brought by the assumption of a single rational economic person and is also an objective requirement to reduce the uncertainty of evaluation in the strengthening of social and public welfare through poverty alleviation and ecological development. Therefore, here, farmer perceptions of PAFCS were treated as observed variables; the perceived value of whether the benefits outweigh the disadvantages was treated as a first-order latent variable; and the willingness of farmers to support the later operation of the projects was treated as a second-order latent variable. In order to better explore the relationship between the observed and latent variables, the following hypothesis is proposed based on the above theoretical model of farmer perceived value.

H1: Perceived benefits have a positive effect on farmer perceived value of PAFCS.

Keller (1993) argued that perceived benefits are all the benefits and gains that an individual perceives to be gained by purchasing a product or service or by taking a certain action. The perceived benefits of farmers can be either direct (such as increased income, access to new technologies, and expansion of social networks) or indirect (such as changes in traditional views and improvements in their production, living conditions, and ecological environment), including both real benefits obtained in the early stage and potential benefits that can be expected. The theory of perceived benefits suggests that the higher the individual's perceived benefits, the higher the perception of value. Integrating the above analysis, this study decomposes perceived benefits into three aspects: perceived economic, social, and ecological benefits, and further proposes the following hypotheses:

H1a: Perceived economic benefits have a positive influence on farmers' perceived value of PAFCS.

H1b: Perceived social benefits have a positive influence on farmers' perceived value of PAFCS.

H1c: Perceived ecological benefits have a positive impact on farmers' perceived value of PAFCS.

H2: Perceived risk has a negative effect on farmers' perceived value of PAFCS.

Dowling and Staelin (1994) and Keh et al. (2002) considered perceived risk as the subjective perception of various objective and potential risk factors. It is the psychological feeling of adverse consequences and uncertainty, usually attributed to several perceptual dimensions such as economic risk, temporal risk, psychological risk, opportunity risk, functional risk, and social risk. Meng et al. (2010) pointed out that risk perception varies from person to person, and factors such as external observation, empirical judgement, fear emotion, and surrounding environment can have an impact on perceived risk. Farmer perceived risk includes not only monetary and non-monetary opportunity costs and the fear of intangible mental costs such as equity protection but also likelihood risk assessment based on similar experience, i.e., situational risk. In the field of public behaviour research, voluminous literature shows that perceived risk has negative impacts on individual perceived value. That is, if the level of perceived risk is lower, the perceived value is greater, and vice versa. Combining the above analysis, this study decomposes perceived risk into three aspects: perceived economic, psychological, and situational risk, and further proposes the following hypotheses:

H2a: Perceived economic risk influences farmers' perceived value of PAFCS negatively.

H2b: Perceived psychological risk influences farmers' perceived value of PAFCS negatively.

H2c: Perceived situational risk influences farmers' perceived value of PAFCS negatively.

H3: Farmers' perceived value has a significant effect on their willingness to support the later operation of a project.

The duration of an FCS project is long and entails pre-forestation, mid- and post-management, and CER verification and certification. Farmer willingness to support the post-operation of an FCS project includes farmers' willingness to strictly comply with the contractual agreement based on economic rationality as opposed to withdrawing from the project due to the opportunity cost that may arise from the long-term change of land use. Also should be included a farmer's willingness, as a socialist, to practise post-management requirements (such as avoiding carbon leakage and safe use of fire) based on the trade-off of non-economic effects (Yang et al., 2016).

Empirical studies have shown that farmers' perceived value of *benefits outweighing the disadvantages* has a positive effect on their willingness to participate in the return of farmland to forest, poverty alleviation through tourism, agricultural land preparation, non-agricultural conversion of farmland, and resource utilisation of agricultural waste, whereas the perceived value of *disadvantages outweighing the benefits* has a negative effect. Accordingly, it is assumed that the perceived value of PAFCS has a significant effect on the willingness of farmers to support the operation of the project in the later stage. In other words, the perceived value of *more good than bad* has a positive effect, while the perceived value of *more bad than good* has a negative effect.

16.2 Investigation and Analysis of Farmer Perceived Value

In order to confirm the validity of farmer perceived value, the researchers ascertained the theoretical hypotheses through an empirical study. This section details the data sources, the survey method, and the overall sample used for the empirical analysis. The samples used for the analysis were gathered from the Novartis Sichuan FCS Project and the Tenchong FCS Project.

16.2.1 Data Description

The data were collected from households in areas of the Novartis Sichuan FCS Projects and the Tenchong FCS Projects. To ensure the rationality of the survey and the quality of the research data, the research team used a combination of Delphi, group discussion, and Likert scale methods to design the questionnaire and made several revisions based on several pilot surveys conducted before the formal survey. The formal survey adopted a combination of typical sampling and stratified sampling methods. First, typical sampling was adopted to select 8 poverty-stricken villages in two counties from among 27 villages in the five national counties of Zhaogjue, Yuexi, Meigu, Leibo, and Ganluo in Liangshan Yi Autonomous Prefecture, where the Novartis FCS Project is located, and three poverty-stricken villages from among the five villages in three townships (towns) where the Tengchong FCS Project is located. Then, 20–50 households were randomly selected for the questionnaire survey, and semi-structured interviews focused on the farmer's intuitive feelings of community changes. A total of 419 farm households were surveyed. Because of the dispersed distribution of the surveyed households, inconvenient transportation in the poverty-stricken areas, and communication difficulties induced by the indigenous language, some data are missing or abnormal. So only 328 valid samples were obtained after centralised questionnaire quality testing, with a sampling efficiency rate of 78.28%.

In the survey, conference talks and individual interviews were also conducted with the responsible persons and managers of afforestation subjects, such as the forestry authorities of the project areas, namely the Sichuan Dadu River Afforestation Bureau and Sujiang State Forestry Farm. The Tengchong and Novartis projects were formally launched in 2007 and 2010, respectively, with a construction period of 30 years, and have progressed from the reforestation stage to the management and operation stage. Both projects have adopted the construction model of *enterprise + government + farmers*. The practices of allocating the carbon sink income and returning all the trees to the landowners at the end of the project are in accordance with CDM-AR standards. A tripartite contract was signed with the farmers involved in the afforestation projects.

16.2.2 Description of the Survey Sample

Among the surveyed farm households (see Table 16.1), male (65.12%) and young adult respondents predominated; the proportions aged 31–40 and 41–50 years old were 25.20 and 31.17%, respectively. The majority of the family members living together were in the range of four to six (56.10%). Those with more than one member working outside the village and those who were or had previously been township (village) cadres were few, accounting for 8.00% and 1.00%, respectively. The education level (mean value 6.38 years) and annual per capita household net income (mean value 3507.18 RMB, about 510 USD) of the sample farm households were significantly lower than the average levels of the region they belonged to; 19.00% of households were poverty-stricken, and the incidence of poverty was significantly higher than the average level of the region. Most of the households have participated in technical training (72.13%) or obtained direct income (92.12%). During the survey, it was found that, first, despite differences in the level of economic and social development among the villages, the general infrastructure lags behind, public services are weak, individual villages are not connected to cement or asphalt roads, and safe water, electricity, and housing have not been completely solved. Due to limitations of education level and language, it is difficult for local farmers to find paid work, and a high proportion of young and strong labourers are left behind. Traditional farming and forestry by-products are still the main source of income for most families. Secondly, under the constraint of limited carbon trading and bank loan funds, there are several reasons for the smooth execution of a project, extensive participation, and benefits to farmers. For example, reforestation enterprises, NGOs, communities, and other stakeholders work closely together; most reforestation plots belong to a farmer's confirmed land, and supporting policies include ecological initiatives by local government, technical training, industrial poverty alleviation projects, and ecological afforestation, ecological compensation funds for grazing losses.

Table 16.1 Basic information of farmer sample

Category	Min.	Max.	Mean	S.D.
Gender (1 = male; 0 = female)	0	1	0.65	0.48
Ethnicity (1 = Han; 0 = other)	0	1	0.39	0.27
Age (years)	18	75	39.28	13.45
Formal education (years)	0	16	6.38	2.65
Participation in technical training	0	9	3.24	1.65
Household net income per capita (RMB/year)	1200	35,000	3507.18	2722.85
Number of family members (persons)	1	10	4.58	1.52
Number of family members working outside village (person)	0	4	0.78	1.13
Whether anyone in the family (had) acted as a township (village) cadre (1 = yes, 0 = no)	0	1	0.09	0.27
Area of land involved in afforestation (ha)	0	1.07	0. 59	0. 32
Whether direct income is obtained (1 = yes, 0 = no)	0	1	0.92	0.28
Whether the household is poor (1 = yes, 0 = no)	0	1	0.19	0.21
Project area (1 = Tengchong, 0 = Novartis)	0	1	0.57	0.19

16.2.3 Farmer Perceived Value

The statistical results show that 72.56% and 69.82% of the farmers (i) thought that the benefits of PAFCS outweigh the disadvantages and (ii) were willing to support the later operation, respectively, showing a clear inclination towards most of the observed variables. This reflects that local farmers perceive both positive and negative effects of project development, and most farmers have a positive attitude of welcoming and supporting the project in general as the project has obvious effects of poverty alleviation. More details are presented in Table 16.2.

16.3 Econometric Analysis of Farmer Perceived Value

Based on the data, the researchers analysed the perceived value of the effect of an FCS project through econometric methods and empirically tested the theoretical hypotheses. The significance of the results of the econometric and empirical analyses is discussed further below.

Table 16.2 Farmer perceived value of the effects of PAFCS

Factor	Investigated items	Mean	S.D.	Agree rate (%)	Result
Perceived economic benefits	Increased external investment PB_1	3.67	1.17	62.20	Agree
	Received carbon trading revenue and government subsidies PB_2	3.59	1.20	61.27	Agree
	Increased labour opportunities for locals PB_3	3.58	1.21	55.79	Neutral
	Increased income from shareholding or transfer of inefficient forest land PB_4	3.61	1.15	61.79	Agree
	Equal access to benefits for the poor PB_5	3.48	1.28	55.49	Neutral
Perceived social benefits	Increased local reputation PB_6	3.73	1.14	64.63	Agree
	Introduction of new ideas and technologies PB_7	3.85	1.13	60.43	Agree
	Enhanced community cohesion PF_8	3.81	1.10	66.16	Agree
	Contribution to poverty alleviation PB_9	3.30	1.22	49.67	Neutral
	Expanded social networks and livelihood skills of community residents PB_{10}	3.81	1.09	66.46	Agree
	Improved forest roads, irrigation, and other infrastructure PB_{11}	3.80	1.10	66.77	Agree
Perceived ecological benefits	Reduced soil erosion PB_{12}	3.70	1.20	56.10	Agree
	Improved habitation conditions $PB1_3$	3.67	1.24	55.18	Agree
	Increased ecological awareness of climate change PB_{14}	3.82	1.19	60.67	Agree
	Contribution to biodiversity conservation PB_{15}	3.81	1.15	62.20	Agree
	Reduced risk of returning to poverty due to natural disasters PB_{16}	3.85	1.20	55.79	Agree

(continued)

Table 16.2 (continued)

Factor	Investigated items	Mean	S.D.	Agree rate (%)	Result
Perceived economic risks	Forest land used for carbon sink afforestation is not cost-effective PR_1	2.13	1.09	10.37	Against
	Difficulties in harvesting and marketing of trees PR_2	2.23	1.30	19.51	Against
	Restriction on traditional grazing methods PR_3	3.77	1.18	62.20	Agree
	Restrictions on traditional fuel wood and Chinese herbal medicine collection PR_4	2.09	1.08	10.67	Against
Perceived psychological risks	Concern that carbon trading revenue will not be available in full and on time PR_5	2.28	1.07	11.89	Against
	Concern that the project will occupy forest land longer than the contract period PR_6	2.41	1.13	17.38	Against
	Concern that the government's support policies will diminish PR_7	2.36	1.15	15.55	Against
	Concern about the cumbersome procedures for handling contract disputes PR_8	2.33	1.10	12.50	Against
Perceived situational risks	Increase in the number and species of wild animals and plants PR_9	3.61	1.24	63.41	Agree
	Loss of native tree species, increase in pest and disease damage PR_{10}	2.36	1.29	19.51	Against
	Increase in the number and scale of forest fires PR_{11}	2.42	1.25	16.16	Against
	Increase in the gap between rich and poor farmers PR_{12}	2.43	1.27	21.04	Against

(continued)

Table 16.2 (continued)

Factor	Investigated items	Mean	S.D.	Agree rate (%)	Result
Farmer's attitude	The advantages of PAFCS outweigh the disadvantages Y_1	3.95	1.30	72.56	Agree
	Willing to support the later operation of forest carbon sink projects Y_2	3.98	1.28	69.82	Agree

Note A 5-point Likert scale was used to classify the perception results into five categories: strongly agree, relatively agree, uncertain, relatively disagree, and strongly disagree, with values of 5, 4, 3, 2, and 1, respectively. Among them, the values of 5 and 4 account for above 60%, indicating agree, 40–60% were neutral, and less than 40% were against

16.3.1 Reliability and Validity Tests and Exploratory Factor Analysis

Using SPSS20.0, reliability and validity tests were conducted on the sample data to obtain the reliability of the indicators. Specific results are shown in Table 16.3. Cronbach's α for each latent variable ranged from 0.82 to 0.93, which were all greater than 0.7; thus, the results met the requirements of stability and consistency questionnaire. Then, two validity tests, the KMO sample measure and the Bartlett sphere, were used to analyse the validity of the data. The results showed that the KMO value of the perceived effect of poverty alleviation was 0.94, and the significance level of the p-value in Bartlett's sphere test was $0.00 < 0.01$, indicating that the data could be used for factor analysis. Exploratory factor analysis was conducted on the observed indicators; the factors were analysed by principal component analysis with maximum variance orthogonal rotation. The factor loadings of each observed variable were between 0.72 and 0.93, which satisfied the standard requirement of multiple factors with upper loadings greater than 0.4. The total variance explained by each factor exceeded the criterion that the cumulative contribution of variance is higher than 60%, which indicated that the factors contained the majority of information in the original data, indicating that the converged effective factors were acceptable.

16.3.2 Structural Equation Model Fit Test

Using AMOS 22.0, the researchers selected the χ_2 test, χ_2/DF(degrees of freedom) ratio, CFI, GFI, Non-normed Fit Index (NFI), and Adjusted Fitness (AGFI) to further evaluate the model fit rate. χ_2/DF was 1.34, which meets the criterion of less than 3. CFI, GFI, and NFI all reached the criterion of greater than 0.9. AGFI was 0.88, which is close to the acceptable range of 0.9. RMSEA was 0.03, which meets the evaluation criterion of less than 0.08. The fit indicators all met the usual criteria for SEM studies, and the structural equation model was considered to have a good fit.

Table 16.3 Results of reliability and validity tests of variables

Factor	Bartlett's test	KMO value	Accumulated S.D. (%)	Cronbach's α
Perceived economic benefits	1295.47(0.00)	0.90	78.57	0.93
Perceived social benefits	963.83(0.00)	0.89	63.50	0.88
Perceived ecological benefits	794.66(0.00)	0.87	67.62	0.88
Perceived economic risk	442.75(0.00)	0.80	65.34	0.82
Perceived psychological risk	749.39(0.00)	0.84	75.68	0.89
Perceived situational risk	743.70(0.00)	0.83	75.13	0.88
Farmer perceived value	519.18(0.00)	0.73	80.51	0.87
Willingness to operate later	367.17(0.00)	0.60	91.12	0.90

16.3.3 Hypothesis Testing and Discussion

Finally, the structural equation was estimated using the method of greatest likelihood estimation. The results, shown in Fig. 16.2, supported all the theoretical hypotheses proposed.

As the figure indicates, perceived economic, social, and ecological benefits had significant positive effects on the perceived value of PAFCS, with path coefficients of 0.36, 0.35, and 0.22, respectively, where perceived economic benefits > perceived social benefits > perceived environmental benefits. This indicates that the farmers had the highest perception of the economic benefits of PAFCS, which were in line with their own personal interests. That is to say, farmers' perceived economic benefits had the greatest positive influence on their perceived value. In terms of perceived economic benefits, the coefficients of the observed variables were $PB_1 > PB_4 > PB_2 > PB_3 > PB_5$. First, the farmers perceived the project as increased external investment most strongly, followed by the increase in income from shareholding or transfer of forest land. The reason for this perception is that there are very few external market players in poverty-stricken mountainous areas and few investments in industrial projects. Also, the majority of the respondents indicated that although their direct income from the project was less than from the ecological compensation for returning farmland to forest, they strongly agreed that they could increase income through the development and utilisation of inefficient degraded barren hills and slopes. Second, the perceived benefit of increased work opportunities for local people was not strong, which is consistent with the fact that the project owner gives priority to employing skilled foreign labourers for reforestation in order to reduce development costs and

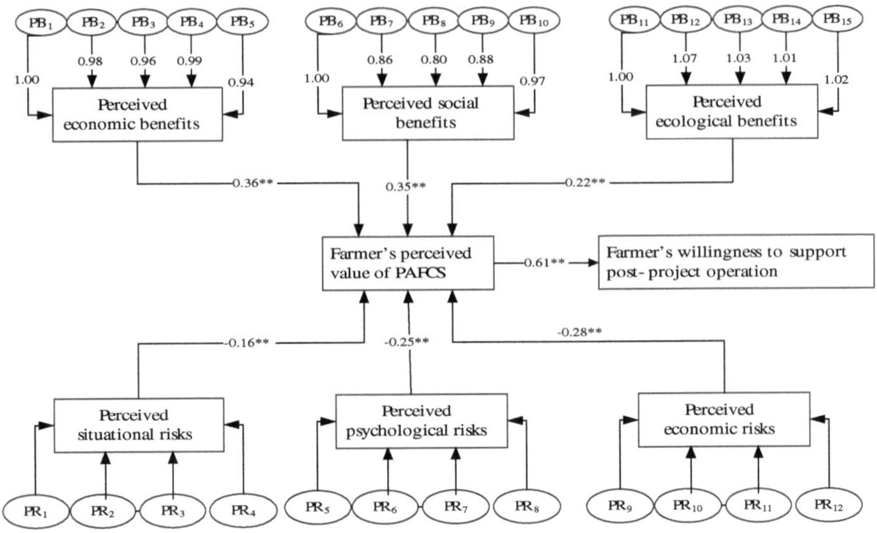

Fig. 16.2 Path coefficients of structural equation model for farmer perceived value of the effects of PAFCS and farmer willingness to participate in post-project operation. *Note* *** indicates $p < 0.001$, ** indicates $p < 0.05$

improve the survival rate of seedlings. Some farmers in project communities are obviously opposed to this. Finally, the perception that poverty-stricken farmers in the project communities had equal access to the economic benefits of participation was weakest, which may be related to the fact that the respondents generally lacked awareness and concern about their priority employment opportunities. In terms of perceived social benefits, the loading coefficients of the observed variables were $PB_6 > PB_{11} > PB_{10} > PB_8 > PB_7 > PB_9$. On the one hand, the farmers' perceived benefits in terms of increased local reputation were the most prominent. The main reason is that, as a new ecological compensation project, it has received wide attention and strong support from all sectors of society, such as governments at all levels, NGOs, research institutes, and publicity media. Hence, more resources are integrated for project promotion and implementation, new opportunities are created for community development, and exchanges and cooperation between farmers and the outside world increased, especially in the sales of traditional agricultural, forestry, and livestock products. Secondly, farmer perceptions of improved infrastructure such as forest roads and irrigation were relatively prominent, which is related to both CDM carbon reforestation project standards that emphasise community functions and the practice of utilising collective earnings to improve backward infrastructure. On the other hand, some project areas have gradually adopted various initiatives to help promote poverty eradication by providing poverty-stricken households with fixed-term cards to participate in technical training. Priority is also given to poverty-stricken households to participate in reforestation replanting, fence construction, management and care work, etc. However, the perception of contributing to local poverty alleviation was weakest. This may be related to the fact that the core objectives in poverty alleviation, such as creating benefits and development opportunities for the poverty-stricken people, were not adequately incorporated into the project's preliminary planning and design, organisation and construction, monitoring and evaluation, and publicity and mobilisation. In terms of perceived ecological benefits, the loading coefficients of the observed variables were $PB_{13} > PB_{14} > PB_{16} > PB_{15} > PB_{12}$ in this order, indicating that the farmers perceived improved habitation conditions as more prominent and less sensitive to the perception of reduced soil erosion. The main reason may be that in the early stage of a project concentrating on pre-forestation and reforestation, macro-scale ecological benefits, such as improved habitation conditions, are more easily recognised by farmers than micro-scale ecological benefits, such as reduced plot erosion.

All three dimensions of perceived risk had a negative impact on the perceived value of PAFCS, in the order of perceived situational risks > perceived psychological risks > perceived economic risks, with path coefficients of −0.28, −0.25, and −0.16, respectively. This result indicates that the farmers' perceived situational risk had the greatest negative impact on their perceived value and willingness to support the later operation of the project, while perceived economic risk had the least. Theoretically, farmers should be most sensitive to the perception of economic risk based on benefit, cost, and risk trade-offs, but the estimation results show that they perceived situational risk and psychological risk more strongly. On the one hand, this is consistent with the reality that FCS projects have adopted a *business-led and government-directed* management model, which reduces farmers' market risk. On the other hand, it also reflects that farmers are relatively sensitive to the perception of near-term, objective risk and tend to ignore potential non-monetary opportunities. In terms of perceived economic risk, the loading coefficients of the observed variables were $PR_2 > PR_3 > PR_1 > PR_4$, indicating that the farmers strongly perceived both the difficulties in forest harvesting and marketing and the restrictions on traditional grazing methods. The reasons could be as follows. First, the project area is remote and mountainous, with inconvenient transportation, and it is difficult to apply for forest harvesting permits and sell trees. Second, most of the project plots are traditional grazing areas, and many respondents felt that the contradiction between forest and livestock was very prominent in the short term and believed that the project's development has squeezed the free grazing space of cattle, sheep, and other livestock and thus increased grazing costs. In terms of perceived psychological risk, the load coefficients of the observed variables were $PR_8 > PR_5 > PR_6 > PR_7$. The strongest perception was the fear of complicated procedures for handling contract disputes. The possible reason is that the respondents considered it difficult to understand the terms and conditions of the contract, and they were also concerned about the difficulty of securing personal rights and interests due to the long project period. In terms of perceived situational risk, the load coefficients of the observed variables were $PR_{11} > PR_9 > PR_{10} > PR_{12}$. Among them, perceptions of the increase in the number and scale of forest fires and the increase in the number and species of wildlife were relatively prominent. A significant number of respondents considered that along with the development of forestry ecological projects such as large-scale artificial forestation, natural forest protection, and return of farmland to forest, the frequency of forest fires has increased, and populations of wild boars, rats, and other animal and bird species have increased rapidly, which has objectively posed a significant hazard to traditional livelihoods, especially plantations. The FCS projects are believed to exacerbate similar risks.

The perceived value of the poverty alleviation effect of FCS projects had a positive effect on the willingness of the farmers to support the later operation of the project, with an effect coefficient of 0.61. This indicates that the perceived value of the benefits outweighs the harms, which affects not only the participation of farmers in the project but also the long-term sustainability of the project operation.

16.4 Summary

This chapter empirically examined the relationship between the perceived value of FCS projects and farmers' willingness to support the project's later operation by using data from samples of two CDM FCS projects. The results showed that, first, FCS is an effective method to achieve win–win results in combating climate change and reducing poverty. The development of large-scale FCS projects has achieved obvious effects of poverty alleviation. In all, 72.56% of the interviewed farmers were in favour of PAFCS projects, and 69.82% were willing to support the later operation of FCS projects. Second, perceived benefits and perceived risks directly affect the perceived value of PAFCS. Specifically, perceived economic, social, and ecological benefits have a significant positive impact on perceived value, with perceived economic benefits having the greatest impact. Perceived economic, psychological, and situational risks have a significant negative impact on perceived value, with perceived situational risk having the most significant impact. Third, the perceived value of PAFCS directly affects the long-term sustainability of a project. Results show that the perceived value of benefits outweighed the perceived value of disadvantages, which had a significant positive effect on farmers' willingness to support the project's later operation, with an impact path coefficient of 0.61.

This chapter has several policy implications. First, enhancing poverty alleviation and regulating market operation should be an important choice for institutional change involving FCS in China. It is essential to accelerate the development and promotion of China's voluntary emission reduction rules to make them in line with international rules and meet the requirements of poverty alleviation at the same time. Thus, poverty reduction potential can be exploited, contributing to international cooperation in addressing climate change. Second, improving incentive mechanisms and compensating for market failure should be an important direction for PAFCS. Investment, finance, insurance, and taxation policies related to industrial poverty alleviation should be integrated. Policies such as forest carbon sink offset, carbon sink mortgage or discount loans, futures trading, etc. should be introduced. Pilot projects should be carried out to stimulate multiple participating entities to participate in project planning and design, organisation and construction, management and monitoring and evaluation, etc. Third, improving farmer perceptions of the benefits gained is an important way to promote a win–win situation in combating climate warming and alleviating poverty. It is important to improve the management of forest land transfer and technical training, labour inputs, and the distribution of timber sales income and to actively guide afforestation entities to employ the poverty-stricken people who are more dependent on traditional agricultural livelihoods to ensure that the interests of farmers in project communities are maximised. It is also important to integrate ecological compensation policies, such as grazing losses, electricity instead of firewood, and technical assistance, to encourage the development of the forest economy and extend the forest carbon sink industry chain. Intensive breeding models could be promoted. For example, captive breeding cattle and sheep could be introduced to reduce farmer reliance on traditional agricultural livelihoods. It is crucial to revise management methods in areas such as wildlife protection, take lawful actions such as remediation of forest fire hazards, and allow hunting of wild pigs to reduce the negative impact of forestry ecological projects on the production and livelihood of regional farmers.

References

Dowling, G. R., & Staelin, R. (1994). A model of perceived risk and intended risk-handling activity. *Journal of Consumer Research, 21*(1), 119–134.

Flint, D. J., Woodruff, R. B., & Cardial, S. F. (2002). Exploring the phenomenon of customer, desired value change in a business-to business context. *Journal of Marketing, 66*(10), 102–117.

Han, C. Y. (2016). Research on the influence of farmers' perceived value on their agricultural waste resource utilization behavior. Ph.D. Dissertation from Huazhong Agricultural University.

Keh, H. T., Foo, M. D., & Lim, B. C. (2002). Opportunity evaluation under risky conditions: The cognitive processes of entrepreneurs. *Entrepreneurship Theory and Practice, 27*(2), 125–148.

Keller, K. L. (1993). Conceptualizing, measuring, managing customer-based brand equity. *Journal of Marketing, 57*(1), 1–22.

Li, H. Q., Li, X. Q., & Hou, L. C. (2012). Study of feeling effects of pro-poor tourism in fragile eco-environment areas of Loess Plateau—Take Guzui village of Shanxi province as an example. *Tourism Research, 4*(3), 1–6.

Li, J., Zhong, L. S., & Cheng, S. K. (2009). A study on residents' perceptions of tourism poverty alleviation effects and participation behavior in ethnic poverty-stricken areas-taking the Sanjiangyuan area in Qinghai Province as an example. *Journal of Tourism, 8*(1), 71–76.

Meng, B., Liu, M., Li, Q. S., et al. (2010). A risk perception theoretical model and analysis of influencing factors. *Chinese Journal of Security Science, 20*(10), 59–66.

Qin, Y. H., Ma, Y. J., & Liu, D. X. (2016). A study on the perceived impact of tourism on poverty alleviation among residents of ethnic poverty-stricken areas-taking Huangshui Town, Shizhu County, Chongqing as an example. *Journal of Southwest University (natural Sciences Edition), 8*, 74–82.

Woodruff, R. B., & Cardial, S. F. (1996). *Know your customer: New approaches to customer value and satisfaction.* Blackwell.

Yang, F., Zeng, W. Z., Zhang, W. K., et al. (2016). Foresters' willingness to participation sustainedly in forest carbon sequestration projects and its influencing factors. *Forestry Science, 52*(7), 138–147.

Zhang, F. Y., & Zhao, X. Y. (2014). Analysis of ecological compensation effects based on farmers' perceptions-taking Zhangye City in the middle reaches of the Heihe River as an example. *Chinese Journal of Ecological Agriculture, 22*(3), 349–355.

Zeithaml, V. A. (1988). Consumer perception of price quality and value: A means-end model and synthesis of evidence. *Journal of Marketing, 52*(7), 2–22.

Chapter 17
Conclusions, Policy Suggestions, and Prospects

This book focuses on two issues: the creation of benefits and opportunities for the poverty-stricken people and the trade-off between FCS and poverty alleviation. At the theoretical level, it explains the essential characteristics of PAFCS, analyses the impact mechanism of FCS projects in poverty alleviation, clarifies the stakeholders and their interests and demands, and delineates the dynamic mechanism of FCS projects and poverty alleviation. Based on this theoretical framework, the current situation and challenges for PAFCS in Southwest China's ethnic areas are examined in depth. Project performance was quantitatively measured both for projects themselves and for community farmers. In this chapter, the researchers provide the main conclusions, policy recommendations, and advice for further research.

17.1 Main Conclusions

First, PAFCS is aimed at the creation of benefits and opportunities for the poverty-stricken people. PAFCS is an emerging form of poverty alleviation in China's participation in the world's actions to address climate change. Its characteristics comprise clear poverty alleviation objects, multiple poverty alleviation subjects, multidimensional poverty alleviation effects, and distinctive policies. Endogenous driving forces include the obligation to reduce emissions and to promote the economic interests and public image of enterprises. Exogenous driving forces include emission reduction standards, government policies, and social demand. It is far from enough to rely on policy guidance for climate change and FCS industry. It is necessary to inject new impetus and vitality to create benefits and opportunities for the poverty-stricken people through moderate government intervention in integrating various poverty alleviation and ecological policies, promoting the synergy of PAFCS subjects, and incorporating poverty alleviation resources and methods. To improve the efficiency and effectiveness of FCS in poverty reduction, the basic principles of *government-led,*

W. Zeng and F. Yang, *Poverty Alleviation Via Forest Carbon Sequestration*, International Research on Poverty Reduction, https://doi.org/10.1007/978-981-99-8191-5_17

enterprise-driven, farmer participation and third-party evaluation should be met. Mechanisms for providing benefits to poverty-stricken farmers should be the focus. Government policies should be market-oriented to ensure the sustainable operation of the projects in order to achieve the poverty alleviation goals.

Second, PAFCS emphasises *alleviating the truly poor and truly alleviating the poor* and promotes voluntary emission reduction rules for FCS in China. The surveys found that PAFCS has significantly developed, though PAFCS continues to be perceived as merely FCS projects or general industrial poverty alleviation projects. On the whole, the pattern of FCS projects has not yet been formed in terms of incorporating poverty alleviation goals into planning and design, certification and registration, organisation and construction, monitoring and evaluation, etc. PAFCS is still in the exploratory stage led by governments. Most policies are introduced as compulsory institutional changes made by local governments to meet the urgent requirements of targeted poverty alleviation and to solve practical problems in FCS development. The institutional arrangement is usually short-term, and the practice is unsustainable. The supply of macro management to promote sustainable PAFCS is insufficient. Normative systems such as FCS standards and methodologies need to be established. These should include elements such as preferential policies (e.g., carbon tax, carbon mortgage, and interest-bearing loans); FCS insurance; FCS interval harvesting; fiscal, tax, financial, investment; forest ecological compensation; and technical assistance, among others. The formulation of China's FCS voluntary emission reduction rules should highlight poverty alleviation according to local conditions, and supply-side reforms of PAFCS-supporting policies are essential to transform PAFCS from focusing only on poverty-stricken areas to targeting poverty-stricken populations.

Third, China's FCS system needs change to strengthen the function of poverty alleviation. The empirical results show that the large-scale carbon sink AR projects in poverty-stricken ethnic areas of Southwest China have achieved alleviation effects, but the poverty alleviation performance of the same FCS project in different regions differs significantly. Nonetheless, 72.56% of sampled farmers believed that the benefits of PAFCS outweigh the disadvantages. The perceived value of PAFCS has a significant positive effect on farmer's willingness to support sustainable operation of the project in later stages, with a coefficient of 0.61. This indicates that strengthening the poverty alleviation function of FCS could enhance the attractiveness of FCS and improve the market share of FCS projects. Doing so would also help reduce uncertainty, as well as transaction costs and the practical risks of projects, by guaranteeing the extensive cooperation and long-term support of farmers in project communities, thus ensuring the sustainable operation of FCS projects. Therefore, the goal of strengthening the function of poverty alleviation should guide changes in China's FCS system. Continuously improving the sense of achievement for farmers in project development is an important strategy to promote win-win situations in addressing climate change and poverty reduction.

Fourth, improving the participation of and sense of gain among poverty-stricken people is an important means of promoting the sustainable operation of FCS projects. The research results show that the willingness of community farmers to continue to participate in the projects is not strong, with 46.35% of the farmers holding a neutral attitude and 24.69% not willing. Factors such as age, area of participating land, household income, degree of part-time farming, project organisation mode, satisfaction with preliminary income, expectation of later income, government support, ease of access to forestry information, and road conditions had significant effects on farmers' willingness to continue participating. To reduce the risk of farmers withdrawing from projects and to ensure long-term sustainable operation of projects in poverty-stricken ethnic areas in Southwest China, it is essential to pay close attention to both the extent and risk of participation of households with a low-income level, small forest land area, and greater dependence on traditional agriculture and to help them obtain participation benefits in a fair and reasonable manner.

Fifth, despite the phenomenon of elite capture, the organisational willingness of community elites to participate in FCS projects remains low. The research shows that the overall extent of elite capture is 0.254. The degree of institutional elite capture is 0.095, the degree of economic elite capture is 0.201, and the degree of traditional elite capture is 0.092, with the economic elite capturing the largest proportion. Meanwhile, only approximately fifty per cent (55.3%) of community elites were willing to organise farmers to participate in FCS projects. Factors such as age, present or past engagement in forestry-related work, main source of household income, perception of benefits, perception of the difficulty of project execution, relationship with villagers, perception of FCS policies, and elite category significantly influenced the organisational willingness of rural elites. The promotion of PAFCS depends on the commitment of not only institutional elites but also extra-institutional elites such as respected community elders, more educated farmers, and clan leaders. PAFCS should exclude the personal interests of community economic elites while preventing elite capture.

Sixth, informal institutions and villager autonomy are also important factors in ethnic regions of Southwest China. The empirical results show that farmers in the Yi ethnic minority project in Liangshan, Sichuan Province, identified significantly more with traditional culture than with modern commercial culture, with an average traditional culture score of 4.05 and an average commercial culture score of 3.48. Integration and separation are the two main adaptation strategies adopted by farmers towards traditional culture and commercial culture. Women, middle-aged people, and those with higher education levels were more likely to be involved in the project. The degree of cultural adaptation of the farmers increased with the size of the area involved in the project. This indicates that it is not wise to rely solely on contract regulation, economic incentives, and administrative means to carry out FCS projects in remote and poverty-stricken areas, especially in poverty-stricken minority areas. The effective integration of formal and informal systems should be strengthened.

Seventh, community function is an essential guarantee for PAFCS. In the analysis of the sample, the overall function of communities was on the rise. Their total function in 2016 was significantly higher than that in 2013, and the social, economic, and environmental levels of communities had improved accordingly, indicating that in the FCS project process, communities play a key role in project promotion. From 2013 to 2016, the levels of the organisation and propaganda function, revenue distribution function, and project maintenance function of communities fluctuated, but the overall trend was upward; in particular, the revenue distribution and project maintenance function improved very significantly. This indicates that the community function will gradually change from organisation and propaganda to project maintenance with the advancement of FCS projects and that the community has both a direct and an indirect influence on the organisation, implementation, maintenance, and benefits distribution of a project.

Eighth, the driving force in FCS is establishing the long-term stability of the project operation mechanism while preventing potential natural and market risks. At the same time, it will ensure that the project promotes regional economic development and poverty alleviation through the *trickle-down* effect. The research results show that the implementation of FCS projects has a significant promotion effect on regional economic development. While, due to their long period, this promotion effect will not be immediate and exhibits a lag, and the longer the project has been implemented, the greater the promotion effect on local economic development. The project promotes local economic development mainly by optimising the local industrial structure, increasing the savings rate of residents, and augmenting the incomes and expenditures of the regional government. Therefore, it is essential to expand the coverage area of FCS projects, increase special investment, guide projects towards ecologically fragile and extremely poverty-stricken areas, enhance local economic development capacity, and realise the organic combination of regional development, ecological protection, and targeted poverty alleviation. In addition, more measurements are needed, including evaluating the effectiveness of the projects in the short and long run, which is indispensable. Appropriately guiding farmers to pay attention to the long-term benefits of projects will enhance their participation willingness, and improving the financing environment will promote sustainable regional development and the *trickle-down* effect in poverty alleviation.

Finally, reducing situational risk is an important tool to promote the development of PAFCS in ethnic areas of Southwestern China. The research results show that the benefits and risks perceived by farmers directly affect their perceived value of PAFCS. Among these, perceived economic, social, and ecological benefits each had a significant positive influence on perceived value, with path coefficients of 0.36, 0.35, and 0.22, respectively; the degree of influence is perceived economic benefits > perceived social benefits > perceived ecological benefits, in order. Perceived economic, psychological, and situational risks each had a significant negative influence on perceived value, with path coefficients of −0.28, −0.25, and −0.16, respectively; the degree of influence was perceived situational risks > perceived psychological risks > perceived economic risks. Thus, it follows that ensuring the maximisation of economic benefits for farmers and actively reducing adverse impacts derived from forestry ecological construction are important ways to improve farmers' sense of access to project development and promote win–win solutions with regard to climate change and poverty reduction.

17.2 Policy Recommendations

As a dynamic, open, and complex strategy driven by the emerging carbon trading market, PAFCS is promoted by the government and assisted by society. The process progresses from starting to piloting, promoting, and improving. Therefore, PAFCS in the ethnic areas of Southwest China should not be confined to FCS projects but also focus on creating benefits and development opportunities for the poverty-stricken people. Their performance should be examined from a combination of macro and micro perspectives.

17.2.1 Establishing a Sound Policy Framework to Promote PAFCS

PAFCS should adhere to the basic principles of *government-led, enterprise-driven, farmer participation and third-party assessment* and contribute to constructing ecological and prosperous countries, implementing rural revitalisation strategies, building China's carbon emission trading market, and promoting sustainable development of the FCS industry. As China is joining international cooperation in addressing climate change, supply-side reforms of PAFCS policy should be oriented to creating benefits and development opportunities for the poverty-stricken people and emphasise both overall regional and individual farmer development. A policy framework is recommended as follows.

First, in line with the principles mentioned above, the balanced FCS standards should be created in the first place, while the market quota of FCS shall be expanded. Only well-designed *rules of the game* can guarantee the realisation of regulatory objectives. Breaking institutional constraints and formulating FCS standards will play a decisive role in PAFCS. As the second-largest carbon emitter in the world, China should emphasise the macro-control role of the government and formulate voluntary emission reduction rules and methodologies that can meet international rules and requirements of targeted poverty alleviation. The market share of FCS should be continuously expanded in China's carbon emissions trading, thereby demonstrating China's responsibility as a large country and enhancing its reputation in international climate negotiations. FCS market transactions should centre on the goal of creating benefits and development opportunities for the poverty-stricken people. Diverse entities should participate in the planning and design, organisation and construction, monitoring, and evaluation of FCS projects. This will lay a solid economic and institutional foundation for enhancing the performance of PAFCS.

Second, macro guidance of PAFCS should be strengthened. PAFCS involves many government departments and market players. Therefore, it is necessary to take into account the differing roles of government and market, break barriers between sectors, and strengthen the government's macro-level leadership and guidance in PAFCS. The concepts of inclusive growth, green poverty alleviation, and win–win solutions should be applied in building the FCS market and industrial development. Firstly, the government's macro-level leadership and guidance on PAFCS should be strengthened. Based on the China National Leading Group on Climate Change and Energy Conservation and Emission Reduction and China State Council Leading Group for Poverty Alleviation and Development, a *Low-carbon Poverty Alleviation Office* should be established. On this basis, efforts should be made to constantly strengthen coordination among departments of ecological environment, economic development, forestry and grassland, poverty alleviation, and other departments. Also, communication and coordination among governments at all levels must be enhanced to establish and improve the macro management system for PAFCS. Secondly, governmental responsibility in poverty alleviation and ecological compensation should be in place. Administrative, economic, and legal means should be established to improve the carbon sink industrial, legal, and technical systems related to PAFCS. Moreover, the government should make full use of administrative, economic, and legal means to promote the FCS industry and the legal, technical, and social participation systems for PAFCS. Also, poverty alleviation resources should be integrated to encourage the poverty-stricken people to participate in FCS projects, while a monitoring and evaluation mechanism of poverty alleviation performance must be established. Most importantly, the FCS developer should focus on solving problems caused by market failure, thus driving the healthy and sustainable development of PAFCS. Thirdly, the government should act in the role of guiding public opinion and the media to publicise the advantages of supporting PAFCS, building public participation platforms and an environment conducive to PAFCS, and optimising the motivation mechanism of PAFCS.

Third, a PAFCS policy system should be established on the basis of pilot tests. The change and transformation of PAFCS are bound to be a long-term, complex, and arduous process; therefore, the government should give full play to its regulatory function, formulate and issue a *Guidance on PAFCS Pilot Programmes*, and clarify the main objectives and key tasks of the pilot programmes as well as their guiding policies on talents, prices, investment, finance, market access, etc. Local governments should be given more room for innovation in the PAFCS system and encouraged to *try out* a pilot programme firstly. If the pilot programmes achieve the expected results in various places, a policy system should be established. FCS projects, executed in areas where the poverty-stricken people live, should be supported in several ways, including registration, recording, approval, etc., with priority given to them with clear, measurable indicators of poverty amelioration. On the basis of the China Green Carbon Sink Foundation and other domestic carbon funds at all levels, a special fund for PAFCS should be established. An approval, registration, execution, monitoring, verification and certification, and CER issuance and evaluation mechanism should be established. Pre-, mid-, and post-evaluations of PAFCS performance by third parties should be also in place.

17.2.2 Policy Recommendations for Promoting PAFCS in Ethnic Areas of Southwest China

The research results show that PAFCS projects have achieved significant progress and that the political, public opinion and practical basis for promoting PAFCS in southwest ethnic areas are already in place. To speed up the promotion of PAFCS in these areas, the needs of farmers must be met based on the resource endowment and the social, cultural, and economic characteristics of the areas, and the PAFCS model should be improved according to local conditions. Furthermore, policy support and interventions are necessary to regulate market operations through corresponding guidance, regulatory, incentive, guarantee, and assessment policies, so as to promote the sustainable operation of FCS projects and maximise the satisfaction of stakeholders. Based on the findings of these studies, the following are policy recommendations for promoting PAFCS in Southwest China in three dimensions: governments, enterprises, and farmers.

17.2.2.1 Strengthen Government Guidance, Focus on Policy Incentives, and Create a Favourable Environment

Give full play to governmental policies, such as China's National Programme for Addressing Climate Change, China Forestry Adaptation to Climate Change Action Programme (2016–2020), the Ecological Poverty Alleviation Work Programme in China, and the Action Plan for Establishing Market-oriented and Diverse Ecological Protection Compensation Mechanisms. Market players should be given more choices in the terms and depth of participation. In addition, project operational models with regional characteristics should be established based on policy incentives, rather than mandatory means. Firstly, local governments should actively integrate PAFCS into their current poverty alleviation work; strengthen the integration of poverty alleviation resources, methods, and policies; and collaboratively promote the practice of targeted PAFCS. Secondly, PAFCS could be incorporated into national key ecological zones and rural revitalisation strategies so as to provide a decision reference and policy basis for supply-side reform of national PAFCS policy and to optimise the institutional and market environment. Thirdly, poverty alleviation funds and their general policies on finance, taxation, finance, and investment should be integrated. Preferential policies should be given to pilot projects, such as carbon sink mortgage and interest-bearing loans, carbon sink forest insurance, and harvesting and financial subsidies. Fourthly, ecological compensation policies such as for grazing loss, electricity instead of firewood, and technical assistance should be introduced to encourage the forest economy and extend the FCS industry chain. Intensive breeding models should be adopted, such as cattle and sheep captive breeding with advanced technologies, so as to reduce farmers' dependence on traditional livelihoods. Fifthly, the transparency of PAFCS should be maintained, and the compliance rate of poverty alleviation participants should be improved. Also, any actions not conducive to the goal of poverty alleviation should be corrected, and the monitoring and evaluation mechanism of PAFCS should be established. Sixthly, an international exchange platform could be set up to publicise successful experiences and practices around China, which could help win the support of society as a whole.

17.2.2.2 Adhere to the Leading Role of Enterprises and Focus on Win–Win Cooperation with the Grassroots of Rural Society

Enterprises are investors, executives, and leaders in PAFCS. The empirical results of these studies show that cooperation with grassroots communities could improve the participation of the poverty-stricken people and maintain the sustainable operation of FCS projects. Therefore, enterprises should take the initiative to take part in social poverty alleviation and promote the participation of the poverty-stricken people, thus increasing their income. Secondly, they should pay attention to local culture, traditions, customs, knowledge, and wisdom. They should encourage the participation of farmers, especially men, the elderly, and those with a lower level of education, a lower level of household income, and a smaller area of forest land. Farmers should have the right to know, speak, suggest, and choose. Thirdly, enterprises should strengthen cooperation with the local government and its institutional elites and actively seek support from extra-institutional elites such as respected community elders, farmers with higher education, clan leaders, and wealthy leaders. Fourthly, they should integrate the institutional arrangements of FCS projects into informal systems such as village rules and regulations so as to secure the mutual support of formal and informal systems. Finally, according to economic and social/cultural differences, enterprises should select practical operation modes for existing and pilot FCS projects.

17.2.2.3 Respect the Will of Farmers and Improve the Participation and Sense of Gain of the Poverty-Stricken People

Community farmers are vital participants, beneficiaries, and judges of PAFCS. The empirical results of these studies show that the perceived value of benefits outweighs that of disadvantages and has a significant positive effect on the willingness of community farmers to support the subsequent operation of a project. Also, improving the participation and sense of gain of the poverty-stricken people is an effective way to promote sustainable operations. Therefore, farmers should be given opportunities to learn advanced technologies, and their initial direct income should be increased, with stable income in the later stages. Local industries should be developed to reduce the dependence of farmers on traditional agricultural livelihoods and to improve the willingness of poverty-stricken households to participate sustainedly in FCS projects (and their sense of gain). Secondly, farmer training programmes should be organised to provide technical skills training and assistance for poverty-stricken farmers with less human, financial, and social capital. Thirdly, forestry organisations, such as forestry share cooperatives, forestry professional cooperatives, and forestry technical associations, should be set up in order to organise poverty-stricken farmers to participate. Fourthly, the conditions of community roads and communication facilities should be improved to facilitate the communication of community announcements and public notices for the planning, decision-making, and implementation of FCS projects, so as to enhance the participation opportunities of poverty-stricken farmers. Fifthly, publicity and communication with community farmers should be enhanced and their opinions and suggestions should be followed up, as social and public opinion are essential to projects.

17.3 Research Outlook

In both theory and practice, there is an inextricable link between FCS and poverty alleviation, but practice in and research on PAFCS are still in the initial stage. Potential topics to be studied are two basic scientific issues of (i) creating benefits and development opportunities for poverty-stricken people and (ii) the trade-off between FCS and poverty alleviation.

First, future research on different types of FCS projects is encouraged. FCS industry development and institutional innovation have always been the two cornerstones of PAFCS. This book reports the systematic study of carbon sink AR projects; research on REDD and REDD + projects was not conducted. Future studies should track the results of international climate negotiations and progress in institutional changes, especially the institutional arrangement of the FCS market in China. It is an important future direction to deepen research on PAFCS based on different types of FCS projects.

Second, future research could focus on the topic of community participation in FCS. PAFCS is a systematic project that requires the concerted efforts of governments, enterprises, communities, and farmers. Effective community participation is an essential means to achieving the goal of PAFCS. This book focuses on farmer participation in PAFCS, but less on the participation of the community itself. It is necessary to investigate community participation models and evaluation systems.

Third, it is also valuable to conduct research on the dynamic process of PAFCS targeting the relatively poverty-stricken population. Identification of the poverty-stricken population and evaluation of poverty alleviation performance are important issues that need continuous attention in PAFCS. Due to our research capacity, data, and time, this research gathered data only from a certain point of time or area, and it replaced the whole poverty-stricken population with the poverty-stricken households in the record. Future research could focus on rural revitalisation strategy and the poverty problems remaining after China has progressed to a moderately prosperous society in all respects, as well as on performance evaluation based on sequential data and panel data in view of the long process of PAFCS projects, which have characteristics of diversity, spatial heterogeneity, time dynamics, and lag of poverty alleviation effects.

Appendix

China Forestry Adaptation Action Plan for Climate Change (2016–2020)

Climate change is a major crisis and a serious challenge faced by humankind in common. As such, it has become a common concern in international political, diplomatic, economic, and ecological arenas. The response to climate change should focus on both mitigation and adaptation. Mitigating climate change is a long-term and arduous task, while adapting to climate change is a more realistic and urgent task. Forestry is one of the areas most seriously affected by climate change and is a key area identified by China for adaptation to climate change. Doing a good job in forestry adaptation to climate change is very important to enhancing the overall adaptive capacity of the country and maintaining its ecological and climate security.

Basic Background

The Situation

The United Nations Intergovernmental Panel on Climate Change (IPCC) has released five scientific assessment reports to date. Six reports were released during the fifth assessment period from 2008 to 2014. Among them, the 2012 *Special Report: Managing the Risks of Extreme Events and Disasters to Advance Climate Change Adaptation* was the first scientific assessment report dedicated to adaptation, showing that climate change has already had widespread and far-reaching impacts on natural ecosystems and on human survival and development, impacts which will be exacerbated by the warming of climate change. *Climate Change 2014: Impacts, Adaptation and Vulnerability*, the second report, confirmed the significant impacts of climate change on socioeconomic systems, natural ecosystems, and human survival and development. Studies show that climate change leads to frequent extreme weather

W. Zeng and F. Yang, *Poverty Alleviation Via Forest Carbon Sequestration*, International Research on Poverty Reduction, https://doi.org/10.1007/978-981-99-8191-5

events and threatens and even irreversibly damages ecosystems, causing significant economic and social losses globally. In 2014, *The Adaptation Gap Report* released by the United Nations Environment Programme (UNEP) stated that the annual adaptation costs for developing countries are estimated to be USD 70–100 billion by 2050. *The United Nations Post-2015 Development Agenda Report* pointed out that human-induced carbon dioxide emissions are the largest contributor to climate change and that adaptation can reduce the risks and impacts of climate change. The international community is highly concerned about adaptation to climate change, and both developed and developing countries are focusing on adaptation as an important aspect of addressing climate change. Developed countries such as Germany, the Netherlands, and Belgium have introduced national programmes for adaptation to climate change. Developing countries vulnerable to the adverse effects of climate change, especially the least developed countries and small island states, have paid particular attention to adaptation to climate change and adopted a series of adaptation policy initiatives.

China has complex climatic conditions and an overall fragile ecological environment, and it is vulnerable to the adverse effects of climate change. Studies have shown that climate change causes changes in temperature, humidity, precipitation, and growing seasons, all of which have real or potential impacts on forestry development. For example, the frequency and intensity of forest fires will increase; the scope of forestry pests and the harm they cause will increase; rare tree species distribution areas and wildlife habitats will shrink; the hydrological resources of wetlands will change, resulting in wetland water shortage, shrinkage in area, and declines in biodiversity and ecological functions; grasslands in Western China may retreat; and desertification and soil erosion may occur on a national scale. In 2008, widespread snow and rain disasters in the south of China caused significant losses in forest resources, reflecting the serious hazards of extreme weather events and highlighting the vulnerability of forest ecosystems. It is increasingly urgent to reduce climate risks and enhance the adaptive capacity of forestry.

Existing Problems

First, China's forestry resource endowment is insufficient. China's forest cover is far below the global average of 31%, the per capita forest area is only 1/4 of the world's per capita, the per capita forest accumulation is only 1/7 of the world's per capita, the wetland rate is lower than the global average of 8.6%, and the per capita wetland area is only 1/5 of the world's per capita. Furthermore, wetland protection pressure is high, and recovery is difficult; hazy days are frequent, sand and dust storms are frequent, and the task of sand control is heavy; fragmentation of the landscape and species endangerment are intensifying, and the need for the protection of biodiversity is very urgent. Ecological fragility is still China's basic national condition; a shortage of ecological products is still the outstanding problem; and forest, wetland, and desert ecosystems are more sensitive to climate change. Second, the foundation for forestry adaptation to climate change is weak. Awareness of climate change adaptation in forestry is generally not high, the capacity is relatively weak, the organisation system is not sound enough, and the number of specialists is relatively small. The work needs to be strengthened.

Policy Foundations

China's government has always attached great importance to climate change adaptation, and it has introduced a series of major initiatives. 2007 saw the release of *China's National Programme for Addressing Climate Change*, which clearly defined key areas and actions for climate change adaptation. 2011 saw the release of the *National Twelfth Five-Year Plan*, which called for actively addressing climate change, strengthening the capacity for adaptation, and formulating adaptation strategies. *The National Adaptation Strategy for Climate Change* issued in 2013 made comprehensive arrangements for adaptation at the strategic level, clarified key areas and tasks, and required the preparation of sectoral climate change adaptation programmes and their proper implementation. *Opinions of the Central Committee of the Communist Party of China and the State Council on Accelerating the Construction of Ecological Civilisation*, issued in 2015, further arranged for adaptation to climate change. In order to thoroughly implement the requirements of the central government and facilitate the work of forestry adaptation to climate change, this action plan was formulated.

General Requirements

Guiding Ideologies

This plan is guided by the Chinese Communist Party's 18th National Congress; by its Third, Fourth, and Fifth Plenary Sessions; and by the spirit of General Secretary Xi Jinping's series of important speeches. It aims to contribute to building an ecological civilisation and Beautiful China, implementing the overall national plan to address climate change and adapting to the strategic requirements of climate change. It centres on scientific afforestation, protection, and management, and it is to strengthen monitoring and early warning, risk management, and team building and to comprehensively enhance forestry adaptation to climate change so as to promote low-carbon development and contribute to ecological civilisation.

Basic Principles

First, it is essential to adhere to the principle of dovetailing with national strategies. The objectives of forestry adaptation to climate change should dovetail with national strategies and plans, highlight the characteristics of forestry adaptation actions, and support national efforts to adapt to climate change. Second, it is vital to adhere to the principle of equal emphasis on adaptation and mitigation. Priority will be given to measures with synergistic benefits in mitigation and adaptation. Third, the principle of avoiding harm by taking advantage of benefits is to be obeyed. The favourable factors brought about by climate change are to be actively exploited, and scientific measures are to be taken to avoid, to the greatest extent possible, various possible risks so as to optimise the exploitation of forestry resources, minimise losses, and promote the sustainable development of forestry. Fourth, it is important to adhere to the principle of active adaptation with a focus on prevention. Monitoring and forecasting are to be strengthened, orderly adaptation targets are to be established, and the ability of forestry to adapt to climate change at all levels is to be improved, from adaptation technology to adaptation policies. Fifth, it is vital to adhere to the principle of promoting broad participation of the whole of society. Publicity of the concept of green low-carbon development and climate change needs to be strengthened to popularise forestry policies and knowledge on climate change adaptation and to raise public awareness. Social participation mechanisms need to be explored to build a good societal atmosphere.

Main Goals

By 2020, the utilisation rate of good forest species will increase to more than 75%, the forest coverage rate will reach more than 23%, forest accumulation will reach more than 16.5 billion cubic metres, the forest fire damage rate will be controlled at less than 0.9 per 1,000, the disaster rate of major forestry pests will be controlled at less than 4 per 1,000, the protection rate of national key protected wildlife will reach 95%, and wetland area will not be less than 800 million mu; more than 50% of sandy land can be treated. In general, the capacity of forest, wetland, and desert ecosystems to adapt to climate change has been significantly enhanced. By 2020, the work of forestry to adapt to climate change will be fully developed, awareness of adaptation will be generally improved, basic capacity will be strengthened, the initial talent team will be established, the basic work system will be in place, and the ability to serve the national work of adapting to climate change will be significantly enhanced.

Focus of Action

First, the Action Plan is to focus on accelerating the protection and use of good genes and vigorously cultivating good seedlings to adapt to climate change. It is also essential to strengthen the investigation, collection, preservation, and utilisation of forestry germplasm resources; strengthen the construction of a forestry seed base; carry out research on and testing of species improvement techniques; increase the selection and use of forestry species; scientifically cultivate strong seedlings with good growth and strong resistance to adapt to extreme changes in temperature and precipitation factors; and improve the supply of seedlings for reforestation and use.

Second, it is to adapt to changes in climatic conditions and to choose the right places for scientific afforestation and greening with appropriate species. According to changes in temperature, precipitation, and other climate factors, it is to adapt to the trend of species transfer to high latitudes and altitudes, optimise afforestation and greening tree species, and choose the suitable season. It is to adjust to local conditions to choose the right tree; improve the proportion of native species and mixed forests; increase the proportion of fire-, drought- (wet-), barren-, disease-, insect-, extreme temperature-, and saline-resistant tree plantations; ensure reasonable configurations of afforestation species and plantation density; optimise the afforestation model; and cultivate healthy forests. In particular, in the work of dry-zone afforestation, whether to plan arbours, shrubs, grasses or a combination of these should be based on local conditions. In these zones, it is important to accelerate the recovery of vegetation and strive to build well-adapted, diverse forest ecosystems.

Third, the concept of natural management is basic to actively promoting multi-functional and natural forest management. Work shall be done to utilise natural forest management concepts and techniques, accelerate the study of climate change adaptation of forest cultivation and business models, and promote sustainable forest management. Forest management plans are to be developed taking into account future climate change scenarios, especially extreme weather conditions. In response to problems such as monocultural forests, unreasonable density, forest degradation, fragile service functions, and other issues, climate change factors are to be combined to carry out scientific forest nurturing and management; optimise forest structure; improve the productivity of forest land, quality, and service functions; and enhance the ability of forests to withstand natural disasters and adapt to climate change.

Fourth, attention shall be paid to strengthening forestry disaster monitoring and early warning and constantly improving adaptive disaster management. Climate change factors are to be considered in establishing a monitoring system to foresee forest fires, forestry pest disasters, and dust storms by using remote sensing and other modern means of forest monitoring to enhance forecasting and early warning capabilities. Research is to be done on the occurrence of forestry disasters to strengthen disaster risk assessment, focusing on the assessment of floods, droughts, snowstorms, freezing rain, typhoons, and other meteorological disasters as well as landslides, mudslides, and other geological disasters and the impact of these on forestry. Disaster prevention and control infrastructure and emergency response capacity are to be enhanced by establishing material and technology reserves, adopting advanced management measures, controlling the scope of disasters, preventing secondary disasters, and striving to reduce losses caused by disasters.

Fifth, efforts shall be made to enhance work in the establishment and management of nature reserves, thereby strictly protecting ecologically fragile areas and related species. Accelerating the establishment and adaptive management of forestry nature reserves is important to work with the aim of establishing a network of nature reserves and species migration corridors and strengthening the protection of both typical forest ecosystems and ecologically fragile areas. Specific work includes improving the monitoring and early warning of wildlife epidemics; increasing the protection of key species; protecting terrestrial wildlife with relatively small populations, fragmented habitats, or severely damaged habitats; and improving the adaptive capacity of both important and rare species under climate change scenarios. Last, it is necessary to strengthen the protection of landscape diversity, promote the construction of forest parks, protect the originality and integrity of natural ecosystems, and build a complete ecological protection network.

Sixth, wetland restoration efforts are to be enhanced for wetland ecosystems to adapt to climate change. The key work lies in carrying out wetland restoration projects and the restoration and comprehensive management of wetlands in key areas, optimising the structure of wetland ecosystems, increasing their area, restoring their functions, and enhancing their carbon storage capacity. It is also important to strengthen the monitoring of wetland resources, increase the protection of biodiversity in wetland ecosystems, and promote assessment of the risk of degradation of wetland functions.

Seventh, attention shall be paid to accelerating the recovery of vegetation in sandy areas and enhancing the ability of desert ecosystems to adapt to climate change. Biological and engineering measures are to be used to promote the construction of the Beijing–Tianjin Wind and Sand Source Control Project and the Sandy Land Closure Protection Zone, and soil erosion and stone desertification control in karst areas are to be enhanced. It is also important to strengthen the protection of species in sandy areas, carry out dynamic monitoring of vegetation conditions and desertification in sandy areas, and accelerate the process of vegetation restoration in these areas. The ability of desert ecosystems to adapt to climate change can be enhanced by enhancing soil conditions, improving the conditions of vegetation renewal, and increasing the vegetation cover of forests and grasses.

Eighth, scientific research on forestry adaptation to climate change is to be strengthened. Research is to be promoted in an in-depth assessment of the sensitivity of forestry to climate change and its risks; the response and adaptation of forest, wetland, and desert ecosystems to climate change; the relationship between climate conditions and forest disasters in historical periods; and assessment tools for forest ecosystem vulnerability and their local application. Also, research is to be promoted in the evaluation index system of forestry's ability to adapt to climate change, the adaptation countermeasures and policy measures, cost-effectiveness, and evaluation of adaptation effects of forestry adaptation to climate change. The research aim is to apply and promote forestry technologies for climate change adaptation that meet China's national conditions, build an adaptation technology system, and inform policy decisions.

Ninth, international cooperation on forestry adaptation to climate change is to be strengthened. Constructively participating in international climate negotiations and IPCC report research, preparation, and evaluation are needed to catch the international process and development trend of forestry adaptation to climate change. China should promote extensive, bilateral, and multilateral practical cooperation on forestry adaptation to climate change and carry out multi-channel, multi-level, and diverse exchanges. Action should be taken to encourage developed countries to provide financial, technical, and capacity-building support to developing countries carrying out adaptation actions, and to use international resources to promote domestic forestry adaptation actions. China should guide and support domestic and foreign enterprises, private institutions, and non-governmental organisations to conduct technical exchanges on forestry adaptation to climate change, and promote practical cooperation.

Guarantee Measures

Organisational Leadership

Forestry authorities at all levels should heighten awareness of the importance and urgency of forestry adaptation to climate change, put forestry adaptation to climate change on the priority agenda, strengthen organisational leadership, establish a sound working mechanism, and implement responsible units. Departmental cooperations are to be strengthened, especially between development and reform, finance, meteorology, and other departments to form a synergy of forestry adaptation action. All localities should develop specific implementation measures according to local realities in order to ensure that the forestry adaptation actions identified in this programme are performed satisfactorily and achieve measurable results.

Policy Support

Forestry authorities at all levels shall incorporate climate change action objectives and tasks into the *Thirteenth Five-Year* overall arrangements for forestry development planning at their level. Forestry adaptation and mitigation work should be organically combined and synergistically promoted. According to national and local planning, it is essential to refine annual construction tasks; develop decomposition and implementation plans; and ensure well implementation, supervision, and inspection. Efforts shall be made to actively explore policy innovation; improve the multi-investment mechanism; mobilise the community, enterprises, and individuals to participate in forestry construction; and strive to build a forestry adaptation to climate change policy protection system. Measures are to be taken to promote the establishment of forestry farmer disaster insurance and explore the adjustment of support disaster logging policy. Additionally, forestry adaptation to climate change is to be financed through multiple channels to maintain the continuity and stability of financial investment and to ensure the funding needs of adaptation work.

Fundamental Infrastructure

Measures are to be taken to establish forest fire warning systems and forestry pest prevention and control systems, increase forest fire prevention roads and equipment, and construct forestry pest detection stations, quarantine inspection stations, and other infrastructure in order to improve disaster response capabilities to provide basic security and to operate the data and information system required for assessment of the vulnerability of forest, wetland, and desert ecosystems.

Publicity and Training

Forestry training courses are to be provided to address climate change and cultivate personnel specialised in adaptation. To increase publicity, the focus should be on mobilising forestry system cadres and workers to popularise climate change-related knowledge and on organising policy lectures to raise awareness of adaptation. Successful experiences in pilot projects of forestry adaptation to climate change are to be summarised and promoted.

Source: *Website of China State Forestry Administration and Grassland Bureau (2016) China Forestry Adaptation to Climate Change Action Plan (2016–2020), Retrieved from* https://www.gov.cn/xinwen/2016-07/22/content_5093660.htm, July 22, 2016.

Work Plan for Ecological Poverty Alleviation in China

This work programme was jointly developed by the China Development and Reform Commission, China State Forestry Administration, China Ministry of Finance, China Ministry of Water Resources, China Ministry of Agriculture, and China State Council of Poverty Alleviation, in accordance with the unified deployment of China State Council Leading Group for Poverty Alleviation and Development. Based on the *Decision of the CPC Central Committee and China State Council on Winning the*

Battle against Poverty and the poverty reduction goals of the *Thirteenth Five-Year Plan* of the China State Council, it aims to give full play to the role of ecological protection in targeted poverty alleviation and eradication and to effectively address ecological poverty alleviation.

Overall Requirements of Ecological Poverty Alleviation

Guiding Ideologies

In the study and implementation of decisions and instructions made in the Nineteenth CPC National Congress, by the General Secretary Xi Jinping on poverty alleviation, the Party Central Committee, and the China State Council, it is essential to practise the concept of *green water and green mountains are golden, lucid water and lush mountains are invaluable assets.* At the same time, poverty alleviation is the fundamental strategy, and poverty alleviation and ecological protection are equally key. Measures are to be taken to increase ecological compensation, vigorously develop ecological industries, innovate ecological poverty alleviation methods, increase effective support for poverty-stricken areas and poverty-stricken people, promote poverty alleviation and related development in poverty-stricken areas, and promote poverty alleviation and sustainable development simultaneously. Thus, people obtain more benefits from ecological protection and restoration and realise a win–win of poverty alleviation and ecological civilisation construction.

Basic Principles

Adhere to the principle of central coordination and local responsibility. On the basis of central coordination, the province is responsible for the overall work, and cities and counties carry out specific work. The relevant departments of the central government are responsible for the formulation of policies, work assignment, and assessment and supervision. The relevant departments of the provincial government are responsible for improving policies and measures and strengthening coordination and cooperation. Relevant departments of municipal and county governments are responsible for the work of ecological poverty alleviation in their administrative areas to ensure that the policies and measures are put into practice.

Insist on the principle of government guidance and diverse subjects. It is important to innovate institutions and mechanisms, widely mobilise forces from all walks of life to participate in ecological poverty alleviation work, broaden the channels of social forces to help the poverty-stricken, and form social synergy. Also, it is vital to fully mobilise people in poverty-stricken areas to protect and restore the ecological environment of their town, enhance initiative and creativity, and promote the spirit of self-reliance and hard work, relying on their own efforts to alleviate poverty.

Adhere to local conditions and scientific development. The relationship between poverty alleviation and ecological protection shall be coordinated by integrating the principle of respecting nature, obeying the laws of nature, and protecting nature throughout the process of ecological poverty alleviation. The relationship between short-term poverty alleviation and long-term development shall be well-handled, focusing on the long term but based on the present, taking into account natural resource endowment, local characteristics, regional economic and social development level, and other factors. Decisions regarding ecological poverty alleviation shall be well-considered in order to promote poverty alleviation and green development.

Adhere to the principle of making effective policy and improving its effectiveness. Precisely target the 14 poverty areas and counties, areas outside the national key counties, and poverty-stricken households, highlighting the areas with deep poverty. Adhere to the principle of problem- and goal-orientation, focus on the poverty-stricken population, strengthen the convergence of poverty alleviation policies, and implement ecological poverty alleviation policies and measures to ensure that ecological poverty alleviation work achieves measurable results.

Objectives of Ecological Poverty Alleviation

By 2020, the poverty-stricken population's income and living conditions will have increased significantly through participation in ecological protection, ecological restoration project construction, and the development of ecological industries. The ecological environment in poverty-stricken areas will improve, the supply capacity of ecological products enhanced, the level of compensation for ecological protection be compatible with the economic and social development, and the capacity for sustainable development be further enhanced. China will strive to set up 12,000 professional cooperatives for ecological construction and poverty alleviation (including 10,000 forestation cooperatives and 2,000 grass and animal husbandry cooperatives) and absorb 100,000 poverty-stricken people to participate in ecological projects; create 400,000 ecological caretaker job opportunities (including 300,000 ecological forest caretakers and 100,000 grassland caretakers); and drive about 15 million poverty-stricken people to increase their income through the vigorous development of ecological industries.

Means of Poverty Alleviation

Income from Participation in Projects

Promote the models of professional cooperatives and villager self-operation for poverty alleviation and forestation and organise the poverty-stricken people to participate in ecological projects. The major ecological projects invested in by the government must absorb a certain proportion of the poverty-stricken people with labour capacity, pay the poverty-stricken a reasonable salary, and increase the income of the poverty-stricken people.

Income Through Ecological Public Jobs

Provide ecological caretaking jobs in poverty-stricken counties with emphases on forest, grassland, wetland, and sandy land care so that poverty-stricken people who meet the requirements of the jobs can participate in ecological caretaking work and realise poverty alleviation at home. In national parks, nature reserves, forest parks, and wetland parks in poverty-stricken counties, priority is to be given to poverty-stricken people with working ability to participate in service and management.

Income Through Ecological Industry Development

Under the premise of strengthening protection, make full use of the ecological resource advantages of poverty-stricken areas; combine existing projects; vigorously develop ecological industries such as ecological tourism, special forest industries, and special breeding and raising industries; establish interest mechanisms through land transfer; share dividends, cooperative operation, labour employment, independent entrepreneurship, etc.; improve the income distribution system; increase asset income; and broaden the channels of income for poverty-stricken people. Under the same quality standard conditions, priority will be given to purchasing forestry seeds and seedlings from poverty-stricken households to increase their income.

Transfer Income Through Ecological Protection Compensation and Other Policies

When arranging subsidies for returning farmland to forest and/or grass, for grassland ecological protection, and other subsidies, priority shall be given to supporting poverty-stricken people in need so that they can obtain subsidised income.

Promotion and Implementation of Tasks

Strengthen the Construction of Major Ecological Projects

Strengthen ecological protection and restoration in poverty-stricken areas, and further tilt to poverty-stricken areas in the arrangement of various major ecological projects and funds. Organise and mobilise the poverty-stricken to participate in major ecological projects and improve the benefit to the poverty-stricken people.

Returning Farmland to Forests and Grasses

The basic farmland retention index shall be adjusted for arable land on steep slopes above 25 degrees in poverty-stricken areas, and the next round of returning farmland to forest and grass in poverty-stricken areas shall start. Priority shall be given to the task of returning new fallow land to forest and grass in 22 provinces (autonomous regions and municipalities) in Central and Western China, and the provinces (autonomous regions and municipalities) should give priority to supporting poverty-stricken counties in need, especially areas of deep poverty. The poverty-stricken counties should prioritise eligible poverty-stricken people. In the selection of tree and grass species, poverty-stricken households shall be guided to develop economic forest and grass species with better economic benefits and adapted to local planting conditions so as to enable poverty-stricken households to receive long-term stable income. Ensure that by the end of 2020, all arable land in poverty-stricken counties that meet the national policy of returning farmland and that have the will to return farmland to forest and grassland do so.

Returning Pasture to Grass

In Inner Mongolia, Shaanxi, Ningxia, Xinjiang, Gansu, Sichuan, Yunnan, Qinghai, Tibet, Guizhou, and other provinces and regions and in the Xinjiang Production and Construction Corps, eligible poverty-stricken counties shall start the project of returning herding land to grassland, and, according to the needs of poverty-stricken farmers and herdsmen, the scale of the annual task of preparing feeding sheds and artificial forage land shall be increased.

Qinghai Sanjiangyuan Ecological Protection and Construction Phase II Project

Deeply promote the protection and projects of forests, grasslands, deserts, wetlands, and lake ecosystems in the Sanjiangyuan area; increase the treatment of degraded grasslands such as black soil beach grasslands; complete the treatment of 2.2 million mu of black soil beach; and improve the productivity of grasslands. Build livestock warming sheds and grass storage sheds for herding households engaged in livestock production, and increase herding income through the development of highland ecological and organic animal husbandry.

Beijing–Tianjin Wind and Sand Source Control Project

Promote the protection and restoration of forest and grass vegetation in 53 poverty-stricken counties (banners) within the project area and sandy land management in key areas; improve the quality and coverage of existing vegetation; curb local sand erosion; arrange for the creation of 3.15 million mu of forests and 60,000 mu of sand fixation projects; and absorb poverty-stricken people to participate in the project.

Natural Forest Resources Protection Project

Focus on the upper reaches of the Yangtze River and the upper and middle reaches of the Yellow River, increase support for natural forest resource protection projects in poverty-stricken areas and the management of public welfare forests through the purchase of services according to law, and create more employment opportunities for poverty-stricken people.

Three Norths and Other Protective Forest Systems

Prioritise the protective forest system projects in 226 poverty-stricken counties within the scope of the Three Norths (including Northwest, Northeast, and North China), Yangtze River, Pearl River, coastal, and Taihang Mountain areas; increase forest management; promote the restoration of degraded forests; and complete the afforestation of 10 million mu. Strengthen the national reserve forest project and the afforestation of 10 million mu of national reserve forests with financial and other social capital in southern areas with suitable sunshine, temperature, water and soil conditions, abundant forest resources, and concentrated poverty-stricken areas.

Key Projects in Soil and Water Conservation

Increase efforts to manage soil erosion in key areas such as the upper and middle reaches of the Yangtze River and the Yellow River, the karst area in the southwest, and the black soil area in the northeast; promote policies and projects in favour of poverty-stricken counties with serious soil erosion included in relevant plans; and accelerate the comprehensive management of sloping land, erosion ditches, and small watersheds. Meanwhile, develop economic forestry as well as fruit and special industries, implement ecological restoration, promote the ecological and economic virtuous cycle, and improve the agricultural production and living conditions in project areas.

Comprehensive Management of Stone Desertification

Insist on poverty alleviation and support for rock desertification treatment projects in 146 key counties in poverty-stricken areas such as the rock desertification area in Yunnan, Guizhou, and Guangxi; Western Yunnan border mountainous area; Wumeng mountainous area; and Wuling mountainous area. Adopt various measures such as mountain sealing, forestation and grass cultivation, artificial afforestation, forest nurturing, and comprehensive treatment of small watersheds and complete the treatment of karst with an area of 18,000 km^2.

Sandy Land Sealing Reserve

In Inner Mongolia, Tibet, Shaanxi, Gansu, Qinghai, Ningxia, Xinjiang, and other provinces (regions) and in the Xinjiang Production and Construction Corps, promote the sandy land sealing protection zone project, with priority for 4.98 million mu of suitable sandy land in poverty-stricken counties, as well as rigorous sealing protection. Increase support for the pilot zones for sand control and sand management in deeply poverty-stricken areas to improve the wind and sand control capacity of the areas.

Wetland Protection and Restoration

Implement a number of major wetland protection and restoration projects in internationally important wetlands and in national wetland nature reserves in poverty-stricken areas to improve the ability of these areas to contain water, store floods, prevent flooding, and purify water. Support poverty-stricken counties to implement wetland protection and restoration, wetland eco-efficiency compensation, and pilot projects for returning wetlands to farmland in order to improve the wetland protection system.

Comprehensive Management of Reclaimed Grasslands in Staggered Zones of Agriculture and Animal Husbandry

Coordinate and promote reclaimed grassland management at the intersection of agriculture and animal husbandry, with priority on poverty-stricken areas. Promote the development of artificial grass and dry quality forage base to improve the vegetation coverage of treatment areas. Increase forage seeding, the use of processing machinery, and agricultural machinery purchase subsidies in order to guide and support the development of grass-fed animal husbandry in poverty-stricken areas. The purpose is to achieve ecological grassland restoration while promoting the quality and efficiency of animal husbandry.

Increase Ecological Protection Compensation

Continuously improve the transfer payment system, explore the establishment of a varied ecological protection compensation mechanism, and gradually expand the extent of ecological compensation for poverty-stricken areas and poverty-stricken people.

Increase Transfer Payments to Key Areas with Ecological Function

The central government will increase transfer payments to poverty-stricken counties in the national key areas with ecological function, especially the 'three regions, referring to Tibet, four prefectures in southern Xinjiang, and Tibetan ethnic areas in Sichuan, Qinghai, Gansu, and Yunnan, and three states, referring to Linxia in Gansu, Liangshan in Sichuan, and Nujiang in Yunnan', and other areas of deep poverty. Expand the scope of policy, improve subsidy schemes, and gradually increase support for ecological protection and restoration in these areas.

Improve Mechanism for Forest Ecological Compensation and Subsidy

Improve the mechanism for the adjustment of financial forest ecological benefit compensation subsidy standards at all levels, mobilise the enthusiasm of relevant interests in forest protection, improve the forest ecological benefit compensation and subsidy policy, and promote more scientific and reasonable compensation standards. Enhance the supervision of forest ecological compensation funds to protect the vital interests of the poverty-stricken people.

Implement a New Round of Grassland Ecological Protection Subsidy and Incentive Policy

In the pastoral areas of semi-pastoral counties in Inner Mongolia, Tibet, Xinjiang, Qinghai, Sichuan, Gansu, Yunnan, Ningxia, Heilongjiang, Jilin, Liaoning, Hebei, Shanxi, and the Xinjiang Production and Construction Corps, implement the grassland ecological protection subsidy and incentive policy and make timely and full payment of grazing ban subsidies and grass-livestock balance incentive funds to herders.

Carry Out Ecological Compensation Pilot Project

Treat the poverty-stricken counties in the national key ecological function areas as the main participating parties in the socioeconomic system. Integrate the transfer payments, horizontal compensation, market-based compensation, and other channels for funds to establish a comprehensive ecological compensation system subject to the local condition. Improve a monitoring and evaluation assessment system with a close combination of ecological compensation fund payments and effective ecological protection to ensure farmers and herdsmen in poverty-stricken areas can obtain due compensation.

Develop Ecological Industries

Rely on and give full play to the advantages of ecological resource endowments in poverty-stricken areas, select industries with special characteristics that are closely integrated with ecological protection and have relatively stable markets, and effectively transform resource advantages into industrial and economic advantages. Support the creation of characteristic agricultural products in poverty-stricken areas and prioritise the products in poverty-stricken areas with the heavy task of poverty alleviation and outstanding effect of driving farmers to increase income in official brand/product evaluation or recommendation. Guide poverty-stricken counties to broaden investment and financing channels, implement capital integration policies, strengthen financial and insurance services, and work to improve the risk resistance of special industries. Cultivate and strengthen ecological industries; promote the integrated development of primary, secondary, and tertiary industries; and establish an interest mechanism between leading industrial enterprises, new business entities, and poverty-stricken people through various approaches, such as share dividends, business assistance, cooperative operation, and labour employment, so as to broaden channels for poverty-stricken people to increase their income.

Develop Ecological Tourism

Promote a sound mechanism for developing ecological tourism and resource protection; increase guidance and support for poverty alleviation via ecological tourism; and strengthen the development of tourism infrastructure in nature reserves, forest parks, wetland parks, desert parks, grasslands, etc. In accordance with the basic law, such as National security law, improve ecological tourism industry standards and establish or improve fire safety, environmental protection, and other regulatory norms. Actively create varied ecological tourism products; promote the deep integration of ecology and tourism, education, culture, recreation, and other industries; and vigorously develop ecological tourism experiences, ecological scientific research, ecological recreation, etc. Advocate smart tourism and low-carbon tourism.

Guide the poverty-stricken people to develop from scattered individual operations to large-scale operations, and provide convenient services for the poverty-stricken people to set up forest (grassland) hotels and engage in the sale and transportation of local specialties. Expand the development of tourism-related planting, breeding, and handicraft industries to promote poverty alleviation and income growth among the poverty-stricken people. Create 50 forest tourism sites with high reputation, 20 forest tourism routes, 30 celebrity towns in forest counties, 10 national forest and forest health pilot bases, etc. In poverty-stricken areas, the number of poverty-stricken people who can increase their income through forest tourism reaches 650,000 households and 2 million people.

Develop Special Forest Industry

On the premise of ensuring ecological benefits, actively develop industries such as woody oilseeds, special forest fruits, fast-fruit forests, bamboo and rattan, and flowers that are suitable for planting in poverty-stricken areas and that have strong market demand and high economic value. Promote the development of standardised production for forest products, standardised production techniques, special forest industries, regional special forest industries, the expansion of industrial scale, and professional management. Focus on the development of primary and deep processing of forest specialties with local and ethnic characteristics, extend the industrial chain, improve storage and logistics facilities, and enhance comprehensive benefits. Give full play to the role of brand, support leading enterprises in developing corporate brands, improve the reputation of special brands, and expand markets for them. Build exhibition and marketing platforms for special forest products in deep-poverty areas, make full use of e-commerce platforms, online and offline integration, 'Internet+', and other emerging means to increase the market promotion of special forest products.

Develop Special Breeding Industry

Based on the bearing capacity of resources and environment, give full play to the advantages of ecological resources such as lakes and reservoirs, forests, and grasslands in poverty-stricken areas; actively develop the forest economy; and promote the compound operation of agriculture and forestry. Vigorously develop industries such as Chinese herbal medicine under the forest, special economic crops, wildlife breeding (cultivation), forest breeding, high-yield forage planting, grass-fed animal husbandry, and special aquaculture, and actively promote the combination of planting and farming. Accelerate the development of the processing industry for agricultural and forestry products, actively develop e-commerce in agricultural products, create pilot projects with special characteristics, establish a production and operation pattern of 'leading enterprises + professional cooperative organisations + bases + poverty-stricken households', and actively guide the poverty-stricken people to participate in the development of planting and farming industry.

Innovative Means of Support for Poverty-Stricken Areas

Carry Out Pilot Project for Ecological Relocation

With the establishment of the national park system, via multi-channel fundraising, promote the ecological relocation of residents living in an ecological area to restore the original ecological environment of the area and to help poverty-stricken people alleviate their poverty. In accordance with the principle of *pilot project first, gradually rolled out*, carry out an ecological relocation pilot project in the Qilian Mountains National Park system pilot (Gansu area) in the core protected areas. Support the relocation of the construction of housing (by purchasing), industrial development and employment arrangements, ecological protection, and restoration of the relocation area. Summarise and promote successful experiences, and adjust to local conditions.

Innovate Ways of Utilising Resources

Promote the orderly transfer of ownership of forest resources and reform forestry rights, such as forest ownership and forest land management rights, among others, so as to broaden poverty-stricken people's income channels. Localities can independently explore the redemption and replacement of commercial forests where harvesting is forbidden within national and provincial nature reserves, and national forest parks and other key ecological zones could be adjusted to render them public welfare forests, so as to realise the greening of society and benefitting the poverty-stricken people. Promote the reform of rural collective property rights in poverty-stricken areas; protect farmers' property rights and interests; transform eligible rural land resources and collectively owned forest resources in poverty-stricken areas into equity in enterprises, cooperatives, or other economic organisations in various ways; promote the demutualisation of assets in poverty-stricken villages and equity in land use rights; and make use of rural resources, assets, and funds.

Promote Successful Models of Ecological Poverty Eradication

Actively explore modes of poverty eradication and income generation through ecological protection, restoration, relocation, industry, and villages. Explore measures and policies on organisation forms, interest mechanisms, and income generation in multi-industry in deep-poverty areas. Summarise and refine good experience models in a timely manner, create ecological poverty eradication models, and actively promote good experiences and practices so as to better protect the ecological environment and help poverty-stricken people achieve stable poverty alleviation.

Standardise Positions in Ecological Management

Research and formulate the ecological caretaker system; standardise selection and recruitment procedures, the scope of care, job responsibilities, and the rights and obligations of ecological caretakers; strengthen team building; and enhance ecological resource management capacity. Strengthen training of ecological caretakers to enhance their management expertise and safety awareness. Gradually increase job opportunities of ecological caretakers for the poverty-stricken, and prioritise deep-poverty areas, key areas with ecological function, and the source area of large rivers. Insist on strengthening the management mechanism of *built by the country, managed by the township and used by the village*, and give priority to families with deep poverty, ethnic minorities, and military veterans.

Explore Methods of Carbon Trading Compensation

Promote the development of the national carbon emission trading market, actively reform the CDM and the trading mechanism of voluntary greenhouse gas emissions, explore valid models for forestry carbon sequestration/sink projects to obtain carbon emission compensation, and increase support for poverty-stricken areas.

Development of Practical Safeguards

Responsibilities at All Levels

China Development and Reform Commission, State Forestry Administration, Ministry of Finance, Ministry of Agriculture, Ministry of Water Resources, State Council Poverty Alleviation Office, and other departments are to strengthen guidance and support, fortify communication and collaboration, and coordinate and promote the work of ecological poverty alleviation, engaging in joint efforts to promote ecological poverty alleviation work. Local governments should refine ecological poverty alleviation work programmes, integrate ecological poverty alleviation (as key work) into the annual work plan, establish key points of annual work, assign and delineate responsibility to various units, identify responsible persons, and task schedule. Ecological poverty alleviation work should be deployed and arranged as key work, and the responsibility and pressure should be transferred from one level to another as appropriate to establish a responsibility system of ecological poverty alleviation.

Increase Investment

Various types of special transfer funds concern people's livelihood, and poverty-stricken areas and people should be further prioritised for government investment. For areas where ecological poverty alleviation is the main poverty alleviation measure, the structure of financial expenditure shall be optimised by integrating funds from all sources, and the focus shall be on ecological poverty alleviation. Innovate means of government funding, standardise the use of government and social capital cooperation and other models, and leverage more resources for ecological poverty alleviation. Prioritise ecological poverty alleviation tasks in deep-poverty areas, and increase support in areas through new funding, new projects, new initiatives, projects for the people, the integration of agriculture-related funds, financial transfer payments, financial inputs, capital markets, insurance institutions, and land quotas.

Strengthen Technical Training

Actively organise technical experts to provide assistance in poverty-stricken areas; increase technical guidance for large ecological industries, cooperatives, and enterprises; and train both local and technical experts who are 'visible, inquiring, and retaining' among poverty-stricken people in poverty-stricken areas. Increase training for grassroots staff and poverty-stricken households; improve their capacity; and enhance their ability to develop themselves, their market awareness, and their risk prevention and control.

Strengthen Supervision and Management

Establish a dynamic management and supervision system for ecological poverty alleviation, coordinate and do a good job in organisation and execution (such as daily scheduling, tracking, and inspection), and implement accounts management and regular scheduling of annual key work. A successful experience can be a good reference for the assessment of the effectiveness of poverty alleviation elsewhere. Ineffective and lagging progress is to be listed for supervision, and effective progress is to be given praise and incentive to ensure the smooth progress of the work programme. Adhere to the principle of *zero tolerance* in the supervision of ecological poverty alleviation funds; improve the use of management methods; increase the number of audits, inspections, and interviews; any problems found are to be seriously investigated and punished. Ensure that funds are publicly audited, and effectively protect the interests of the poverty-stricken people.

Source: *Website of China State Forestry Administration and Grassland Bureau, Retrieved from* http://www.forestry.gov.cn/main/72/content-1071264.html. January 18, 2018.

Postscript

This book is a monograph under Policy Research on Promoting PAFCS in Ethnic Areas of Southwest China (project number 15BJY093), chaired by myself and funded by the National Social Science Foundation of China (NSFC). First of all, we are grateful for the support of the NSFC, Social Science Literature Press, and Springer Nature, as well as for the excellent work of the translation team led by Chengzhi Long and Jing Lin.

Global warming has brought significant impacts on human survival and socioeconomic development and has directly or indirectly exacerbated poverty. Countries around the world are working together to address the two major global challenges of climate change and poverty, and China is also vigorously implementing the strategies of targeted poverty alleviation and poverty eradication, ecological development, and rural revitalisation. This book is targeted at people who care about, participate in, and promote research and practice related to PAFCS and who contribute to the fulfilment of the Paris Agreement in China. We hope to provide reference and inspiration for promoting FCS to benefit more poverty-stricken people and to assist in the win–win situation of climate change and poverty reduction in China and the world.

We have received a lot of guidance, selfless help, and support from many people and organisations, including but not limited to Huang Chengwei of CICP; Professor Wang Sangui of Renmin University of China; Professor Zhang Qi of Beijing Normal University; Professor Dong Wenyuan of Southwest Forestry University; Professor Chen Yong of Sichuan University; Du Shouhu, Guo Xiaoming, and Zhang Kejun of Sichuan Academy of Social Sciences; Professor Xu Chenghong of Southwest University of Finance and Economics; Professor Krishna P. Paudel of Louisiana State University; Deputy Director Bao Jianhua, Chief Engineer Luo Jianguo, Division Chief Zhou Gupeng, and Ren Tongwei of Sichuan Forestry and Grassland Bureau; Division Chief Jing Shigang and Cai Hong of Sichuan Provincial Department of Science and Technology; Wang Jingdong of Sichuan Rural Science and Technology Development Centre; Deputy DirectorLiu Weijia, Division Chief Zhang Bing, and Xue Bing of Sichuan Provincial Bureau of Poverty Alleviation and Migration Work; Deputy MayorJiang Xuemei of Tengchong City; Deputy Director Xie Wu of Forestry Bureau; Deputy Director Zhang Yong and Zhao Yan of Xishuangbanna Dai Autonomous Prefecture Forestry Bureau; Deputy Director Yang Limin of the Forestry and Grassland Bureau of Liangshan Yi Autonomous Prefecture; Bureau Chief Deng Lin, Deputy Director Su Xianwen, Liang Xiucheng, Wang Huaipin, and Gong Qiaoyuan of Dadu River Afforestation Bureau of Sichuan Province; Secretary General Ling Lin and Senior EngineerTang Caifu of the Sichuan Greening Foundation; Dr. Zhao Mingshi of the Sichuan Nature Conservancy; the leaders and staffs in Yunnan Mengxiang Bamboo Industry Co. Ltd; Zazuo Forestry in Guizhou Province, Yuexi County, Meigu County, Leibo County, Zhaojue County, Ganluo County, and three nature reserves in Yuexi Shenguozhuang, Ganluo Ma'anshan and Leibo Mamize, Li County and Mao County in Aba Tibetan and Qiang Autonomous Prefecture, Sichuan Province, and Pingwu County in Mianyang City, etc. I would like to take this opportunity to express my sincere gratitude. I would also like to take this opportunity to thank my team, the faculty, and students of the Southwest Poverty Reduction and Development Research Centre (SPRDC), for their hard work in conducting household research, data collection, data processing, and report writing. Most importantly, special thanks also go to the local community farmers who create and manage carbon sink forests in remote and poverty-stricken mountainous areas and who contribute their land, labour, and time for the future of all mankind with their very limited livelihood resources. Also, we have a responsibility to involve them in the governance of global climate change and in sharing the benefits in a fair and reasonable manner.

PAFCS is a cutting-edge topic in practice as well as in research on global climate change, forest carbon sinks, ecological compensation, and poverty alleviation. As the FCSes themselves are still at the pilot and exploration stage, researchers face the difficulties of insufficient knowledge and practice in PAFCS, as well as language barriers, challenging transportation, and high research costs because the farmers are scattered in the ethnic regions of Southwest China. In spite of all these difficulties, our research has produced preliminary results, which is most gratifying to me. In this regard, we continue to believe that strengthening the function of FCS in alleviating poverty can enhance the attractiveness of the FCS market and reduce the uncertainty of FCS industrial development. This function is also a major guarantee of reducing the risk and transaction costs of projects as it helps win extensive cooperation and long-term support from farmers in project communities and ensures the sustainable operation of FCS projects. We firmly believe that although there are still differences in the positions and specific initiatives of countries around the world in addressing climate change, and the Paris Agreement signed in 2016 is not perfect, the general trend of human cooperation to address climate change and protect the biosphere will persist. This presents a great opportunity and a serious challenge for theoretical research and practical work on PAFCS. We will continue to collaborate, side by side, with colleagues from all walks of life to produce results and promote the win–win situation of climate change and poverty alleviation. I would like to dedicate this book to all those who have been working hard and with a low profile for the cause of global climate change and poverty reduction.

MIX
Papier aus verantwortungsvollen Quellen
Paper from responsible sources
FSC® C105338

If you have any concerns about our products,
you can contact us on
ProductSafety@springernature.com

In case Publisher is established outside the EU,
the EU authorized representative is:
Springer Nature Customer Service Center GmbH
Europaplatz 3, 69115 Heidelberg, Germany

Printed by Libri Plureos GmbH
in Hamburg, Germany